SOUND AND NOISE

SOUND AND
NOISE

A Listener's Guide to Everyday Life

MARCIA JENNETH EPSTEIN

McGill-Queen's University Press
Montreal & Kingston · London · Chicago

© McGill-Queen's University Press 2020

ISBN 978-0-2280-0388-5 (cloth)
ISBN 978-0-2280-0449-3 (ePDF)
ISBN 978-0-2280-0450-9 (ePUB)

Legal deposit fourth quarter 2020
Bibliothèque nationale du Québec

Printed in Canada on acid-free paper that is 100% ancient forest free (100% post-consumer recycled), processed chlorine free

This book has been published with the help of a grant from the Canadian Federation for the Humanities and Social Sciences, through the Awards to Scholarly Publications Program, using funds provided by the Social Sciences and Humanities Research Council of Canada.

Funded by the   Financé par le
Government   gouvernement   Canada Council   Conseil des arts
of Canada   du Canada   Canada   for the Arts   du Canada

We acknowledge the support of the Canada Council for the Arts.

Nous remercions le Conseil des arts du Canada de son soutien.

Library and Archives Canada Cataloguing in Publication

Title: Sound and noise : a listener's guide to everyday life / Marcia Jenneth Epstein.
Names: Epstein, Marcia Jenneth, 1951– author.
Description: Includes bibliographical references and index.
Identifiers: Canadiana (print) 20200290576 | Canadiana (ebook) 20200290630 |
    ISBN 9780228003885 (cloth) | ISBN 9780228004493 (ePDF) |
    ISBN 9780228004509 (ePUB)
Subjects: LCSH: Noise. | LCSH: Noise—Social aspects. | LCSH: Noise—
    Psychological aspects. | LCSH: Sound. | LCSH: Hearing. | LCSH: Listening.
Classification: LCC RA772.N7 E67 2020 | DDC 363.74—dc23

This book was typeset by True to Type in 10.5/13 Sabon

For Irvin,
whose passion for science lifted mine.

Contents

Tables and Figures

Foreword

For the past forty years, I have researched and written on the impact of loud sounds and noise on our health and well-being and more recently have focused on the importance of quiet time in our lives. Although there was some interest in the field of noise when I first began exploring the area, interest has grown immensely in recent years. That is why *Sound and Noise: A Listener's Guide to Everyday Life* is of greater interest and importance to more readers today than it would have been years ago. When I started writing about noise, the people interested in this issue were academics who wrote about their research in specialized books and journals. For research to have a greater impact, it must reach the general public and be presented in a way that can be understood by individuals unfamiliar with the technical language of those specialized publications. Marcia Epstein's book is successful in giving research on sound and noise this impact.

While Dr Epstein does draw on information from the research of medical and environmental scientists on noise, she also adds the perspectives of musicians and journalists to her discussions. She presents noise as a "multifaceted subject in itself, with connections to history, philosophy, anthropology and culture." Viewing noise in this way, from the vantage points of a variety of disciplines, makes the book especially appealing. *Sound and Noise* will be appreciated as much by professionals who have studied noise for years and by members of other academic fields as it will by members of the general public.

Readers will learn about the effects of loud sounds on hearing and the impact of noise on overall mental and physical health, including adverse effects on children's learning. Dr Epstein wants us to acknowledge that noise is a public health issue and that it should be recognized as a pollutant that is as dangerous to our health as are air and water pollution. Although we may understand that loud sounds and noise are harmful to hearing – this issue has received considerable attention – this book guides us through the growing body of research that has linked noise pollution to mental and physical stress, damage to the heart and circulatory system, reading deficits in young children, and increased errors in the workplace.

Yet, sound cannot be viewed only as unpleasant and annoying. Dr Epstein introduces the work of acoustic ecology and its soundscape studies, work that not only examines the noises around us but also explores the welcoming sounds in our environment. Recent soundscape studies include soundwalks through the quieter areas of cities and small towns where participants are given the opportunity to tune in to the sounds around them.

Dr Epstein encourages us to go beyond spreading the word on the dangers of noise, and to use the knowledge about the dangers of noise to take action to lessen its harmful effects. She views this activism as a social justice issue, particularly for those living in high-density, low-income urban areas where the "detrimental effects of ambient noise are most severe." As a long-term advocate of lessening noise in our environment, I was pleased that Dr Epstein directed her readers to become personally involved in reducing noise. In this book, she has provided the information and knowledge that will allow us to move forward with wisdom toward a quieter and healthier environment.

This book shares a knowledge of sound and noise that will serve to enhance lives, and I hope, as Dr Epstein does, that its readers will be encouraged to share this book and its valuable information with others so that we can better understand the dangers of noise and the benefits of the beautiful sounds that surround us.

Arline Bronzaft

Preface in Mid-Pandemic

At any other time, this moment would be held within a suburban rush hour – somewhat muted by its distance from the downtown core but characterized by a variegated soft whooshing roar interspersed with the rumbles of trucks on a nearby freeway. Today, there is quiet. The pandemic keeps us all at home and usually indoors, with businesses and schools closed. Traffic is sparse and sporadic, while aircraft have been almost absent for nearly two months. Bird calls, often relegated to an indistinct background, are easily audible now: crows and magpies, sparrows and robins and (*dee-dee-dee-dee*) chickadees; now and then a gliding hawk.

I listen while reading about the sounds of other locked-down places, thanks to colleagues in the World Forum for Acoustic Ecology (WFAE) who share their carefully observed soundscapes in emails and recordings.[1] These recordings include birds in Australia, England, and the Czech Republic; street protests in Brazil; and someone opening a creaky gate to walk down a traffic-free street in Rome on a Saturday afternoon. There are neighbourhood gratitude demonstrations (applause, whoops, clackers, bells, clanging pots, car horns, a trumpet, a cello) every evening, supporting healthcare workers all over Canada. Just as we now travel virtually through videos, we can eavesdrop on everything from wilderness to urban centres.

Reports come in, as well, about the calls of urban birds being less loud than usual as they adjust to a quieter environment, no longer straining to be heard above human soundscapes. Human listeners interpret the conversations of trees and wind, recognizing the dialects of leaf shape and density.[2] Ambient sound level readings taken in major cities show nighttime levels during the day.[3] Quiet may be hard to find indoors as children are home and music, news, computer games, and videos are in constant play: gaps left by cancelled activities get filled by whatever raises spirits, a practice with historical precedent. Literature about the plagues that occurred in local waves in Europe from the fourteenth to the seventeenth centuries shows that music was considered crucial for keeping people from despair: along with funerals and masses for the dead and processions of chanting penitents, there were sprightly dances and folksongs described in stories but not preserved in notation. I imagine flutes and drums, bagpipes, fiddles, frenzies of dancing to keep fear at bay.

Sound makes us aware of larger worlds than the ones we can see. So do silences: as field recordings from Europe and North America grow in number, large areas of the world are currently voiceless for lack of recording equipment or of computer and internet access, or because people lack sufficient health or time for sharing experiences. We wait for a "new normal." Will it bring more awareness of our connections to each other and the natural world? Will we hear our everyday lives more clearly?

Marcia Jenneth Epstein
Calgary, Canada
May 2020

Acknowledgments

Sound and Noise: A Listener's Guide to Everyday Life has been shaped by several influences. One is the recent proliferation of concern about noise and hearing loss on websites and in the media: innumerable sources on the Internet now warn of the hazards of earbuds or document the legal regulations for hearing protection in industrial workplaces, but few actually explain the science involved. Even fewer explore the personal and cultural decisions that underlie the persistent attractions of loud noise in spite of its hazards. Another motivation was the need to make available the most recent general knowledge about auditory anatomy and function, which has been recently advanced by the use of technologies for scanning and imaging the brain: in this I have attempted to build on the foundations set so clearly and thoroughly by Jones and Chapman (1984) and many others. Finally, I wish to honour the early insights in the work of R. Murray Schafer and his followers, who formulated a pragmatic social philosophy of acoustics that interlaces the scientific aspects with history, art, and a deep appreciation of the natural world.

There are frustrations inherent in research on topics whose search terms can be ambiguous. Because both terms are used metaphorically in popular and scientific literature, Internet searches using "sound" and "noise" are doomed to attract huge numbers of ads for recorded and live music, audiology clinics, noise abatement companies, and "sound" investment schemes, as well as (ironically!) innumerable studies of statistical irrelevance in data searches. The scope of the topic also required that I become familiar with a great many potential subtopics, not all of which turned out to be suitable to the focus of the book. Time for exploration and sifting, in the absence of research assistance, was crucial. Thanks are due to the former Faculty of Communication and Culture (now the

Department of Communication, Media and Film) at the University of Calgary, which granted me six-month sabbaticals in 2008, 2012, and 2016 to enable periods of unchallenged focus on the initial research, development, and then revision of the manuscript.

Among the people who provided assistance at various stages, several merit special mention. Stephan Bonfield helped with reviewing research literature in neuroscience and provided a second pair of eyes on parts of the manuscript. Several of his suggestions helped substantially with the development of chapters 5 and 6. Thain Nicol took on the task of formatting initial chapter files, maintaining the early stages of the burgeoning bibliography, and knowing just what to do to convert working files between operating systems without losing the endnote anchors. Janine Chan, whose knowledge of the *Chicago Manual of Style* greatly exceeds mine, joyously accepted the momentous task of assisting me with formatting the bibliography and ferreting out URL and citation errors. Janine's talented colleague Aaron Swanbergson helped with edits to the reference citations and endnotes.

At the University of Calgary Library system, Susan Beatty tracked down some obscure online citations when they eluded me. At McGill-Queen's University Press, Jonathan Crago shepherded the manuscript through peer review, revisions, and a grant application while graciously responding to a spate of questions. Kathleen Fraser provided expert assistance with formatting changes; Barbara Tessman applied keenly honed editing skills to the final version. I am grateful for their aid and advice, and especially for their ability to make a challenging process into a gratifying one.

My desire to change public perception about the importance of soundscapes and auditory awareness links me to a growing community of scholars, researchers, broadcasters, and activists whose work has nourished mine. They include Eric Leonardson, Hildegard Westerkamp, Bernie Krause, and Jim Cummings, acoustic ecologists par excellence, as well as urban planner and architect Antonella Radicchi, who has provided collegial correspondence and supportive enthusiasm for my work. Arline Bronzaft, whose anti-noise activism in New York City has led to many effective bylaws, graciously agreed to write a foreword. The late Dr Ursula Franklin's extraordinary grasp of the intimate human side of science also remains an inspiration.

Immensely helpful colleagues from the Alberta community of acoustic engineers include Daryl Caswell, Teresa Drew, Richard Patching, and Jess Roy, all of whom responded several times to my requests for advice about technical terminology and provided general comments on parts of the manuscript. Terry Avramenko, Monica Fercho, Dana Negrey, and Dawn Stewart shared concerns about community noise problems that sharpened my awareness of the differences between theory and practice in noise abatement law and of the variations in individual sensitivity to sonic ambience. I am grateful also for the diligent work of several undergraduate students from my initial "Introduction to Acoustic Ecology" course at the University of Calgary in September 2007: Javan Bernakevitch, Deanna Cameron Dubuque, Camille Diaz, Marko Gregorsanec, Mark Knibbs, Michelle Mobarrez, and Wendy Nadon on noise and community issues around the Calgary Spy Hill asphalt plant; and Eva den Haan, Lauren Dickie, and Heather Ross on open-plan office workplace soundscapes.

Bylaw Officer Bill Bruce and Detective Stephen Johnston of the Calgary Police Service provided information and inspiration for discussions of community noise issues and bylaws. Carole-Lynne LeNavenec of the University of Calgary's Faculty of Nursing collaborated with me on a course to introduce nurses and nursing students to the basic concepts of music therapy, an experience that led to initial questions about the effects of noise on healing processes.

My cheering section – local colleagues and friends who helped sustain me through the long process of questioning, researching, writing, and editing – includes Tania Smith, Roberta Rees, Sandra Evans, Michelle Perkins, Brenda Forsey, and the extraordinary members of the Calgary Renaissance Singers and Players.

Deepest gratitude (in memoriam) follows Irvin Shizgal, best friend and husband, for keeping a series of computers in optimal working order, offering periodic reminders that it was time to stop working and eat something, and just being there.

Abbreviations

ADHD   attention deficit hyperactivity disorder
ASTM   American Society for Testing and Materials
DAP   digital audio player
dB   decibel
dBA   A-weighted decibel
dBC   C-weighted decibel
EEA   European Environment Agency
EHS   electromagnetic hypersensitivity syndrome
EMF   electromagnetic frequency
EMRP   European Metrology Research Program
HL   hearing loss
HVAC   heating, ventilating, and air conditioning
HZ   Hertz
ICAO   International Civil Aviation Organization
ICBEN   International Commission on Biological Effects of Noise
ICU   intensive care unit
kHz   kilohertz
LFN   low-frequency noise
LRAD   long-range acoustic device
NICU   neonatal intensive care unit
NIHL   noise-induced hearing loss
OITC   outdoor-indoor transmission class
OSHA   Occupational Safety and Health Administration (US)
PAD   personal audio device
PLD   personal listening device
PMP   personal music player
PTSD   post-traumatic stress disorder

STC	sound transmission class
SHA	Sight and Hearing Association
TTS	temporary threshold shift
WCB	workers' compensation board
WFAE	World Forum for Acoustic Ecology
WHO	World Health Organization

SOUND AND NOISE

Introduction

As I write, a window slightly open to the secluded suburban street below, a car horn sounds a repeated staccato alarm. It goes on for perhaps a minute, then stops. Almost immediately afterward it starts again, and stops, and starts. This becomes a pattern in the late summer afternoons for the next two days. A short circuit, I think, from across the street. Annoying. Something wrong with someone's car. Finally, on day three, the window fully open in hot weather, I decide to walk across the street to investigate the source of the noise, a van parked in front of the house directly across from my home office window. The recently arrived occupants are at this point unknown. I ring the doorbell, knowing the culture of the neighbourhood to be benign. A young woman opens the door, revealing a glimpse of the source of the noise: a child, perhaps a year old, with her car keys – complete with panic button and automatic door control – in his hand. I smile and we chat for a while, then I mention gently that the noise is disruptive. "Oh!" she says, surprised, "Sorry!" The next afternoon the sound starts up for a few seconds, stops, starts, stops. It does not continue.

The incident of the child with the car keys is indicative of several cultural norms in my neighbourhood. The first is the secluded suburban parent, sometimes at a loss to keep an active baby amused – but that is a subject for a different book. The second is the degree of disconnection among households in North American cities, but that is not the focus here either, at least not yet. The third is awareness of noise. I was surprised, that summer afternoon, that the young mother was able to ignore the repetitive stimulus immediately outside her door. Was she intentionally tuning it out? Was she plugged into a media player before opening the door and unable to hear the horn even though she responded to the doorbell? Was she running a vacuum cleaner? Did she have mildly impaired hearing? I had no indication that any of these conditions applied. All I knew was that the thought that the sound might bother neighbours came as a surprise to her, and that her level of tolerance for noise was different from mine. Of course it was: she was living with a toddler and I wasn't.

We become accustomed to ambient levels of domestic noise, whether the chortles, shrieks, and wails of small children or the intermittent soft clacking of computer keyboards. In this, we follow a pattern set by the evolution of our species and shared with other mammals. Familiar acoustic backgrounds, the everyday noises to which we are subjected on a regular basis, are relegated by the brain to a state of inattention in order to avoid fatiguing the nervous system with constant alarms. Our species evolved in the natural soundscapes of African forest and plains; loud with wind, rain, thunder, water, the calls of birds and insects, the earth-borne thunder of vast herds of hoofed animals – but not with traffic, aircraft, factories, leaf blowers, or rock concerts.

We learn accommodation to noise from infancy, with the repetitious sound stimuli to which we expose children – toys, television, listening devices. It continues with the teen who grows up loud, encased in sound by means of earbuds. We take our listening habits into adulthood: raised loud, we think nothing of exposing ourselves to the din of traffic, commuter trains, noisy factories or offices, and a whole host of sonic environmental conditions, from combustion engines to cellphone conversations. Heating and cooling systems, the mechanics of modern comfort, form a constant drone in the background of our indoor lives. We learn to ignore them, but our ears, brains, and bodies nonetheless continue to react.

What would happen if we were to listen with intention?

LISTENING UP

Take a moment, this moment, to listen to your surroundings. Closing your eyes will help. Right here and now, what do you hear?

- How many natural sounds are perceptible? Human sounds? Music? Machines?
- Have you chosen what you are listening to, or is it so much a part of your inescapable surroundings that you just haven't noticed it?
- How many layers of sound can you distinguish? What are their sources?
- What feelings, images, or memories do they evoke, if you let them?
- What can they tell you about this moment, this place?

What you have just experienced is a soundscape. It is a perceptual frontier of philosophy and neuroscience, but it is also as old as human awareness. In fact, far older. Sound, whether perceived as an auditory stimulus or a tactile one, is a vital sensory clue that all animals use to evaluate their surroundings. Human cultures, as well as individuals, have responses to it ranging from highly articulated awareness to near obliviousness. If you are reading this book in North America, the culture that surrounds you is likely to be extremely visual, and not particularly aware of the soundscapes it produces. Culture, and therefore language, helps to determine what is actively heard. The European heritage of the Western world has not yet developed a strong linguistic toolkit for the study of soundscapes. In a cultural vocabulary full of perspectives, overviews, outlooks, viewpoints, standpoints, aspects, prospects, panoramas, maps, and microscopic examinations, where are the clicks, crunches, reverberations, and resonances?

This book is about how to rediscover such knowledge through attention to sound, hearing, and conscious listening. The acoustic ecology and sound studies movements, two related branches of scholarship that are now examining sound as both ephemeral environment and a component of human culture, provide maps for previously unexamined territory. In their wake, the principles of a focused auditory culture have been taking shape since the 1970s, with new perspectives appearing steadily. Research publications related to sound and noise appear in technical journals of engineering, medicine, and environmental studies every month, as does information for the public on websites and blogs and in popular magazines. Book-length explorations of scholarly theories written or edited by philosophers, psychologists, anthropologists, and sociologists are now approaching questions of noise as sensory phenomenon and social signal.

Some of the recent literature on the cultural significance of noise is concerned with urban soundscapes as carriers for intercultural communication. While thoroughly in favour of this perspective for its insights as well as its potential to encourage active listening in urban environments, I am concerned here more with the biological and emotional burdens of loudness as it permeates everyday life in a variety of ways: industrial machinery, traffic, media, advertising, insistence on entrenched points of view. Amid the obvious attractions of loudness as cultural expression, celebration, protest, and mapping of territory, I want to encourage developing and embracing space in homes, workplaces, parks, schools, hospitals, and public venues that allows access to restorative doses of relative quiet, and to promote the building of dialogue on a foundation of listening.

The effects of noise on health are given considerable attention here, as are the social and legal complexities of noise annoyance in communities. Salient focal points – or riffs and cadenzas – include the effects of ambient noise on children, noise as both a unifier and divider of communities, the implications of loudness as a cultural preference, and the uses of music as noise. Here you will find reference to discoveries in physics and medicine, but also to questions of ethics and aesthetics, designed to enhance awareness of personal choices and – perhaps – the negotiation of decisions and policies pertaining to community soundscapes, including the following:

- What are the conditions under which noise is not simply a nuisance, but rather – or also – a clue to conditions in a given environment and to the culture that shaped it?
- How do noisy environments affect the ability of children to learn?
- Is the trend toward use of portable acoustic devices – personal media players and smartphones – a cause of fragmented attention, an attempt to insulate personal space from noise over which we have no control by replacing it with a controlled soundscape, and/or a search for personal meaning in a depersonalized mass culture?
- What are the rights and obligations associated with community soundscapes as an acoustic commons?

Many more questions are raised in these pages; relatively few have definitive answers. Far more research and considerable creativity – as well as attention from scientists, media, and the public – will be needed. Technical and legal solutions are crucial components of successful noise control, but so is speculative thinking, around questions such as the following:

- How might soundscapes contribute to individual, social, and cultural communication and healing?
- How can the harmful effects of urban noise be mitigated, abated, or eliminated? What might our future sound like if the issues presented here are given optimal attention?

The stereotypical association of peace with quiet provides another area ready for inquiry. Noise calls attention to itself and already has a history of meticulous, if not comprehensive, study; silence has been relegated to a very quiet corner, used as a control in studies of the effects of music and noise but not deemed worthy of a lot of attention as part of the soundscape. If noise is a problem, could quiet be a solution? Medical studies on the effects of meditation and examinations of silence as a spiritual practice suggest some directions for further investigation. Others are called into focus by the legal and ethical perspectives on quiet as a human right, essential to health and civil society, and the potential applications of quiet and of conscious listening as tools for conflict resolution.

KNOWLEDGE TRANSFER:
MOVING FROM THE ARCANE TO THE FAMILIAR

The majority of reliable information about noise as an everyday occurrence has until fairly recently been limited to specialized professional journals about civil engineering and public health. These have had limited accessibility to the public, as a result of both restricted subscription services and technical vocabularies. At this time, the Internet and social media are starting to dispense morsels of information about noise as a public nuisance and a threat to auditory health. Most of them are quick summaries of already summarized research – generally valid but lacking links to any substantive evidence. My intention here is to open the box by including perspectives drawn from the work of journalists, musicians, and social scientists along with the data-backed investigations of medical and environmental scientists, in hope of enabling wider access to evidence and thereby encouraging both professional and community activism. Since the detrimental effects of ambient noise are most severe in high-density and low-income urban districts worldwide, such activism carries an element of social justice.

The work of making technical information accessible to the public is called knowledge transfer. It involves reading a wide scope of technical literature, recognizing which elements of its content can lead to accurate and relevant general observations, and effectively translating the observations into information that will be of interest and use to people without advanced specialist degrees. It's considered by publishers to be a risky business: specialists may critique the work for being insufficiently detailed; the public may stay away from it in droves because there's too much detail and not enough narrative. In trying to find a balance, I have probably erred at times in both directions, but I hope that the benefits of the transfer will still be received and put to use.

A NOTE TO GENERAL READERS

You, the reader of this book, might be someone sensitive to noise, or simply curious about it. You might be a specialist in a field that does not study sound directly, but you suspect that you would benefit from additional knowledge about it – as an engineer, an educator, a city planner, an architect, a lawyer, a bylaw officer, a musician, a physician, a nurse, or a hospital administrator. Perhaps you are a student aspiring to one of these professions and seeking answers to questions that intrigue you. You might as easily be a parent concerned about your child's attention deficit diagnosis or your teenager's listening habits; you might equally be the teen whose ears ring for hours after attending a concert or the one whose love of wilderness hiking motivates putting the smartphone away and listening to all those birds. General readers, students, and professionals will all find useful information here; facts and perspectives that both sound alarms and build harmonies. Both theoretical analysis and pragmatic advice are offered, so listen carefully to what speaks to you and apply earplugs, at least for now, against the rest.

If you wish to follow your own explorations of the subject with this book as a base camp, or to check the validity of particularly startling assertions, references to sources and endnotes for each chapter are provided. If you wish to ignore the scholarly paraphernalia, you will not be disadvantaged by doing so, although you might occasionally miss some amusing comments and hidden gems of information in the endnotes.

The writing, which modulates between descriptive and lyrical tones according to need, is grounded in speech and shaped with the mind's ear. Listen as you read.

A NOTE TO SPECIALIST AND ACADEMIC READERS

My intention here is to bring attention to such issues as noise-induced hearing loss, medical impacts of ambient noise, effects of workplace noise on productivity, and the functions of noise in unifying, and dividing, communities. This is done through reference to a very large – but also incomplete – literature reporting the results of research undertaken primarily from the 1990s to the present. It is incomplete because new discoveries constantly follow from new developments in technology and comprehension, but also because many aspects of human perception and processing of sound are not yet fully understood. I hope that better public knowledge of the issues and the research will lead, eventually, to more research and to better methods for facing the challenges of public policy, urban and mechanical design, and health care that are implied by the literature.

The sources cited in the endnotes and compiled in the bibliography represent a small sampling of what is available from databases in numerous fields of study, including acoustic engineering, educational psychology, health sciences, communications, urban planning, anthropology, sociology, music therapy, and musicology, chosen for their relevance to the issues and questions discussed.

I have not included the vast quantities of scientific literature on the acoustic conditions of underwater environments, some of which are at this time more carefully studied than many land habitats. My intention is to focus on a series of issues about human responses to noise without removing them from their larger context, so occasional references are made to other species.

Priority for inclusion in the citations and bibliography has been given to articles available in open-source academic and scientific journals rather than proprietary ones; items from popular press and blog sources are sometimes referenced as well. Entire research histories for specific subtopics have not been included because they can be found in the sources cited here. Citations refer primarily to the most recent and comprehensive or representative articles available; thus, literature reviews were often chosen over reports of specific experiments. In some cases, a search for the most up-to-date articles during the final revision of the manuscript did not reveal anything more recent than what was already listed in a prior search a few years earlier. Research budgets being what they are, several promising areas of investigation have thus far languished in their preliminary stages. Therefore, many cited articles are to be regarded as directional signposts toward further research rather than definitive and conclusive reports.

1

Life without Earlids

The general acoustic environment of a society can be read as an indicator of social conditions which produce it and may tell us much about the trending and evolution of that society.

Schafer 1977c, 7

Noise leaves no physical mess to clear up. If it were a garish color and had a powerful stench, would it have been taken more seriously earlier?

Coates 2005, 642

Listen:
This is a guide to living with sound, and especially with noise.

Not constantly fighting battles against it, not inflicting it on others, not hiding it, or yourself, under more layers of sound. This is about being a listener, whatever else you also are.

This is about how you listen and what you hear, as well as how you hear and what you listen *to*; this is about having a dialogue with the sounds around you, which carry powers to harm and to heal. This is about the pain and the pleasure, the power and the danger, the science and the art of sound. This is about the cities, the medicine, the music, the children, and you.

Let us begin, then, by looking at listening.

Consider the photos and videos you have collected that record your life, calling forth memories and giving others a glimpse of what your experience has been. We can see the past as a series of images, whether recorded or held in memory, and the pictures can summon memories of sound. Recall that birthday party or wedding: whose cheering and chattering voices were present in the crowd? What words of comfort rang most true for you at a grandparent's funeral? What raucous and glorious call crowed out with the first breath of your firstborn child?

Pictures summon the sounds of special days and moments. We can hear the past even more directly when a recording has been taken of a cherished concert, a family celebration, or a gathering of friends; when a song evokes a vivid moment or a special person in our lives. But what about the unremarked continuity of days with ordinary schedules and repeated activities, the days when everything familiar is humming along unnoticed?

What is the everyday sound of your life? Do you seek quiet or excitement, and do you find them? Is your acoustic background birdsong, rain, a river, and wind? The giggles and screams of children at play? Traffic, mechanical hum, mechanical roar? Non-stop music? Now that you're paying attention, how do you feel about it?

Now consider whether the sounds you hear every day are noise. Can you concentrate on something else while you hear them? Are your neck and shoulders and breath relaxed and your thinking clear? Can you ignore the sounds? Are they within your control? If you answer 'no' to any of these questions, you are experiencing some degree of noise. In some cases, like the sound of the aircraft that carries you on a business trip or the truck that picks up the garbage every week, you may be grateful for its presence despite the annoyance it can produce. In others, like the construction site across the street or that furiously barking dog next door sounding off *again*, you might recall that the word *noise* shares its linguistic root with *nausea* and *nuisance*.

While noise has many definitions, the most salient one is unwanted or uncontrollable sound. In the absence of earlids[1] to block the reception of noise, we either stop listening or intentionally cover what we don't want to hear with a layer of controlled music or another electronically generated stimulus. Both reactions are illusory, however, because the sound to which we do not listen is still heard. The human auditory system – intimately linked with the nervous system, and thus with the function of organs, glands, and the immune system – processes the sounds we hear regardless of whether we consciously consider them to be worth listening to. Studies conducted on human and animal subjects confirm that hearing is active even during sleep (Nielsen-Bojlman et al. 1991; Perrin et al. 1999), and that two-thirds of dream activity may include auditory content (Velluti 2002). Thus, our hearing is alert even when we're not. This is because evolution is not a rapid process, and it is thought that humans today have essentially the same systemic responses to noise as early *homo sapiens* and even their tree-dwelling primate cousins. Surround prehistoric humans with a constant barrage of modern urban mechanical noise and they would be in a constant state of alarm, interpreting roaring machines as predators, endangered both by the unrelenting stress and by their inability to distinguish noises that signal immediate danger from the background din.

We are less susceptible than our early ancestors because we grow up with a background of mechanical roars and crashes as well as the soundtracks of broadcast urgency: we learn that they are "normal." Normal does not mean entirely safe, however. Our refined biological alarm systems can be in constant overdrive as a result of ambient urban and mechanical soundscapes, as they contribute to a host of stress-related insults to health: anxiety, muscle tension, headaches, insomnia, irritability, fatigue. Like the noise, these conditions have become "normal."

Noise has many manifestations: it can alert us, annoy us, attract or repel us. It can carry information or submerge it in a sea of distractions. It can be loud, repetitive, abrasive, startling, or just at the threshold of hearing and impossible to identify. It can stimulate, soothe, or infuriate. Noise is a form of communication: it tells us that something is in motion, vibrating and moving the air around it into waves. When the waves of air reach our ears, the delicate components of the auditory system vibrate as well, and we hear what the vibrations are telling us: thunder, traffic, siren, children, the wind.

HEARING AND LISTENING

Hearing involves both perception of sound and its interpretation. For example, we might ask:

- What is the nature of the vibration reaching my ear: how loud, fast, close, complex is it?
- How long does it last?
- What is causing it? Can I see or otherwise identify the source?
- Can the sound – or what it represents – be dangerous?
- Do I like the way it sounds? Do I want to hear more of it?
- How does it make me feel? Does it remind me of anything?
- What does it signify?

All of these aspects of perception, from the physical to the cultural, are processed by the auditory system. Much of what we hear, however, does not reach consciousness. Hearing encompasses both the conscious act of listening and the unconscious processing of sound without attention. In the former category are such acoustic phenomena as speech and music, which carry meaning. In the latter category is the ambient noise of your daily existence, the natural and mechanized soundscapes that surround you.

We listen actively to a minority of what we hear. Hearing is a physical process, set in motion by sound waves striking the resonant membrane of the eardrum, named for its resemblance to an instrument of percussion. It happens whether or not attention is paid, although the extent to which the hearer is aware of what has been heard depends partly on physical properties – acuity of hearing, proximity to the sound source, and amount of background noise and other distractions – and partly on such subtle factors as degree of attention, attitude toward the source of the sound, and its social context. Attention is paid particularly when sound is loud, intrusive, abrasive, or attractive. Advertisers on television and radio are well aware of this, knowing that the commercial that annoys you will still cause you to remember their product. Sound will also grab your attention when it carries emotional content, which is why music can have such a strong impact. Recite the lyrics of a favourite song and you might not be impressed by them; restore the music and they will carry you to a place where every syllable can take on the resonance of profound truth.

"Tuning out," like listening, can occur with or without intention when a sound is repetitive, overly familiar, mildly annoying, or associated with unwelcome messages. Classic examples come from communication patterns within families: the child so absorbed in play that a parent's call goes unattended, the spouses whose expectations of familiar content in conversation mask the actual exchange of information. Habitual patterns of activity reduce attention, as well: if you are a frequent flier, you probably listened closely to your first and second presentations of aircraft safety features, but not to the forty-seventh repetition. Even more than the soundscapes of human interaction, machines in the home or workplace create a constant dull roar that discourages auditory attention both by fatiguing nerves and by establishing a familiar background to repetitive activity. Maintaining an alert state in such conditions, actively scanning the repetitive mechanical hums for significance, would quickly exhaust the listener. Mechanical sound patterns become significant when they change or cease, indicating that attention must be paid to a malfunction, but as long as they are constant they are usually heard without awareness.

Because communication as well as survival depends on listening, distinguishing meaningful sound from ambient noise is deeply important. What is heard without attention may register physically but produce neither reaction nor retrievable memory. Listening, by contrast, is defined by attention. It is an act of respect, a recognition that important information will result from the sound being heard. Whether the sound source is speech, music, or noise, conscious listening brings it to the foreground of awareness and allows us to derive both meaning and memory from it. Along with visual observation, listening is essential to acculturation. All human cultures, as well as other mammals and birds, teach their offspring through sound, whether their communication is conscious speech or instinctive call. Human infants practise an enormous range of vocal sounds before they are able to speak; squealing and gurgling, chortling and cooing, roaring and grunting, playing with vocal capabilities by imitating the rhythms and shapes of actual speech. The range is gradually reduced to match the sounds their parents and siblings make, because language in its earliest form grows from the set of sounds that merit attention from caregivers. Since communication is not limited to speech, however, small children also imitate animals, machines, and anything else they hear: noise is an important part of their world and becomes a means for understanding it. Noise is important, but it can also endanger health in ways that are both obvious and subtle.

THE OTHER AIR POLLUTION

The World Health Organization (WHO), founded in 1948 as a branch of the United Nations, lists noise among its responsibilities. Along with climate change, air pollution, unsafe water, parasites, and epidemics, noise is considered a global health issue. Major WHO surveys conducted in the 1990s found that the populations of "developing" countries endured the greatest risks of stress-related reactions and damage to hearing because of traffic noise. Urban street corner noise of 100 decibels,[2] loud enough to cause hearing loss in hours or minutes, depending on proximity, was recorded in major cities of India, Pakistan, Argentina, and Brazil (Schwela 2001); shopkeepers, rickshaw drivers, and traffic police were particularly prone to hearing loss as a result. Residents exposed to these levels would be at somewhat less risk for hearing loss, but greater risk for noise annoyance, chronic stress, and high blood pressure.

Nor is the "developed" world immune. According to European Environment Agency (EEA) statistics published in 2014 and referring to 2012, at least 130 million people in Europe – where the most thorough investigations have been conducted – were regularly exposed to traffic noise above 55 decibels, the WHO guidelines threshold for continuous day and night exposure (Berglund, Lindvall, and Schwela 1999). This is believed to contribute to 1,629 heart attacks per year in Germany, the first European country to be studied in such detail (Bartz 2017). A major pan-European study of the effects of urban noise levels on health (Haralabidis et al. 2008; Jarup et al. 2008) even found a correlation between urban ambient noise and respiratory illnesses, a finding also suggested by a separate Spanish study of hospital admissions (Linares et al. 2006). Is the correlation of respiratory illness with noise a coincidence, because cities with dense traffic also have polluted air? Or could the illnesses be connected with prolonged stress and its effects on the immune system?

We can finally supply definitive answers to such questions: evidence gathered over nearly three decades shows clearly that noise causes stress and stress causes illness, and that the world is getting noisier. Rising urbanization is the principal cause; globally, the urban population is now a majority for the first time in history. Another contributor is the voracious appetite of growing middle classes in the developed world for new mechanical devices designed for convenience in home maintenance and recreation. Power mowers, leaf blowers, weed whackers, motorcycles, snowmobiles, powerboats, jet skis, and all-terrain vehicles, products of twentieth-century Western ingenuity, have brought industrial-strength noise to burgeoning suburbs and shrinking wilderness areas. Car stereos hit the market in the 1970s, followed by boom boxes in the '80s and car alarms in the '90s. Ambient noise is now reaching consistently harmful levels in large cities: traffic in New York, the noisiest of North American cities, routinely produces levels between 70 and 85 dBA, in the range considered potentially hazardous with long or frequent exposure. Even restaurants now produce background noise from machinery, music and conversations that will damage the hearing of staff and potentially that of regular customers, with levels in Los Angeles recently measured at 80 to 90 dBA.[3]

Worldwide increases in urban density and in the volume of road and air traffic, the spread of manufacturing economies to formerly agricultural areas, and the mechanization of agriculture itself, as well as rising demand for heavily amplified entertainment, all contribute to a growing problem that impacts "First" and "Third" World nations differently. Where ambient urban noise is regulated by laws, and the laws are actually enforced, control is possible as long as financial resources are available to solve problems associated with infrastructure like roads, rail lines, airports, and factories. Where resources are stretched and laws unformulated or difficult to enforce, sonic air pollution has stronger effects. The everyday noise levels of North American and European cities, regulated by enforced law, can be a full 20 decibels lower than those in cities in South and East Asia where such laws are still being planned or formulated. In some cases, the latter report levels high enough to gradually damage the hearing of residents living near major roads and factories simply through daily exposure (Singh and Davar 2004; Chauhan 2008; Al-Mutairi, Al-Attar, and Al-Rukaibi 2011).[4]

While public knowledge about risk is growing thanks to websites developed by health-promotion agencies and noise-abatement firms, as well as occasional mentions in mainstream news sources,[5] noise is not yet given a prominent place on the list of common risks to health. Smoking, substance abuse, air and water pollution, and processed foods laden with salt, fat, sugar, and chemicals all rank higher as risk factors and their associated risks are better reported. Noise is so much a part of common experience – and has, until recently, been so routinely ignored – that the WHO guidelines seem at first reading to be an overreaction. This is because our species' ability to ignore sound even when surrounded by it leads us to assume that it isn't worth noticing, and that no harm can result unless the impact is very loud and sudden. As numerous studies cited in subsequent chapters will show, this is not the case.

Noise is a public health issue. Hearing impairment is the most studied effect, because the damage is both measurable and easily attributed to the acoustic environment, but less obvious relationships between noise and health are also significant. Sleep disturbance, chronic stress conditions, depression, anxiety, and medical conditions involving the heart and circulatory system can be caused or aggravated by exposure to excessive noise and even by prolonged exposure to levels that are not excessively loud. The effects may be blatant or subtle, immediate or slow to develop, but they are measurable and well documented. They are not specific to any particular class, age, gender, ethnicity, or cultural group: noise is a universal cause and amplifier of stress, which affects physical, mental, and social functions.

Noise affects productivity in learning and work. When classrooms are loud, attention, reading development, problem-solving skills, and memory are affected. When noise presents a distraction or covers the sound of verbal instructions in the workplace, errors and accidents result. Noise affects mood as well as productivity for both physical and mental tasks: noise is an occupational health issue, whether the workplace is a factory or a classroom.

Noise affects economies, too. Studies have been conducted on the costs of noise-induced hearing loss (NIHL). It's official: NIHL cases present the greatest economic burden in workplace loss of hours and cost to employers in North America.[6] The majority of cases of NIHL occur in low-income countries,[7] but economic prosperity does not insure against them. An Australian study estimates that, in that country, noise is responsible for 37 per cent of hearing loss cases from all causes in adults (Kurmis and Apps 2007). A 2008 article in the *New Zealand Medical Journal* reveals that new cases of NIHL reported to the government's Accident Compensation Corporation doubled between 1995 and 2006, resulting in cost increases of 20 per cent per year. The vast majority of claimants were middle-aged males employed in agriculture, fisheries, or industries.

Noise affects individuals, but also families and communities. The WHO recognizes "environmental, residential [and] domestic noise" as significant contributors to social stress and conflict. Noise is a potential trigger for aggression, and thus for disputes within families, neighbourhoods, and entire communities (Dzhambov and Dimitrova 2014), as well as feelings of helplessness in children (Maxwell and Evans 2000). Traffic volume, airports and flight paths, construction sites, playgrounds and sports fields, restaurants, clubs, bars, parks, beaches, and public festivals can all polarize opinion about acceptable levels of noise. Interruption of quiet leisure activities – conversation, reading, listening to music or media, taking naps – is a typical component of noise annoyance, and when outdoor noise overwhelms indoor activity, tempers can flare. Quality of life issues are involved. So are political decisions, measurement standards and processes, and laws.

Noise is an issue that can force communities to think of themselves as communities, with individual preferences in music and recreation contextualized within a framework of collective cost, benefit, and compromise. Since one individual's entertainment can be another's idea of auditory torture, disagreements about rights and freedoms occur within households and neighbourhoods. The key to successful resolution of complaints and disputes is public awareness of the issues and of the options for controlling noise. The key to developing more humane soundscapes in our cities and transportation routes is the combination of expert knowledge and public will that results from such awareness.

In view of the ubiquitous impacts of sound – including speech and music as well as noise – on health, communities, communication, and everyday life, several questions are worth raising:

- What do we know about human responses to sound?
- What are the influences of nature, culture, physiology, and psychology on our responses?
- Why are some sounds pleasing while others annoy us?
- How can annoyance and physical harms be reduced?

Current knowledge is incomplete but growing rapidly as scholars and scientists investigate a broad range of issues about natural and mechanical sources of sound, physical and emotional responses, and cultural attitudes. The first step in such investigations is understanding the physical properties of sound and the anatomy of the human auditory system, which is the subject of chapter 2.

Sound Science

Noise:
1a. Sound or a sound that is loud, unpleasant, unexpected, or undesired.
1b. Sound or a sound of any kind: The only noise was the wind in the pines.
2. A loud outcry or commotion: the noise of the mob; a lot of noise over the new law.
3. Physics: A disturbance, especially a random and persistent disturbance, that obscures or reduces the clarity of a signal.
4. Computer Science: Irrelevant or meaningless data.

http://www.thefreedictionary.com/noise

It is important to remember … that ears do not "get used" to loud noise. As the League for the Hard of Hearing notes – they "get deaf."

Goines and Hagler 2007, 290

The process of hearing begins with the disturbance of air: clapping your hands will produce the effect quite well. As your hands move toward each other, air molecules are compressed between them. When the hands strike together, the molecules are set in motion, forming a wave that moves outward from the place of impact by bumping against the next layer of molecules and the next and the next after that, like the circular ripples formed by a pebble thrown into a pond. But if there is no receiver, there is no sound: the waves of air dissipate into silence. If an ear – human or otherwise – is present within the radius of the sound wave, that wave is perceived as sound. Thus, the putative Zen Buddhist riddle – *If a tree falls in the forest when no one is there, does it make a sound?* – is actually a quirky question in physics as well as a goad for moving beyond anthropocentric thinking. The tree certainly displaces air while falling. If no human is present, is there a receiver? A forest is densely populated: all animals and birds, and some insects, perceive and respond to sound.[1] Thus, sound has a dual identity: as acoustic wave in the realm of physics and auditory sensation in the realm of physiology.

WHAT WE HEAR
(EVEN IF WE'RE NOT LISTENING)

Because sound begins with the transmission of a signal, let us begin with the acoustic properties common to all sounds. What we define as sound falls into particular spectra of loudness and rates of vibration, but our perception is further influenced by factors external to the signal itself. These include how far away it is, how insistent (a matter of frequency components as well as loudness), what is producing it, whether it is familiar, what other sounds accompany it, and whether we are paying attention to it. Each aspect of a particular sound interacts with the others in complex and sometimes unexpected ways, but all hearing begins with the acoustical properties of all sounds.

Loudness/Volume and Sound Intensity/Pressure

The first quality that you are likely to notice about sound is how loud it is. *Loudness*, also called *volume* in the vocabulary of music and the recording and broadcasting industries, is actually the amount of pressure exerted by a sound wave on air, its consequent impact on the eardrum, and its subsequent conversion to signals that activate the auditory system and the brain. If the pressure is extremely low, the sound may be hard to hear distinctly unless its ambient background is very quiet. If the pressure is high, it will grab attention. If extremely high, the pressure can produce pain and damage the delicate nerve cells of the inner ear.

Sound intensity or *pressure* – which creates what we perceive as loudness – is measured by a sound level meter and represented by the decibel system, which has two scales of measurement: *A-weighted* (dBA) and *C-weighted* (dBC).[2] The A-weighted scale is most commonly used, because it is calibrated to approximate the range of human hearing and adjusted for sensitivity to the higher frequencies of 500–10,000 Hertz.[3] Decibel readings are given as a number and usually followed by the abbreviation dBA or dB(A). The notation *L aeq*, sometimes including a range of numbers, often follows the decibel abbreviation. It indicates a measurement of sound energy averaged, or equalized, over a specified period of time and converted to the decibel scale; the number averages the "peaks" and "valleys" with any concurrent steady tones. Decibel ratings increase logarithmically rather than numeri-

Table 2.1
Average decibel readings, selected activities/settings

Sound	dBA
Whisper	20
Quiet bedroom	30
Quiet conversation	40
Average office noise	50–60
Residential traffic	60
Shopping mall food court: lunch hour	70–75
Busy street traffic	75
Smoke detector, power lawn mower, train	80–90
Rock concert	90–120
Snowmobile, motorcycle, leaf blower	100
Jet engine takeoff	120–155
Threshold of hearing damage	*82–90 (depends on length of exposure)*
Threshold of pain	*90–140 (depends on individual sensitivity)*
Threshold of immediate damage	*140 (no safe exposure)*

cally. This sometimes causes confusion for the general public, and even for engineers who are not specialists in acoustics. An increase of 3 dBA represents a doubling of sound intensity, not just three units of increase, although the ear unassisted by technology is likely to perceive a doubling of intensity only when it reaches a difference of 10 dBA. Because high frequencies are perceived as louder than low ones, the decibel scale also adjusts mathematically for frequency differences. Octave band analysis, which accounts for frequency as a component of loudness, may also be used to refine measurements designed to determine the potential for annoyance.

The measurements in table 2.1 show how the A-weighted system is applied to measurement of sound intensity, perceived as loudness. Such estimates of loudness, along with actual readings of sound energy, are often used as the basis for municipal noise bylaws, but measuring sound for legal purposes is not as exact a science as it might appear. Since the equipment and software used in measurement systems are constantly being improved, recent measurements may be more accurate than ones taken five years ago or more. Also, more than one mathematical scale can be used to interpret readings, resulting in controversies within the noise abatement industry.[4] In addition, there is no mathematically precise way to measure the *perceived* loudness of a sound – as distinct from its actual level of sound intensity – since human hearing and sensitivity to sound vary from one individual to the next.

The numbers given in table 2.1 are intended to provide estimates for average noise emission by common activities and machines, but actual loudness readings will vary with the age and model of each machine measured and with the hearer's proximity to it, as well as the equipment used for measuring. Annoyance levels will vary with the person listening, and can relate to their hearing acuity and sensitivity, their mood and personality, their physiology and susceptibility to illness, their social habits and stress level, the time of day, the weather, and even whether the listener has had any training in music.

Frequency/*Pitch*

Another component of sound that affects its ability to cause annoyance is frequency, a measure of how high- or low-pitched the sound is. The designations *high* and *low*, as well as the term *pitch*, come from the vocabulary of music, and have been in traditional use since the nineteenth century, but they are subjective terms and can be somewhat misleading. In physical terms, frequency is actually a measure of how fast the sound wave is vibrating, as measured in Hertz (Hz) or cycles per second. Thus, what musicians call a "high" or "low" note is actually a "fast" or "slow" one: the faster the rate of vibration, the higher the frequency of the note and the higher the pitch.

Outside of an acoustics lab or an audiology clinic, sounds never contain just a single frequency. All natural, musical, and mechanical sounds are composed of a spectrum of frequencies that influence each other's vibratory patterns, giving the sound its characteristic quality. Annoyance related to frequency can be caused by extremes – the high-pitched whine of a buzz saw or the low-frequency thudding of pile-driving equipment, for example. Human hearing perceives the frequency range of speech, between 1 and 4 kilohertz (kHz; 1000–4000 Hz), with the greatest sensitivity. Mechanical noise emitting frequencies in this range is most likely to cause annoyance, and noise-induced hearing loss typically occurs first in this range. Disturbance based on frequency may also be caused by particular combinations of secondary frequencies – called *partials* or *overtones* – that set up discordant vibratory wave patterns, or even by unpleasant associations with the sound. Dentists' drills are a prime example of all three: they contain high-pitched frequencies, discordant wave patterns, and unpleasant associations. Frequency is not usually mentioned by existing laws on noise annoyance, which concentrate on loudness.

Proximity

Measurements of sound pressure and its effects depend on proximity to the source of the sound, and this is another way in which accurate readings taken in relation to a noise complaint can be problematic. If you are bothered by the sound of the leaf blower four houses down the block and the people living next door to it are not, your noise annoyance complaint is less likely than the next-door neighbours' tolerance to carry weight with noise bylaw officers. What might influence whether or not you complain in the first place are such factors as the time of day, your hearing acuity and that of your neighbours, wind velocity and direction, terrain, the building materials in your neighbourhood (i.e., the degree to which they reflect and amplify the offending sound), and even your degree of acquaintance with the offending neighbour. In general, noise is likely to be more annoying the closer you are to its source, but geographic and weather conditions can complicate the science because sound waves do not move in straight lines. They can bounce off surfaces and ricochet across streets, getting amplified by hills and buildings and swallowed by valleys and vegetation. Conditions indoors also depend on proximity to the source: take, for example, the steady roar of the ventilation system in your workplace. If it forms a steady distant background to other sounds, you probably don't notice it, but if the vent is next to your workstation, the noise may be enough to interfere with your ability to concentrate on your work.

Duration

How long a sound lasts, and how long you stay close to its source, can determine whether it affects you. Some people become accustomed quickly to continuous sounds and are able to ignore them; others do not. Sleep can be disrupted by sounds relatively low on the decibel scale: the hum of a furnace or air conditioner, a distant train, a partner's gentle snoring. In physiological terms, a loud, random, non-continuous sound is the most alarming to the nervous system, and the hardest to ignore. This principle forms the basis for the acoustic design of many alarm clocks and warning sirens, which broadcast a loud discordant signal with regular brief interruptions so that each new burst of sound reactivates the basilar membrane of the inner ear – translator of sound signals to the brain – and startles the listener into paying attention.

Masking and Fidelity

Continuous noise, even at low levels, can cause a problem when it blocks reception of desired and useful sounds. If your child attends school in a building with loud ventilation, the teacher's voice might not be consistently audible. Multiple conversations and music in a crowded restaurant may keep you from hearing what the people at your table are saying. Of course, this principle applies for high levels of loudness as well: if you drive with the car stereo turned up high, you may not hear the ambulance behind you, and thus may ignore it for precious seconds until your attention shifts from the music to the siren and your brain interprets the siren as a signal to pull over. This phenomenon is called *masking*, and is related to the notion of *fidelity* in soundscapes. *Fidelity* is a descriptive categorization, not a precise measurement, of how many competing sounds are present at a given time and place. It is high when few sounds are present and each is easily and clearly audible, or when one sound is noticeably louder than others. It is low when so many sounds at similar levels of loudness are present that none can be easily distinguished, or when loudness levels of informative signals are too low to permit them to be heard easily over ambient mechanical noises. When masking is present, fidelity declines. This is an issue in architectural acoustics for educational institutions, theatres, and concert halls, and also in workplaces where worker productivity and safety require attention to auditory signals.

Fidelity of the outdoor soundscape is also a factor in determining noise annoyance. Your gas-powered lawnmower or leaf blower is likely to bother your neighbours if you run it against a high-fidelity background of gentle breezes and birdsong because it will be noticed immediately by its contrast to these ambient sounds. If you turn it on next to a noisy construction site, it will be less noticeable as a distinct sound because the soundscape is already loud.

Having examined the components and effects of sound transmission – the noise and how we react to it – we can move to the system for reception, with the following questions. What is the process of hearing? How does it affect our reactions? And do we all hear the same things in the same way?

WHAT IS LISTENING, WHAT IS HEARD?

Hearing is central to our lives and our sense of well-being. What we hear, whether we listen or not, affects us. Do we notice when hearing happens, or is it an unconscious act? And what, exactly, makes hearing happen?

We might remember school biology courses that taught the basic anatomy of the ear. We remember that we have ear drums, and inside our heads somewhere there is an auditory nerve that conducts messages to the brain, but how is hearing itself carried out? First, remember that listening and hearing are two different activities. Listening is an active participatory function, or at least it should be. Hearing is largely a passive activity, essentially a series of mechanical steps taken by an entire system to give us fidelity of signal transmission from the external environment to a body for resonance reception. The signal is then transmitted to and interpreted by the brain, and only then does it make sense to us. Sound waves are received passively; understanding what they mean requires attention and interpretation.

Consider fish for a moment. They receive vibrations through their bodies from persistent wave action. They swim guided by the vibrations and by dislocation of the water along their lateral lines. Fish have refined neural receptive fields that approximate sound wave reception, but instead they interpret vibration as tactile signals to guide their swimming responses. This mechanical vibration system converts the information received into a behavioural response that signals the fish to swim toward its familiar habitat, companions, and food sources, or away from danger. The human auditory system, which perceives vibration within its audible spectrum as sound, functions in similar ways.

HOW WE HEAR (AND HOW WE DON'T)

Hearing acuity varies from person to person; thus, there is a subjective element to the perception of any sound because of subtle differences in listeners' experiences. Those differences are dependent on culture, past experience, and individual preferences, but physiology must also be added to the list: we all hear things a little bit differently because our ears, auditory pathways, and brains show tiny physical differences. Ear canals vary in length, shape, and diameter from person to person and even – very slightly – between left and right ears. They also vary considerably from species to species. Dogs and cats hear quite differently from humans because of differences in outer-ear construction as well as brain function: their auditory systems evolved to focus on the high-pitched calls of rodent prey, far above the frequency range of human hearing.

Like a mechanical sound system, the human auditory system consists of units for receiving and amplifying sound. The visible part of the ear, the *pinna*, acts as a funnel to direct sound waves into the ear canal. Its convo-

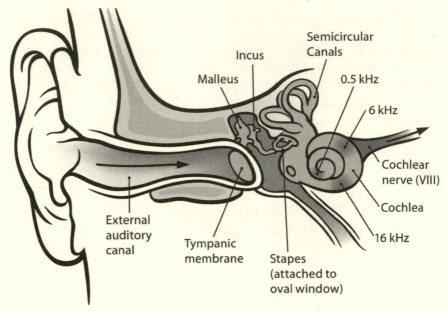

2.1 The human ear and auditory canal

luted shape in humans and other primates maximizes the collection of sound from a stationary position; many other mammals are able to swivel their ears, but we turn our heads to increase the reception of sound, and our brains compensate for the difference in reception time between the two ears. Sound waves are also carried by the bones of the face, which differ considerably in size, shape, and thickness within our species. These differences cause subtle amplification of some sound frequencies and damping of others, influencing the tones of our own speech and singing as well as our auditory perception of the vocalizations of others.

From the pinna, the sound waves move into the ear canal located within the skull and strike the flexible *tympanic membrane* (the eardrum), causing it to vibrate (see figure 2.1). Attached to the membrane is the first of three tiny adjoined bones, the *ossicles*: the *malleus* (hammer), the *incus* (anvil), and the *stapes* (stirrup), so called because of their shapes. The ossicles are essentially amplifiers: wave pressure increases with the action of each, and is regulated by tiny muscles in the middle ear. The last of the three, the stapes, has a flat terminal surface that regulates the entrance to the inner ear, a membrane-covered opening called the *oval window*. The stapes ends in a flat little plate of bone; its motion, rocking against the window, transmits the airborne vibration of the sound wave to the *peri-lymph*, a fluid medium that fills the inner ear.

Your inner ear changes the many incoming timbres (sound colours) and amplitudes (volume levels) from vibrational pressure to neural signals, so that your brain can interpret a bewildering array of sounds striking your eardrums every millisecond. This happens by means of the spiral-shaped *cochlea*, the active organ of the inner ear and translator of auditory signals to the brain. Inside its spiral, the basilar membrane divides the organ into two elongated chambers. A duct running along one of them, the *organ of Corti*,[5] is lined with layers of microscopic columnar *hair cells*, so called because they resemble tiny hairs in shape, although they are far smaller. Once inside the cochlea, sound waves move the hair cells, waving them back and forth at different angles. These inner ear cells are tremendously sensitive to a wide variety of timbres and a broad range of amplitudes, from the faintest to the loudest. Each hair cell is fragile, especially at its tip, where specialized proteins link with the tips of other nearby hair cells, binding them together into bundles. The bundles wave very slightly, bending with the pressure of each sound wave sent to them from the ossicles. When the hair cells move, they transform their mechanical energy into neural energy, sending the sound signal on to the next important leg of its journey – the auditory nerve. But in order for the signal to travel along the pathway from the auditory nerve inside your brainstem all the way into the brain with perfect fidelity, each of those hair cell bundles must work properly.

Now imagine the inner ear and its 15,000 to 20,000 cochlear hair cells lining the basilar membrane, all linked together in bundled small groups. You need every one of them because each bundle bends at a different frequency: high sounds bend cells on one end of the membrane and low sounds bend cells on the other. The loss of even one of these bundles means a diminution of reliable hearing. Temporary hearing loss occurs when hair cell bundles are bent excessively or flattened by the rush of extreme sound pressure in sustained or sudden loud sounds. Sound at high volume will often flatten cells at the frequencies your inner ear is trying to process and will damage the delicate connecting synapses that enable the cells to transmit information (Liberman et al. 2016).[6] When this occurs, it can take time for the "hairs" to straighten back to a normal receptive position. Although the tips can regenerate after a period of quiet time, usually 24–48 hours (Oishi and Schacht 2011),[7] when exposure to loud sound is prolonged, frequently repeated, or extreme, the bundled hair cells will remain flattened and can never recover (see figure 2.2). This

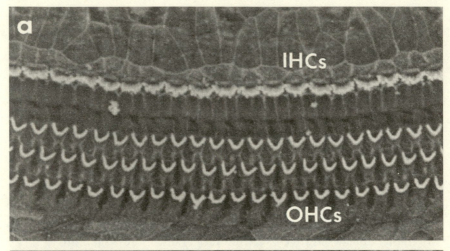

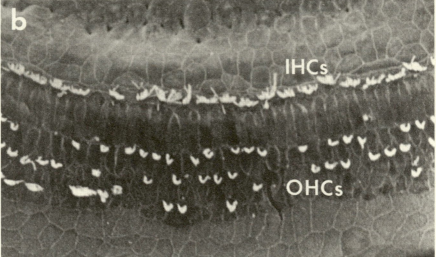

2.2 Damaged cochlear hair cell groups (b)

condition is referred to as noise-induced hearing loss (NIHL), which affects not only the rows of hair cells but the surrounding tissue. All the structures in the inner ear are affected. Such deformities happen even in the slightest circumstances of partial hearing loss. As tips flatten, fray, or break, the result is permanent loss for the relevant frequencies, often those falling within the range of speech. Although some fish, amphibians, and birds can regenerate damaged hair cells, mammals – including humans – cannot do so (Stone and Cotanche 2007). Currently, researchers have made some progress in attempting to regenerate hair cells in animal experiments (Ryan 2000), but there is as yet no miraculous cure to reverse hearing loss caused by hair cell damage.

THE MANY CAUSES – AND EFFECTS – OF HEARING LOSS

Human hearing, when measured in optimum conditions – for example, a child living in a quiet rural area – perceives a range of frequencies from approximately 20 to 20,000 Hz, and a range of loudness measuring upward from one decibel. As we age, the spectrum for both narrows gradually, with perception of the highest frequencies and the lowest levels of loudness declining first, owing to the stiffening of inner ear membranes; this type of age-related hearing loss is known as *presbycusis*. Eventually, with advanced age, hair cells along these membranes die off; this is *sensorineural hearing loss*. Recently, the Hearing Health Foundation in the United States estimated that hearing loss affects over 477 million people worldwide (Hearing Health Foundation 2019). According to the National Organization for Hearing Research Foundation, 80 per cent of all cases are sensorineural, and those affected "have enough sensorineural hearing loss to affect their ability to communicate. This form of hearing loss is particularly devastating because it affects the ability to discriminate between sounds and understand complex combinations of sounds, including speech."[8]

The underlying mechanisms behind such hearing loss are still not well understood. While it is certain that the accumulation of "low-level damage due to noise and other insults" (Gates and Mills 2005; Wong and Allen 2015) is a contributor to age-related hearing loss, such loss has also been found in lab animals who have spent their lives in silence (Sergeyenko et al. 2013). Aging affects hearing acuity to some extent in everyone, regardless of their experience with acoustic environments, but exposure to noise makes it happen far earlier and to a much greater extent. This latter finding is crucial, because hearing loss may be a precursor to even more serious conditions. One of them is semantic dementia, which involves losing the ability to understand the meaning of words: neuroanatomical changes are broadly shared in both conditions (Mahoney et al. 2011). A study at Johns Hopkins University also found that seniors who were diagnosed with significant age-related hearing loss showed an increased tendency to develop Alzheimer's disease and other forms of dementia (Lin et al. 2014), due to shrinkage of brain tissue. Patients with impaired hearing, whether age-related or NIHL, have sometimes been found to have "lost more than an additional cubic centimetre of brain tissue each year compared to those with normal hearing" (Lin et al. 2014). This suggests that treating hearing loss as soon as possible might have the potential to delay or even mitigate other potentially damaging consequences to the brain.

Damage to cochlear hair cells can also be caused by factors other than noise or aging. High fevers can induce inflammation at sufficient levels to damage hearing; they can even burn the hair cells, causing total deafness, sometimes only in one ear. Certain drugs and medicines also have been known to reduce hearing, as have anesthetics; the effects may be temporary in some patients but permanent in others.[9]

Along with flattening, stiffening, or inflammation of the cochlear hair cells, loss of fidelity in hearing can be related to inflammatory diseases of the ossicles, such as otitis media. Long-term problems resulting from ear infections can develop in the middle ear, often when fluid is persistently present: children are especially susceptible to this condition. If left untreated, it can result in permanent damage. Even if the damage is not permanent, perceived noise can be very disturbing to sufferers well into adulthood. Noise can also be difficult for children who have particularly sensitive hearing, as the smaller construction of the passageways of the auditory tract increase the capacity for amplification of sound.

If there is insufficient blood supply to the middle ear, the patient can become deaf and the auditory nerves can atrophy, as happened to the composer Ludwig van Beethoven (Mai 2006). Had Beethoven been alive today, much of his hearing loss could have been treated, although probably not perfectly or perhaps to his satisfaction (Traynor 2011).[10] Hearing aids, in spite of their many and remarkable improvements, still cannot provide complete compensation and often are effective only in certain acoustic conditions. Hearing aids do not cure damaged hair cells, but they do provide relief for many patients by amplifying the sound environment. Unfortunately, this can make hearing more difficult in some circumstances because they amplify every sound: imagine taking a book to a coffee shop and hearing every crackling page-turn as well as the amplified chatter of other customers and the growl of the espresso machine. Adaptation to the devices depends on the individual, the type of hearing loss they experience, and the devices' specifications.

Hearing damage is widely under-reported, largely because its management with visible devices carries the stigma of disability. Many who experience it deny its effects and learn strategies to hide their condition from employers, co-workers, family, and friends. In doing so, they may endanger their jobs through loss of productivity and themselves through accidents. A 2009 survey of auto insurance records in Quebec, focusing on 46,000 male industrial workers exposed consistently to workplace noise over 100 dBA, showed that their risk of serious collisions increased by 31 per cent if their hearing had been impaired by noise. Even those whose hearing was not measurably impaired by their exposure to occupational noise were at 6 per cent greater risk of major car accidents than the general population (Picard et al. 2008), possibly as a result of fatigue. As well, workplace accidents were more likely to happen to workers with NIHL, and risk was particularly high among young workers.

DO I HAVE NIHL? HOW CAN I TELL?

One of the first indications of hearing loss, whether noise-induced or age-related, may be increased difficulty in hearing conversation, especially in noisy and crowded circumstances. NIHL is not the only possible cause: room acoustics, as well as the frequency ranges and resonances of individual voices, can influence perception. The need to ask people to repeat themselves can be caused by *signal masking* – the result of a noisy and reverberant room blocking your ability to focus on a person's voice – rather than hearing loss. However, when background noise interference with conversations is consistently present in different acoustic conditions and gets worse with time, it is usually a sign of NIHL (Bess and Humes 2008; Oishi and Schacht 2011). If you have been exposed to frequent periods of loud noise and have not taken measures to protect your hearing, you may be suffering from NIHL. Given that approximately 10 per cent of the world's population deals with hearing loss, and that about half of them have NIHL, you are in good company (Daniel 2007; Oishi and Schacht 2011). Unfortunately, because the early stages of NIHL are often mild (15–20 dBA reduction), the condition can go unnoticed in everyday life until it becomes disabling.

Even short-term exposure to loud noise, with twenty-four to forty-eight hours' recovery time, has been shown in experiments with mice to accelerate age-rated hearing loss. In other words, NIHL intensifies the effect of aging on your ears even after they recover from transient exposure to prolonged loud noise (Humes 2005). Even more alarming is that, through exposing mice to noise starting at a very young age, researchers have demonstrated that NIHL is cumulative over time. The implications of this for youth using personal music players (PMPs) are enormous. As NIHL researchers Naoki Oishi and Jochen Schacht (2011, 236) state bluntly, "the self-inflicted recreational noise damage of the current generation might exacerbate age-related hearing loss and diminish quality of life in the future."

The best way to determine whether your hearing has diminished is to visit an audiology clinic. Whatever your age, knowing whether your hearing is compromised will enable you to prevent additional damage by turning the volume down on your audio devices and using earplugs in extremely loud surroundings.

TINNITUS

For many people on the threshold of NIHL, the first meaningful warning is a mild and intermittent form of tinnitus, the most common and least severe form of damage to hearing caused by noise. Tinnitus is medically defined as a sensation of high-pitched ringing in one or both ears when no actual sound is present, although it can also manifest as whining, buzzing, clicking, hissing, or whistling sensations. It results from over-stimulation of auditory nerves, which consequently fire in the absence of actual sound waves. You may have experienced it temporarily as you walk out into a quiet area after attending a rock concert. If you work in a noisy place, it might be a part of your commute home. Tinnitus is easy to dismiss as long as it's temporary. If it becomes permanent, it can impact quality of life. In either case, it may be a warning sign for the early stages of NIHL.

The condition is often associated with depression and anxiety. While in some cases these can be a result of pre-existing emotional conditions (Zirke, Goebel, and Mazurek 2010), in others tinnitus can exacerbate these problems by interfering with sleep, communication, and the ability to experience daily enjoyment in life, all due to the unrelenting nature of the internal sound. Brief episodes of tinnitus, lasting seconds or minutes, can be caused by ear or sinus infections, allergic reactions, migraines, or blows to the head, but persistent episodes are usually a result of workplace noise, abundant enthusiasm for loud music, or a combination of the two. Eventually the condition can impair the ability to enjoy music at all.

When William Shatner and Leonard Nimoy, actors in the original television series *Star Trek*, made the episode "Arena" in 1967, about an outpost that was under attack, the special effects team set off simulated mortar explosions in close proximity to where the two were standing. Both suffered some permanent hearing loss, and Shatner was diagnosed with severe tinnitus. He reported that at one point he felt that he could no longer live with the condition. Shatner underwent a regimen of treatment called tinnitus retraining therapy, which provided some remediation and now, as a spokesperson for the American Tinnitus Association, he offers help to those who experience the same debilitating effects.[11]

In recent years, the brain has been the focus for research on the internal cause of this annoying affliction. Feedback from the brain's auditory cortex through pathways that reach all the way to and from the inner ear are said to modulate the circuit pathway for hearing. Research has suggested that the problem with tinnitus lies in the auditory pathway's own "dysregulating" of the incoming sound signal, leaving behind residual noise that results in a whining, high-pitched, sustained tone (Leaver et al. 2011 and 2012; Eggermont 2013). The brain itself may undergo change when tinnitus is present: the hippocampus can show anatomical change due to the stress and depression associated with it (Salvi et al. 2011), although it is not yet known whether the depression or the tinnitus is directly causative of the change. Not surprisingly, another effect is sleep loss. In large patient studies that compared the rating of sleep by tinnitus sufferers and a control group, participants with tinnitus reported a higher level of sleep disturbance. The study found that sleep loss was not due to the whine of tinnitus; instead the research demonstrated that patients were acutely sensitive to *external* noises in the middle frequency range. A US study conducted at the Oregon Tinnitus and Hyperacusis Treatment Clinic found that treatment of insomnia was "bi-directionally related" to reducing the impact of tinnitus itself: in other words, reducing the intensity of tinnitus improved sleep while treating the insomnia made the tinnitus less disturbing (Folmer 2000; Folmer and Griest 2002).

Millions experience tinnitus daily in some form, and their numbers appear to be growing across all ages and demographics. With 70 million suffering from the condition in Europe and over 50 million in the United States, it is small wonder that the search for treatments has intensified in recent years (Zirke et al. 2013; Hall et al. 2015; Martinez et al. 2015). At one time, cognitive behavioural therapy and relaxation techniques were recommended as the only effective means for coping with its psychological effects. Today, the preferred prescription is tinnitus retraining therapy, a form of habituation therapy. The procedure is simple: the patient wears headphones for several hours each day, masking the sound in the head with low-level "white noise," a balanced blend of engineered frequencies that produces a soft neutral whooshing hiss. Natural sounds such as running water or soft music may also be used, with or without headphones. For some patients, hearing aids can provide relief. Others try a white-noise machine with pillow speakers at night, or even a fan, humidifier, or air conditioner in the bedroom to mask the internal sound.

For many, the process of living with noise includes learning to live with tinnitus. Some sufferers whose tinnitus is chronic but mild report becoming accustomed to it and adjusting their interpretation of ambient sounds, speech, and music to ignore it as much as possible. For those with more severe cases, finding a reasonably successful treatment may require experimenting with several different methods. The first step in that journey should be a visit to an audiology clinic for diagnosis, which begins with a series of tests to measure hearing acuity, speech recognition, and middle ear function, as well as identifying specific effects unique to the individual's case through computerized recordings of inner ear function. New technology even allows the audiologist to measure activity in the cochlea in order to match the internal tone and loudness of the tinnitus: finding the correct tone to match what a patient hears internally can be a moment of revelation and emotional release. Diagnosis leads to advice about treatment options and counselling on how to reduce the stress caused by the condition. Although actual cures are not available, treatments can significantly improve everyday life, speech comprehension, and perception of music.[12]

3

Studying Sound

The visual is the known – we have ways of dealing with it, talking about it and studying it. The auditory is the unknown, the unfamiliar, the new – it is the stranger knocking at the door threatening to disrupt the world.

Pinch and Bijsterveld 2004, 637

The home territory of soundscape studies will be the middle ground between science, society and the arts.

Schafer 1994, 4

Sound Studies continues to emerge as an expanding discipline involving many concentrations and discourses. From musicology to anthropology, histories of media and cultural practices, to performance and vocal studies, the range is dynamic and also highly suggestive.

Labelle 2010, xix

What is the history of human reactions to noise, and how has that history been preserved? Such questions require speculative interpretation as well as – or sometimes instead of – measurement, and are the concern of scholars in fields outside of traditional science: social history, sociology, anthropology, literary and culture studies, environmental studies, and even art history. The process of studying sound was, until the advent of electronic recording devices, a challenging one: it still is, considering the number of intricate variables presented by both source transmission and recording procedure. Even today, accurate recording requires planning and precision as well as appropriate equipment. Historians seeking information about ambient sound from the centuries before such equipment became available have to rely on scraps of evidence. Aside from poetic or epistolary mentions of birdsong, parish bells, battles, storms, merchants' calls, market-square gossip, passing carriages, and children's games, there are few direct references to sound and even fewer to noise.

Before the industrialization of the Western world, European writings about sound did not specifically consider noise. Ancient documents that mention sound are primarily medical or philosophical, concerned with descriptions of the ear's structure or with the metaphor of civic harmony. Medieval treatises related to sound focused mainly on music, considered to reflect the mathematical science of proportion and to provide a formula for cosmic harmony as well as civic and physical health: the motion of planets, systems for good government, and maintenance of balance in thought and action were all connected with the concept of harmonic relations. Their authors, concerned more with symbolic meanings of music than with its actual sound, hardly mentioned noise, which represented chaotic conditions considered unworthy of systematic study. In the writing of medieval Christian theologians, noise was the voice of demonic forces. While heaven was said to be filled with the exquisite singing of angels praising God, hell was believed to echo with roaring, screaming, and the braying of devils tormenting the damned.

Ironically, it may have been noise that led to an early theory about the acoustical properties of music. The rhythmic ringing of blacksmiths' hammers, iron striking iron, was supposed to have inspired the Greek philosopher Pythagoras to describe the ratios of musical intervals – the basis for what we now call harmony – in the sixth century BCE. The legend was repeated in texts for centuries afterward, giving primacy to the disciplined structures of music over the disordered conditions of noise. The story illustrates the dual identity of music as sound and mathematics: the vibrational rates of sounded notes in the proportions 2:1, 3:2, and 4:3 produced the intervals considered consonant (the octave, fifth, and fourth) in ancient Greece; other ratios produced dissonances.[1]

Unlike music theorists, composers of the late medieval and early modern periods sometimes considered noise to be inspiring. Imitations of ambient noise occur in written scores of music from the fourteenth century onward: bird calls, battles, hunting parties, and street vendors provided late medieval and Renaissance composers with material for complex vocal counterpoint in songs. Later works from the seventeenth through nineteenth centuries for keyboards and orchestras incorporated acoustic references to bells, clocks, songbirds, hens, military artillery, and thunder; the best-known orchestral examples today are Beethoven's Sixth Symphony (*Pastoral*) with its rainstorm and Tchaikovsky's *1812 Overture*, complete with recorded cannon-fire (provided by real cannons in Tchaikovsky's time). Still, these musical references are islands in a sea of relative pre-industrial silence about noise. Urban and industrial growth, of course, changed both ambient soundscapes and music: booming symphonies, the rise of jazz, amplified rock, and electronic pop – as well as musical imitations of trains, typewriters, and traffic – announced and incorporated the growth of modern mechanized soundscapes.

With mechanization, noise itself became a subject of attention. Urban and even rural ambient noise levels rose with the development of railways and factories in the nineteenth century. By the early twentieth century, a substantial literature of complaints existed, and scientific investigations began conveying information by and for engineers tasked with measurement and abatement. Most published information focused on selected, targeted milieus where noise tended to harm hearing, and on specific environments such as factories and airports that are consistently noisy. Investigations of the effects of noise on health followed the world wars, as soldiers returned from battlefields with damaged hearing. Social implications were given attention somewhat later by urban planners and architects, alert to the soundscapes associated with rising population density and ubiquitous machinery.

The study of sound as a multifaceted subject in itself, with connections to history, philosophy, anthropology, and culture, is now gaining a significant presence among academics and journalists thanks to the investigative methods of two primary schools of thought: acoustic ecology and sound studies. Both schools include professional scholars, sound recordists, composers, musicians, multimedia artists, architects, environmental scientists, and a host of talented amateur observers. Their work is gradually generating a broad – and often speculative – history of noise. Paradoxically, much of the information about the subject has been conveyed by visual sources.

ACOUSTIC ECOLOGY:
LOOKING FOR NOISE AND HEARING LANDSCAPES

Substantive information about ambient noise and its effects before the twentieth century comes primarily from occasional mentions in literary works – novels, essays, poetry, and letters – and from legal statutes designed to regulate offensive noise in cities. All of this information was preserved visually through writing, or through drawings and paintings of urban and workplace activities. Photography, invented in the 1820s and refined to an accessible form in 1839, pre-dated the earliest and least reliable form of sound recording by some fifty years, during which it also contributed glimpses of evidence about lost soundscapes.

Our current knowledge about historical sound is based primarily on assumptions of similarity. We "know" what nineteenth-century roosters sounded like because their living descendants, identical or closely comparable in physique and instinct, are presumed to sound the same. We "know" the sounds of nineteenth-century trains less well because their structures, materials, and speeds have changed significantly. If we have a sense of their sound, it's taken largely from films set in the time they were in use, with soundtracks based on a combination of surviving specimens in working order, reconstructions, and imagination. It was not until the 1970s, with easily portable recording and broadcasting technologies firmly established and widely available, that a specialized literature about environmental sound found its niche in academic investigations outside of formal sciences and inside the public imagination. It began in Canada, with the work of composer R. Murray Schafer and the concept of acoustic ecology.

In *The Tuning of the World*, his first book on the subject, Schafer gives a concise and lyrical account of sonorities that were altered over time by changes in technology and building materials, commenting on the actions of water mills and blacksmiths, the nocturnal attention given to hearing before the invention of electric light, the richly varied resonance underfoot of wood-plank or cobblestone streets in comparison with the "drudgery of asphalt and cement" (Schafer 1977c, 59). His earliest observations were drawn from a combination of his own experiences, the memories of community elders, and visual evidence in written documents and paintings. Schafer and his followers began by looking for noise. They went on to listen intensely to their own surroundings and create maps that documented the human, avian, mechanical, and weather activities producing the ambient sounds of specific locales. These were soundscapes, as powerfully characteristic of places as their terrains and built infrastructures, their regional cuisines, or their plant and animal species.

The west coast of Canada provided rich nourishment for the initial growth of acoustic ecology. The Vancouver Soundscape Project, launched in 1972 from Simon Fraser University by a group of graduate students and sound recordists under Schafer's direction, created brief sound files to represent Vancouver's soundscape: harbour boat traffic, cathedral bells, murmurs of water. These files pale by the standards of later technologies; new, edited recordings were taken in 1996 from the city, as well as the islands that dot the British Columbia coast. Several of them feature the recorded sounds of water, wind, birds, boats, and/or traffic integrated with composed instrumental or electronic music. This merging of art with nature, and musical structures with noise, advanced Schafer's initial plan of sensitizing composers to natural sounds.[2] At the same time, attention to soundscapes as evidence of environmental conditions began to gain traction as a subject for study, giving evidence of the density of bird and insect populations, the travel of marine mammals, and the effects of human incursion into wilderness areas. What resulted was an alliance of composers and sound artists with environmental researchers skilled in technological methods for recording and measurement.

The primary organization for adherents from both backgrounds is the World Forum for Acoustic Ecology (WFAE), founded in 1993 and now coordinating affiliate groups in North America, Europe, Mexico, Japan, and Australia. Its members conduct research, map and describe local soundscapes, advocate for the preservation of wildlife refuges, organize regional and international conferences, maintain educational websites, and teach the techniques of soundwalking.[3] The WFAE keeps a growing reference list of central literature linked to its library website,[4] and a journal, *Soundscape*, that publishes relevant articles as well as conference reports. The insistence of the acoustic ecology movement on avoiding professional hierarchies makes its work accessible to both artists and scientists while keeping the amateur and student components of its membership engaged. The goal is to make awareness of sound environments familiar to both professionals and the public in order to promote knowledge of and advocacy for auditory health and "unpolluted" natural soundscapes.

Today, acoustic ecology is a respected branch of environmental studies as well as a path to artistic expression. Its focus has changed significantly from Schafer's original desire to improve urban soundscapes by banishing mechanical noise and recorded "moozak" – his term for the commercial music industry's invasion of public spaces. While the aim of critiquing public indifference to ambient sound is still active, most practitioners today focus more closely on teaching active listening skills, evaluating urban or architectural design for its acoustic properties, investigating the effects of highway traffic noise on wildlife, or documenting natural soundscapes before they are changed by human activities. A central canon of defining literature and composed music includes works by composers and educators Barry Truax, Hildegard Westerkamp, Pauline Oliveros, and Andra McCartney. Psychoacoustician Brigitte Schulte-Fortkamp conducts pioneering research for the WHO on the effects of noise on urban populations, acoustic designer Mai-Britt Beldam investigates the efficacy of noise-reducing features in hospitals, and environmental psychologist Arline Bronzaft has published copious cautionary research about urban noise levels; all three continue to advance the scientific aspects of the field as it impacts public health. Wilderness ecologist Bernie Krause has archived decades of recordings documenting the gradual impoverishment of wild soundscape diversity. His work is now recognized as a model for research in environmental sciences.

Acoustic ecology's younger sibling, sound studies, was born in twenty-first century academia from an alliance of literary studies with history, psychology, anthropology, and sociology. It seeks to comprehend the ways in which human responses to sound interact with culture and identity, filling a middle ground between the arts and sciences that have shaped its predecessor.

SOUND STUDIES:
PHILOSOPHIES, TECHNOLOGIES, MUSIC AND MEDIA

The primary mission of sound studies is to explore what might be called the social sciences of sound: its relation to history, economics, communications, linguistics, and technological developments. Unlike the acoustic ecology movement, it is firmly established as a field of formal academic study, first practised by historians, anthropologists, and scholars of literature.

Pioneering work began well before the field was actually defined. French economist Jacques Attali published his foundational work *Bruits* (*Noise*), an examination of the political uses of music and of how governments use it along with noise and broadcasting to gain power, in 1977. It took a few decades, and the influence of the acoustic ecology movement, to sharpen a focus on what could not be seen. Several subsequent academic publications moved sound studies forward – Peter Szendy's lyrical essays in *Écoute: une histoire de nos oreilles* (*Listen: A History of Our Ears*, 2001); Jonathan Sterne's study of recording technology and its historical significance, *The Audible Past* (2003); Michael Bull and Les Back's anthol-

ogy *The Auditory Culture Reader* (2003); aural historian Emily Thompson's
The Soundscape of Modernity (2002); and Mark M. Smith's anthology *Hearing History* (2004) – and provided a broad launching pad for its wide diversity of subject matter.

Sound studies scholarship now concentrates on an expanding variety of topics, many of which merge traditional fields of study like music history and sociology with new perspectives based in critical studies of race, ethnicity, gender, disability, social class, demographics, and/or technological change.[5] Such topics are typically considered from the standpoint of "voice" – the ways in which Western culture defines what is and is not "worth listening to" – with regard to speech, protest, music and its distribution, and the determination of historical validity. Sound is often regarded as a cultural marker that represents or enhances beliefs and social movements, as the 1960s in the United States can be evoked by the protests inherent in "We Shall Overcome" or "Blowin' in the Wind," or current anxieties about climate change by the recorded sounds of hurricanes or wildfires in news reports. Other examples come from speculative reconstruction of historical soundscapes and their significance as evidence with respect to everyday life in the past: in addition to numerous film soundtracks, this category is represented by the work of scholars who mine literary works for mentions of ambient noise as well as environmental sound and music. An early example is Bruce R. Smith's *The Acoustic World of Early Modern England* (1999), which surveys sixteenth- and seventeenth-century literature and theatre – with an emphasis on Shakespeare – in pursuit of an understanding of how sound was perceived by the playwrights and actors who worked intimately with it and the audiences that received their work.

Like acoustic ecology, sound studies is by its nature an *interdiscipline*,[6] not suitable for fitting exclusively into the traditional academic silos of humanities/social sciences/sciences. Since knowledge derived from all three divisions is necessary to many aspects of the field, it is sometimes practised by teams of researchers who share their specialized expertise in order to cover a wider scope of interpretation than a single specialist could: an example might be the collaboration of a music historian with a film historian or urban sociologist. Each will notice a different set of details in the material they study and apply a different set of methods for gathering information: the merging of observations and methods will usually enrich the results. The field is not yet well established in academic programs, but some courses and a few graduate programs can be found in a small number of European and North American universities.[7] Like the literature of sound studies, programs are diverse. Most focus on either environmental or cultural studies approaches, sometimes blurring the distinctions between the two fields.

An intriguing general introduction to the broad scope of sound studies today is provided by the blog *Sounding Out!*, which includes 176 categories for studying sound, including advertising, deafness and disability studies, game theory, Indigenous studies, performance, queer studies, rural space, urban studies, voice.[8] Commentaries on event- or place-specific soundscapes, regional or cultural speech idioms, animal vocalizations, sci-fi movie soundtracks, and the ethics of broadcasting can all be found in posts dating from the summer of 2009 through the present. The blog provides a broad survey of many of the ways that sound – as noise, music, speech, or silence – influences our perceptions, experiences, and social norms, as well as how culture shapes our ability and inclination to listen.

Many of the blog's posted articles are concerned with the symbolism of "voice" as a representation of cultural respect. While some "voices" are heard, others are not, as the histories and protests of minority groups are promoted or suppressed by dominant cultures. Commentators also apply these perspectives, adapted from the field of critical studies, to music, film, and broadcasting, asking what subjects are covered, what styles are promoted, what aesthetic decisions are considered acceptable or not, what is ignored. As an example, consider an article by critical studies scholar Reina Alejandra Prado Saldivar that discusses the concept of "sonic brownface" – that is, the casting of white actors as Hispanic characters in film, recognizable as such by stereotypical speech patterns. Prado Saldivar's critique of the practice is activist, arguing that Hispanic characters have been portrayed in American films as foreigners with exaggerated accents when speaking English, despite the identity of many Hispanic people as US citizens of long standing. "Sonic brownface," she notes, is analogous to "blackface" portrayals by white actors in early twentieth-century films, although its effect is more sonic than visual (Prado Saldivar 2013).

Its adherents tend to regard sound studies as the next generation of a rapidly growing field that began with acoustic ecology and quickly surpassed its parent, encompassing new technologies in sound production, recording, and analysis – as well as the methods associated with critical studies – in order to move beyond the model advocated by Schafer and his followers: that is, a nostalgia for quieter, more natural soundscapes and a fixation with white Anglo-Canadian culture as normative, to the neglect of Indigenous and immigrant communities (Akiyama 2015). Purist practitioners of acoustic ecology also insist on maintaining a separate identity for their field because its basis is as much scientific as sociological, depending on on-site recording and measurement as well as theoretical interpretation. The implied conflict seems to be a necessary stage of growth for any new field of study differentiating itself from prior models, but it is ultimately counterproductive, largely because each group can benefit from the other's approaches.

As sibling interdisciplines, both fields are gradually opening to degrees of collaboration. Acoustic ecology is moving out of a stance exclusively opposed to noise and toward greater attention to cultural attitudes about sound. Sound studies has recently expanded its focus to include aspects of acoustic ecology, as exemplified by the adoption of soundwalking, which is designed to increase awareness of the elements that constitute environmental soundscapes, into the education of film sound designers.

Historical studies of urban culture, social values, and infrastructure are also benefiting from the methods of both fields. A recent example is provided by a study on the mapping of sound in nineteenth-century Madrid, in which musicologist and cultural historian Samuel Llano identifies parallels between districts associated with excessive noise in the city and similar mapping of unpleasant smells and social transgressions. Llano's conclusion is that mapping was, and is, far from a neutral and value-free technique: as recent alternatives to the Mercator global map projection demonstrate, it carries considerable evidence of social attitudes. The foremost of these attitudes in nineteenth-century Europe was condemnation of poverty and the aggregate poor: district mapping was used to advance social engineering that equated poverty (associated with noise, smells, dirt, and illness) with crime and "moral disease" (Llano 2018).

The analysis of historical urban noise maps alongside contemporary documents revealing social attitudes and municipal laws suggests one important new trajectory for sound studies: finding insights into past cultures by means of techniques rooted in acoustic ecology. Another is growing out of an alliance with cognitive psychology: the use of an electroencephalogram (EEG) headset to measure the responses of pedestrians to urban soundscapes and compare them with reported reactions (Thibaud 2014). The object of such studies is to provide urban planners with verifiable details about how people react to urban environments, and how to improve the process of planning.

It is likely that similar integration and expansion of acoustic ecology and sound studies will continue, with practitioners producing work relevant to both. As both fields gain further recognition, as more graduate students are trained and more enthusiasts undertake significant projects, understanding of the history of ambient sound and its implications for human culture will grow to illuminate more areas of knowledge.

4

Thunder, Rooster, Hammer, Hum

Although some occupations such as coppersmithing and blacksmithing were noisy, the number of people exposed to potentially damaging noise would have been limited to the metalworking craftsmen and apprentices within the workshops ... The noise from water-powered grinding mills was significant, but the workshops and mills were islands of noise within an otherwise agricultural environment, which meant that a limited number of people were exposed to the sounds.

Thurston 2013, section 4

Modernity, it seemed, and seems, disturbs the peace. Large factories, steam locomotives, industrial whistles and bells, then the sewing machine and the phonograph, the machine shop and the telephone, the ringing cash register and the elevated train, the automobile and the subway, the truck and the machine gun, these were hardly epiphenomenal to modernity: they were of its essence.

Schwartz 1995, 6

As the capitalist economies of the Western world fuelled a race for economic dominance in the nineteenth and twentieth centuries, new inventions for transportation, manufacturing, and communication grew to ubiquity and were supplanted by even newer ones. Noise was established as a fact of life, present in multiple-unit housing, in clerical offices, on factory floors and city streets – but how loud was it? And what of pre-industrial history? Were ancient Romans awakened by chariot traffic under their windows at night? Were medieval knights deafened by the clashing and ringing of armour and swords? Did the clatter and clop of carriages on the streets of seventeenth-century London cause residents to complain? What sounds might have produced annoyance, and was noise-induced hearing loss a problem at all? Searching history for evidence of life without earlids – or earbuds, or earplugs – raises a new series of questions about noise and how it affected people. With a combination of research and informed speculation, some of them can be answered.

Outside of occasional literary descriptions and sporadic complaints, the acoustic backgrounds of everyday life went largely unremarked until they became a focus of scientific study in the twentieth century. Mentions of noise annoyance are rare and usually concerned with urban trade and transport or with street musicians. Outside major cities, pre-industrial soundscapes in the Western world were relatively quiet by our standards. They could certainly distract, annoy, or disturb the peace, but they were not likely to damage hearing or physical health outside of close proximity to smithies, metalworking districts, or battlefields.

Ancient hunter-gatherer settlements were situated within natural soundscapes. Their inhabitants were not at risk for compromised hearing unless they were consistently located close to large waterfalls or lived long enough to develop *presbycusis*, the gradual loss of hearing acuity with advanced age (Thurston 2013).[1] Later agricultural societies in the ancient world had greater population density and a soundscape inclusive of concentrated herds of browsing animals: noise was more layered and pervasive than in smaller hunter-gatherer settlements, but not particularly loud by modern standards. Given that, for centuries, the vast majority of people lived in rural communities or on isolated farms, noise was not considered to be an environmental hazard except in the case of certain occupations exposed consistently to percussive actions or grinding machinery: primarily metalworkers, blacksmiths, millers of grain, and miners. As cannons and guns gradually replaced bows and swords as weapons of choice – between the fourteenth and seventeenth centuries in Europe – soldiers and gunship sailors joined the ranks of the occupationally hard of hearing (Thurston 2013). Compromised hearing was considered to be a part of their job, an annoyance to be accepted as a condition of employment. It was not until the Industrial Revolution, initiated by the invention and growing use of steam-powered engines, that additions could be made to the list of workers at risk: factory and textile workers who had to spend long workdays surrounded by multiple machines clattering and echoing in enclosed spaces, as well as railway workers.

What do we know about reactions to noise prior to industrialization? Written descriptions in historical law and literature suggest that urban noise caused by construction, manufacturing, and traffic has always been problematic for those living close to it, and that noise annoyance – distinct from noise-induced hearing loss – has been present for many centuries. In 720 BCE the Greek city of Sybaris enacted regulatory laws against the noise caused by industrial workshops – presumably the hammering of metal and cutting of stone and wood by blacksmiths, masons, and carpenters – as well as prohibiting urban residents from keeping roosters.[2] Ancient Rome, which prohibited delivery wagons and chariots within the city walls in the daytime because of traffic congestion, further regulated them at night, when the sounds of hooves and wheels on cobblestones kept residents awake. Nonetheless, interruption of sleep may have been a common event in parts of the city.[3] As the growth of cities through the medieval and early modern periods brought increasing numbers of people into close proximity, urban districts were zoned for quiet and noise: blacksmiths and other metalworkers had to set up shop outside of residential areas, as did carpenters and stoneworkers.

Rural life remained quiet. For most rural people, roosters, dogs, and babies may have produced the loudest sounds encountered on an everyday basis. One soundscape historian estimates that the loudest sounds encountered on average in rural areas of seventeenth-century England were thunder, cannon fire, and church bells (B. Smith 1999, 53–8). Of these, only the bells occurred regularly, and their impact depended on distance from the bell towers. Until the Industrial Revolution sent trains chuffing and shrieking across agricultural land, farm life did not disturb anyone's hearing or peace of mind, unless there was a grain mill or a smithy nearby.

Even in cities, a literature of published noise complaints can be traced back only to the eighteenth century. While residents of London at the time complained in letters to the newly fashionable newspapers about the din made by street vendors, drunks, dogs, and busking musicians underneath their townhouse windows (Sanborn 2009), the complainants apparently came from a small group of privileged gentlemen who made a living by writing for newspapers, a circumstance that casts some doubt on the ubiquity of the noise. Still, growing populations in all cities raised the level of noise coming from crowds, wagons, carriage wheels and horse hooves, as well as buskers and drunks, in the cobblestoned streets. So did the rising density and height of their residences, which resembled canyons in their ability to send the street noise echoing upward. In the course of the nineteenth century, new factories came to dominate the soundscapes of manufacturing districts and towns along with the outskirts of cities. Steam engines used for manufacturing and transportation shattered the quiet of rural and coastal areas as well as factory districts. By the early twentieth century, complaints against noise were appearing regularly in letters to newspapers: the American anti-noise campaigner Julia Barnett Rice, for example, counted intrusive steamboat whistles on the Hudson River from her family's New York mansion and encouraged public complaints about the noise.

The urban soundscapes of local transit, crowds, and neighbours also caused complaints. Objections to noise occurring in streets and audible in residences constituted the majority of official "quality of life" complaints in New York City in the 1890s. The issue remained prominent a century later: according to the website for New York City Tenants Services, a police hotline opened in August 1996 to deal with municipal complaints unrelated to emergencies reported that 43 per cent of total complaints received were about noise. The offending sounds included "shrieking subways, thudding pile drivers, roaring boilers, noisy neighbors, and raucous restaurants" (Locker and Bronzaft 1996).

A SPECULATIVE HISTORY OF AMBIENT NOISE

The work of the acoustic ecology and sound studies movements has brought attention to the concept of ambient noise as an element of culture as well as nature. Searching written sources for references to weather conditions and human activities brings recognition of continuity as well as change. The tone and volume of the language spoken by children depicted in a seventeenth-century painting might be quite different from their twenty-first-century counterparts, and the structures and materials of the houses on their street would almost certainly have reflected sound differently from the ones in a modern neighbourhood, but we can approximate the auditory content of their play because they are shown throwing balls, running with dogs, chasing each other, and playing toy drums and flutes. Such investigations and interpretations are the work of historians, historical novelists, and playwrights as well as soundscape researchers.[4] They represent documented knowledge, but also the work of the same sort of auditory imagination that enables small children and their parents to give confident – if undocumented – voice to prehistoric or entirely fictional monsters.

Some activities and settings have always been noisy: battles and arguments, sports, construction work, festivals, heavy traffic, and crowds. We can infer their levels from knowledge about population densities and urban design, even in the absence of written information. Combining the known with the speculative, we can fill out the daytime soundscape of that seventeenth-century painting with the knowledge that the town's streets contained horses, wagons, carriages, street vendors, and shops and their customers, as well as children and dogs. Let us begin an exercise in imagination, then, with some of the earliest communities formed by our species and their interaction with other forms of life, and follow the timeline forward to the present, surveying activities that range from preparation of food to participation in warfare. A good historical or environmental imagination will serve as the requisite time machine.

Soundscape 1:
A Community of Hunter-Gatherers
in a Temperate or Tropical Climate

Our encampment is in a clearing surrounded by dense forest or jungle: leaves rustle or clatter, rain roars or hisses or patters. Birds call incessantly, animals intermittently. Small children scamper, jump, giggle, shriek, chatter, play games with clicking sticks and stones. Older ones have work to do, finding food near the encampment: picking plants, digging roots, shooting arrows at small prey, carrying rustling deadfall branches for firewood. Women pound, thump, grind, and scrape with stone or wooden tools, preparing food and hides or fibres, chatting and singing as they do so, pausing to call to a child or feed an infant. A successful hunting party bursts through the forest into the clearing, heard by the others before they are seen. After the enforced quiet of stalking prey and avoiding predators, they are free to announce their return with calls of greeting, boasts and teasing, and ritual songs. One man, lightly wounded, groans; another laughs as he hoists a squealing child onto his shoulders. Cooking fires fed with damp wood crackle and snap.

Most or all of this activity takes place outdoors, to the accompaniment of rain or nearby water and bird calls. The loudest ambient sounds? Birds, rain, and sometimes children. We are alarmed as well, on occasion, by thunder or by a predator's snarl close enough to freeze the blood, by arguments or by the shouts and grunts and thuds of a dispute settled with fists.

Privacy is unknown, except by intention. Every moment, waking or sleeping, is spent in the company of family and tribe members whose voices are as familiar as the birds and the rain. Songs are dense with melodic information, as the forest is with birds and with rustling, sighing, clattering, pattering leaves, and rain. Where vision beyond the settlement is limited, drums or voices call signals through the forest.

Soundscape 2:
A Community of Hunter-Gatherers
in an Arid or Arctic Climate

The keynote is wind. Dangerous weather – storms of dust, sand, or snow – is signalled by changes in the sound and feel and direction of the wind; favourable weather by its lesser intensity or absence. Human activity in the settlement is much like that in the tropical setting, but its resonance is muted and masked by the wind. The thuds and crunches of food preparation are almost constant. Because tree cover is less dense, bird calls are transitory rather than constant, and animals call at greater distances.

People retreat into caves or tents or huts in harsh weather, and in winter sleep more than their tropical counterparts, huddled together so that the breaths and sighs and snores of the group are a large unconscious conversation. Ice creaks, squeaks, and groans in winter, clashes and sloshes in spring. Large prey animals moving in vast herds create distant thunder, small ones scurry and click in grass and underbrush. Hunters and predators move in near silence, the sound of their movements masked by wind and rustling grass or the hiss of moving sand. Because of the open terrain and the sounds of wind, visitors or invaders approaching the community may be seen before they are heard.

There are periods of relative quiet when only wind is heard. Songs, like the wind, have limited melodic scope and long phrases.

Soundscape 3:
Early Agriculture, Fishing, and Herding

From a fenced enclosure at the edge of the settlement comes a chorus of goats or cows in the morning, and the sound of milk hitting wooden pail or metal bucket. A dog huffs and pants and patters just outside the door of a shelter, whether hut or tent, while chickens add their voices and the agitated flapping of their wings. There are sounds of field and garden in each age and season: digging sticks, hoes, ploughs, scythes, rakes, and birds – and children chasing and shouting at birds to keep them from the ripening grain.

The harvest is rhythmic, accompanied by songs or drums, or both, to keep workers from tiring. Then there is the dull sound of stone on stone grinding grain; fires crackling in the open or enclosed within ovens for baking and roasting; the clatters and clicks of clay or metal or wooden cups and bowls and jars. If the culture has developed wheels, their creaking and clattering on hard ground or paving stones is loud, but birds and nearby animals and children are often still louder. So are harvest festivals, with drums and flutes, and singing aided by a sense of celebration and fermented beverages.

If the village is on a seacoast or a major lake or river, the soundscape is based on the motion of – or upon – water itself, as well as the creaking of boat hulls at rest. Oars or paddles, and leaping fish, split the water. Humid air carries sound better than dry, so that conversation and song are amplified by still water, though masked by water in motion. The collective hauling of nets heavy with fish is coordinated by rhythmic chanting or singing; so is rowing or paddling for long distances.

In herding cultures, made possible by large expanses of open grassland, wind is again a prominent feature of the soundscape. At signals of danger or seasons of migration, the thuds and thunder of hooves predominate. From the back of a horse a rider hears wind, their mount's breath and their own, the creaking of saddle and harness, the aural arithmetic of a quadruped's steady gait. At times of extreme weather or warfare, unusual and more violent noise disrupts the familiar soundscape.

Soundscape 4:
The Ancient Mediterranean

An ancient city, hub of commerce and government: perhaps Alexandria, Jerusalem, Athens, Constantinople, or Rome. Enter the marketplace where sights, scents, and sounds are as rich and complex as the patterned rugs for sale or the spicy stews simmering at the food-vendors' booths. Voices patter and clash as vendors announce their wares, customers haggle over prices, news and gossip are exchanged, children exclaim and implore. Street musicians add their voices, drums, flutes, and the plangent commentary of reeds and strings. Professional public orators, trained in the skills of vocal projection as well as rhetoric and logic, exhort, encourage, or chastise passers-by. The density of urban life gives rise to early forms of noise annoyance: market wagons with metal-rimmed wheels rumbling along stone-paved roads, residences crowded close enough to echo and amplify sounds from the streets, groups of late-night revellers singing and quarrelling their way home.

In the city, the noise of human activity provides a constant daily ambience. Nights are more quiet, except in imperial Rome, where traffic is so congested that market wagons, by law, can enter the city only at night. The soundscape of the countryside surrounding the city walls is still dominated by weather, birds, and animals, as well as the sounds of rivers or streams and the rising technology of water wheels.

There are sounds specific to specialized groups in the culture, as well. One of them is a trained military: the rhythm of synchronized marching, the clanking of armour, the clashes and thuds of sword or spear and shield, ring out along with the ubiquitous shouts, screams, and groans of the practice yard or battlefield. Ceremonial trumpets and horns signal and accompany armies, but also religious festivals. Places of worship, another site of specialization, may contain the clamour of festivities, the quiet of contemplation, or both.

Rural residents wake each morning to a chorus of roosters, songbirds, sheep, goats, and cows, and proceed through the day with the thudding and clanking of tools, with the plodding of animal feet and the symphony of their voices, with rustling of barley or wheat or grass. In this they continue the patterns of their ancestors: as urban soundscapes change, rural ones remain consistent.

Soundscape 5:
Late Medieval Europe

The urban marketplace is much the same as it has been for centuries, but it is regulated and kept aware of time and spiritual obligation by the ringing of enormous cast bronze church bells, designed to be audible throughout the extent of a parish. The city is loud, at times, with the clatter of wagons and carts over cobblestones, amplified by buildings that include a second storey or more and overhang the narrow streets. In the fourteenth and fifteenth centuries, mechanical clocks and the towers that house them become symbols of civic pride, competing with church bells and steeples for dominance of the urban soundscape. If a substantial city has a population as devout as it is wealthy, it may be authorized to build a cathedral. Construction in wood and stone brings the sounds of saw and hammer, mallet and chisel. Guild foremen shout; labourers banter and grunt; wives tend crackling cooking fires, clank and scrape simmering pots, and speak in gentle voices to toddlers or reprimand them harshly; crowds gather to watch and comment as construction progresses. Wagons pass constantly, carrying supplies and removing rubble. The process of building may go on for decades, with two or three generations involved, inuring locals to the soundscape.

Travelling along a stone or dirt road outside the city, one hears hoof-beats at various speeds and timbres, wagons rumbling and creaking, groups of people chatting or singing as they walk, birds, a chorus of sheep in the distance. The trades necessary to the maintenance of manor and village life – the bakers, brewers, millers, weavers, carpenters, leather workers, blacksmiths – contribute their distinctive pounding, pouring, grinding, tapping, hammering, clacking, clanging to the soundscape close to their workshops. The noisiest jobs are held by blacksmiths, along with miners and millers. Mills for grinding grain or stone are powered by water, wind, or animals, the thudding and creaking of the cogs and dull gnashing of the grinding stones accompanied by the slap and squeak of the waterwheel, the thumping of the windmill, or the clopping of oxen hooves. In the thirteenth century, gunpowder is introduced into Europe from China and used in fireworks, the loud bangs stoking the excitement of celebratory crowds. By the fifteenth century, its use in weapons is well established, and soldier gradually joins the list of occupations that can be hazardous to hearing.

Silence is cherished and cultivated in the expanding monasteries of late medieval Europe. Some monastic communities discourage or forbid speech for most of the day, communicating by means of hand signals except when concentrating their voices in the distinctive formulaic melodies of liturgical chant or the graceful cadences of spoken prayer. Others, like the Dominican and Franciscan preaching orders of the thirteenth century, specialize in public speech, employing the rhetorical thunder of trained orators to refine the faith of the urban public. In this, they carry on a tradition that predates them by centuries and was first developed by the ancient pagans whose beliefs they despise, the public orators commenting on politics, religion, and social customs of Greece, Rome, North Africa, and the Middle East.

Soundscape 6:
Early Modern Cities and Towns
and Their Surroundings

Population density increases, and noise with it. Wagons transporting goods to market are larger, louder, and more frequent; city streets resound sharply with the clatter of wagon wheels as well as the hoof beats of carriage horses transporting prosperous citizens at a brisk trot. The noise of neighbours – their trades, their children, their cooking, their conversations and quarrels – intensifies because there are more of them. The buildings in which they live, taller now and partitioned into separate rooms, may contain a number of unrelated families or one multi-generational clan with its servants. Street musicians and vendors selling food or announcing their services – knife grinders, pot menders, chimney sweeps, traders in used clothing – punctuate the urban soundscape and provoke complaints. So do the calls of night watchmen, paradoxically charged with announcing their presence to reassure sleeping residents that they are providing protection.[5] Small-scale industries – spinning, weaving, printing – are beginning to move from home production to workshops as they are mechanized and concentrated in particular districts of the city. Mechanization increases the noise levels but also isolates them in particular districts.

Military soundscapes, dominated for centuries by the clashing and thudding of swords and the twanging and whistling of massed archery, now commonly involve cannons and guns of various sizes. Military encampments are noisy, close-quartered transient tent cities occupying the countryside with populations that can number in the thousands. Soldiers, officers, messengers, cooks, food vendors, armour menders, horse grooms, blacksmiths, bone-setters, prostitutes all contribute to a soundscape punctuated by marching, drills, shouting, hammering, quarrelling, and laughter.

Soundscape 7:
The Industrializing World
of Northern Europe and North America

Mechanization and industrialization in the late eighteenth and nineteenth centuries create a spectrum of new noise. Steam-driven pumps used in mining are modified to produce the steam engine, which is used to drive trains and steamboats. The large-scale factory is born and, with it, relentless industrial noise as manufacturing is mechanized. The industrial worker, often a displaced subsistence farmer forced by economic necessity to move to a city or factory town for salaried work, lives in precarious conditions and works from before sunrise until past dark, his – and increasingly her, as the textile trades are mechanized – ears constantly assaulted by rattling clanking screeching grinding clashing crashing engines, gears, presses, pumps. By the end of the nineteenth century, it is well known that workers in certain trades – locomotive engineers and boilermakers as well as blacksmiths and soldiers – suffer hearing loss as a result of occupational noise.

As the population of urban industrial workers expands and the conditions in which they live and work worsen, protest movements grow. Chants, songs, and marching feet resound in city streets; clashes with factory guards or police produce shouts of indignation and alarm, screams, the thudding of clubs against muscle and bone, gunfire, clanging alarm bells, milling hooves, fire hoses turned on crowds.

Noise annoyance, well established in urban centres, arrives in the countryside as well. The building of railroads exports a level of disturbance that generates a sense of excitement about the future and its possibilities for greater knowledge and greater wealth, but also protests about the noise. Train and steamboat whistles add a new layer of piercing sound to everyday life in the areas between cities and ports. Factories, and the newly built manufacturing towns that provide them with workers, appear with alarming speed and alter the rural soundscape as well as its terrain; the sounds of construction are followed by those of manufacturing processes and dense populations living in cramped and poorly maintained residential districts. In such neighbourhoods, sleep is punctuated by the sounds of machinery, the speech of night-shift workers, the fragmented songs of drunks returning home, the coughing of neighbours whose lungs have been compromised by the poor air quality both inside and outside the factories, the dripping of a leaking roof.

Music also becomes louder during the Industrial Age. Instruments, as carriers of technological change, proliferate and sprout mechanical devices – keys, valves, and geared peg-boxes – as well as heavier construction to enable louder sounds. The first fully mechanical instruments – barrel organs, then player pianos – are developed. Orchestras, bands, and choruses are larger and consequently louder; public concerts and the massive concert halls that contain them change composed music from an elite entertainment to nourishment for the masses. As technological change and the rising wealth of the manufacturing class erode the economic and political dominance of monarchy and aristocracy, music moves from private to public patronage. The meeting of new industrial wealth with desire for active participation in the magical realm of music engenders amateur choral societies, and the sound of pianos becomes more common as they are welcomed into the homes of the middle class as well as the wealthy.

The next acoustic revolution impacts communication and transportation. By the end of the nineteenth century, early technologies for amplifying, recording, and broadcasting sound are available. First comes the clacking of telegraph keys, then the phenomenon of remote voices heard over the telephone and radio, then the amplification of voices by the microphone. The removal of sound from its immediate source opens possibilities for vicarious auditory experience through recording and broadcasting, but also threatens a loss of intimacy with natural soundscapes.

Soundscape 8:
Modern North America

Increasing mechanization of industries for both wartime and peacetime production produces a sharp rise in ambient noise throughout the twentieth century. Over the course of the century, the automobile gradually colonizes the North American soundscape: road traffic becomes a constant source of ambient daytime noise in cities and along highways. The aircraft industry, which expanded dramatically during the Second World War, is reconverted to post-war civilian use, and skies as well as highways become a source of reverberant machine noise as the desire for recreational air travel grips the public. During peacetime, many factories are geared to the production of household appliances, and domestic soundscapes increasingly hum with the fans of refrigerators and furnaces and the swishing of washing machines. In many places, urban populations retreat, if they can afford it, to the newly invented and highly coveted quiet of the suburbs, where the demands of yard work and home maintenance gradually give rise to power mowers, power tools, snow blowers, and weed whackers. Fashionable new sports drive new passions for noise, as motorcycles, powerboats, and snowmobiles feed a growing market for speed in outdoor recreation, while a culture of youth fed by marketing interests demands technologies for producing louder music.

New technologies for communication and entertainment revolutionize the culture of listening. Radio and then television change the definition of community from the live interactions of people living in proximity to vast populations sharing the same broadcast information and entertainment, regardless of location. The concept of intentional background sound – the radio left on all day, the television turned on at night to provide the illusion of community for the individual or the nuclear family isolated at home – becomes normalized and provides a formative ambience for children. Shopping malls serve up a constant soundtrack of commercial and commodified music designed, at first, to be inoffensive and later to serve as in-your-face brand advertising.

Soundscape 9:
The International Information Age

We arrive at the present, with ambient acoustic backgrounds that will be familiar if you are reading this in a city almost anywhere. Outdoors, the dominant element of the soundscape varies from wind punctuated by traffic to perpetual traffic, with overlays of aircraft, commercial music, and possibly commuter trains above or below ground. Indoors, it is a combination of several layers of appliance and ventilation fans with electronically generated music or speech and live conversation. The soundscape of the technologized world hums at us constantly in multiples of 60 cycles per second in North America, 50 in Europe and Australia. Computers, copiers, scanners, and fax machines, with the fans that keep them from overheating, generate a constant drone in the background of indoor life in office workplaces. Electronic entertainment, no longer communal, dominates the soundscape of each room in a suburban house whether it issues from speakers or earphones. Closed airflow systems controlled by fans, fluorescent lighting, furnaces, air conditioners, and household appliances all contribute layers of noise to the domestic soundscape. Much of it – even the carefully chosen music on the MP3 player that accompanies daily activities – occurs below the threshold of our attention, largely ignored because it is always present. Unobtrusive background noise that does not vary, whooshing or rumbling softly and steadily away, can still affect us because the brain and nervous system register its presence. It can produce feelings of contentment when the furnace clicks on in January, but if you make an effort to concentrate you might find subtle tensions accumulating in your neck and jaw and shoulders, or perhaps a subtle shift in your breathing pattern. If the background sound is loud and/or high pitched, you might start to feel irritated and tired. These are signs of noise annoyance, a common condition that takes many forms, and that is the subject of the next chapter.

Why Noise Annoys

Auditory perception intrudes into thought much more directly than visual phenomena.

Ihde 2007, 212

The physical characteristics of sound are not sufficient to understand why particular sounds came to be defined as noises or why private problems of noise became public ones. These questions require acknowledging transformations in the ways people listen to sounds and their cultural meanings.

Bijsterveld 2008, 24

Because the auditory system in humans and other mammals has evolved to distinguish informative and alarming sounds from familiar ambient patterns, sudden sounds catch our attention while repetitive patterns do not. Sudden noise can motivate and energize physically, producing a jolt that sends us into high alert and arouses us to greater activity. Its ability to interrupt sleep can be put to good use: just as our prehistoric ancestors were startled out of sleep and into adrenaline-fuelled motion by thunder or a predator's roar, we wake and begin the day with the acoustic caffeine of alarms from clocks or smartphones, successors to our very long history of songbirds, roosters, and hungry children. Loud noise is an attractor of attention, consciously used to gather crowds, to advertise, to control and shape both individual and collective attitudes and emotions. It can release emotions or set up a wall of distraction that shields us from feeling much of anything. It can be imposed or chosen as a weapon or a shield. If the sudden burst of noise that motivates action becomes repetitive or continuous, however, it will produce a state of panic: imagine trying to work and sleep in the presence of a buzzer alarm that can't be turned off.

Now imagine something far less loud, like the air conditioning at your office or the traffic from the intersection a block away from your home.

There's no need to imagine this one; you probably experience it every day. It doesn't send you into a panic because it isn't loud enough to activate a startle response as an alarm would, but perhaps you find yourself with tension in your neck, a faint sensation of headache, a tendency to feel irritable with your co-workers or family, a feeling that the day has been stressful even when no particular incident has made it that way. If the noise pauses, or you walk into a room where it can't be heard, do you feel a sensation of relief as your shoulders relax? If so, your mood and your physical responses may be influenced by noise annoyance.

TUNING OUT

What noise? Oh. I didn't notice. It's there all the time.

Because it activates the potent biochemistry of stress that interferes with normal sleep and digestion, a prolonged state of irritation is detrimental to health. Evolution has provided us with a defence: the human nervous system compensates for ambient noise by "tuning out" any sound that is constant enough to become familiar, as long as it isn't extremely loud. This may include rain, traffic, indoor ventilation systems, the voices of family members, the construction project down the street that has been there since last April, and sometimes even the alarm clock. We don't notice hearing them because we're not listening. If we did, the ability to focus on whatever we're doing would be compromised. The prehistoric ancestors who developed the ability to tune out ambient noise needed that ability to focus on their hunting and gathering in order to live long enough to become our ancestors. The crack of a branch would startle them because there might be a leopard in that tree, but the constant rustling of leaves was safe to ignore. To their conscious minds, and to ours, familiar noise is not worth noticing. To our nervous system and all the subtle physical and emotional reactions that it perceives and regulates, the noise to which we don't listen is still clearly heard. The noise that we tune out has effects on our mind and body.

IT'S THERE ALL THE TIME –
IT MUST BE HARMLESS, RIGHT?

The ability to ignore noise reduces the stress that results from distraction, but that's a mixed blessing because noise is a carrier of messages about the immediate environment. The ambient noise of traffic is annoying to residents along a busy street, who may tune it out or cover it with music or broadcast media. At the same time, it is a constant source of vital information to the drivers, cyclists, and pedestrians on the street itself.[1] Ignoring noise can even be an immediate threat to safety, as it was to the early human who ignored the sudden creak of a branch as the leopard leapt. Without a conscious effort to listen to environmental cues, the earbud-wearing pedestrian or cyclist who mediates the soundscape with an MP3 player risks failing to notice the approaching car or train. While that poses a danger only to that individual, the problem of noise posing a distraction is compounded when a constant ambience of loud noise is present in everyday life for large numbers of people. That circumstance, which poses a hazard for entire populations, is the norm for most people living and working in cities unless they have the means to afford effective soundproofing. The nature of the hazard is dual: a chronic stress response to the ambient noise and an inability to access the calming and restorative properties of quiet natural soundscapes.

The hazard is greater for some people because the ability to ignore noise is not available to the same degree to everyone. A minority of the population is deemed "noise sensitive" and may find it difficult or impossible to ignore auditory signals and to tolerate loud soundscapes. It is not yet known whether this phenomenon occurs in all cultures or only in industrialized ones, nor whether it can occur in all mammals, but what is known is that people with *hyperacusis* – heightened sensitivity of hearing – have lower thresholds for painful and panicking responses to noise and that their heart rates vary more in reaction to noise when compared with the general population (Soames Job 1999). True hyperacusis can result from any of several possible causes: genetic expression, diseases of the auditory system, some forms of autism, fibromyalgia or chronic fatigue syndrome, or trauma to the acoustic nerve. Milder sensitivities can result simply from short ear canals or from states of general anxiety in which all senses are on high alert, and children may be at particular risk for these. One study on hyperacusis suggests that people with extreme sensitivity can suffer migraines, depression, and middle-ear malfunctions as well as persistent fear of harm resulting from exposure to noise (Andersson et al. 2005). While no causative mechanism for hyperacusis has yet been found, the condition is not trivial. The painful and emotional reactions reported are quite real and often resistant to treatment, although relaxation therapies, "pink noise"[2] recordings, and soft gentle background music can be effective in some cases.[3]

For extremely sensitive individuals, adaptation to the soundscapes of urban life may be impossible. It is easy to dismiss their concerns and resign ourselves to the assumption that excessive noise is part of the price we must pay for progress and prosperity. Where there are choices about where to live, and some control over the activities of neighbours and wider communities, some of those who suffer from hyperacusis or simply dislike noise will have the means and the resolve to change their work and residences in search of quieter soundscapes. In the absence of such options – that is, for most people – public awareness of the problem is a more appropriate response than dismissal or resignation, because we are all – the sensitive and tolerant alike – affected by it. Ears, brains, and nervous systems react to noise even when we stop listening.

Individual reactions vary considerably. Age, gender, hearing acuity, life experience, occupation, economic status, and culture of origin can all influence conscious and expressed responses, but there is also a range of unconscious internal responses that transcend such divisions. They include the electrical and chemical reactions in our brains, nerves, and muscles as well as the activity of the endocrine system that regulates the production of hormones, which, in turn, influence both physical and emotional states. Such reactions are involuntary and may occur regardless of our mood or our aesthetic judgment of the soundscape. In this context, the following questions can be posed:

- Are there universal aspects to noise annoyance, characteristics of noise that remain irritating regardless of personal or cultural reactions?
- If noise can be an irritant, what makes it so? Are there particular acoustic properties that irritate the human nervous system? What are the actual characteristics of annoying noise?
- Does the perception of noise as annoyance relate to the musical characteristics of noise?
- Do "natural" and "unnatural" noises differ on the scale of annoyance?
- Does resemblance to human vocalization make a sound more pleasing and less annoying?

In the absence of large-scale cross-cultural or cross-species studies at this time, some of these questions cannot be definitively answered, and others are simply theoretical. Definitive cross-species studies are even more difficult than comparisons of human responses, of course. We cannot get descriptions of reactions from animals and can rely only on observation of their behaviour in response to a stimulus. Differences in interpretation of animal behaviour further complicate the matter, but the questions are worth raising because they define the boundary between individual reactions that depend on the sensitivity of the listener and universal ones that occur in all cases, even when the listener is not human and therefore not capable of articulating opinions about the noise and its source. A negative response based on opinion gives one type of information; a spontaneous startle reaction gives another. Both are important.

WHAT NOISE ANNOYS?

We know from research done by scientists and engineers in the noise abatement industry that there *are* particular properties to universally annoying noise, specific acoustic and auditory conditions that will produce a sense of irritation in anyone with healthy hearing who encounters them. These have to do with the acoustic properties of the sound, its perception by the auditory system, the conscious associations it evokes, and its unconscious effects. Extremes of loudness, duration, and proximity are most likely to cause annoyance. A frequency range between 2,000 and 5,000 Hz – equivalent to the highest keys on a piano – can also be irritating.

Urban traffic, office machines, and home appliances and broadcasting devices, especially in combination, now routinely produce sufficient levels of ambient mechanical noise to annoy sensitive individuals and to raise additional questions about public health. Reactions to noise are socially, as well as physically, determined. Performers, athletes, and politicians all thrive on crowd noise, and the factory or construction project that disturbs a community may also provide its residents with employment. Reactions to intrusive sound can depend on the sensitivity of the listener, or on subjecting the sound to a process of analysis to determine whether it is acceptable or simply unavoidable:

- Have I heard this before?
- Is it part of a regular pattern or something unexpected?
- How often, and at what time of day, does it happen?
- Is it happening for a good reason, and do I know what the reason is?
- What is causing it? Can I see or otherwise recognize the source?
- Do I have any control over the cause? Does anyone?

Because lack of control increases the potential for annoyance, the boom box on your neighbour's balcony – playing a song you like but playing it outside your control – can move from pleasure to nuisance, especially on the nineteenth repetition. Because what registers as an annoying sound also depends on the sonic ambience that backgrounds it, the unfamiliar motorcyclist roaring down your quiet suburban street to wake you at 2 a.m. (*unknown source; no apparent reason*) is potentially more annoying than a siren (*known source; there's a reason for it*) or thunder (*known source; no control possible because it's weather, not human activity*), although all three sounds have equal capacity for disrupting sleep.

Expectations about the soundscape also play a significant role in noise annoyance. Urban residents who move to the countryside expect undisturbed quiet and may be unwilling to tolerate levels of noise well below what they experienced in the city. The urban motorcycle may be replaced as a source of annoyance by the rural rooster. New communities that advertise peace and quiet as part of their appeal are subject to construction noise as additional houses are built, then to the sounds of traffic and lawn maintenance. The types and levels of familiar ambient noise in established communities change as children grow, economies and prevailing cultures change, and new technologies are adopted. Whether such changes are perceived as deterioration or improvement may depend on individual attitudes, but also on cultural ones: one person's nuisance is another's normalcy. Reactions to the backyard party across the street will vary according to taste in music and beliefs about privacy as well as tolerance for noise. To parents, the sounds made by active children on a playground are usually welcome, but the same sounds may be deeply distressing to someone who has lost a child or has tried unsuccessfully to have one. The factory that produces an incessant roar in the background of your daily life may be consciously accepted if it also provides you with an income, but it can still produce annoyance on an unconscious level as well as physical reactions.

Finally, there is the aesthetic dimension. *Is the sound pleasing or irritating? Beautiful or ugly?* This is the most subjective category and partly dependent on individual taste, although some general conclusions apply: the sound of a flowing brook is more likely to be rated as pleasing than that of a jackhammer, even when tastes in ambience vary considerably. Familiarity – which is known to be a significant factor in musical taste – may be operative in responses to soundscapes as well. Just as the music you heard in childhood is likely to make you feel secure as an adult,[4] the ambient noise of your earliest home shapes the standards you set for tolerance of noise. People who grow up and remain living near railway lines, highways, or major waterfalls may even be comforted by the noise emanating from them, given sufficient distance and moderate levels of loudness, and may have difficulty sleeping in consistently quiet surroundings. Beauty, it seems, resides in the listener's ear as well as the beholder's eye – or does it? Are there any properties of sound that make it inherently attractive or offensive?

CATS, DOGS, MACHINES, AND MUSICIANS

The study of music – as well as the technical vocabulary of physics – can provide additional tools for describing and categorizing noise, especially with reference to its aesthetic properties. The terms *concordant* and *discordant*, derived from harmonic theory, refer to the tendency of certain combinations of sounds to produce pleasant or relaxing sensations (concords), while others produce restlessness or tension (discords). The musician's perspective considers rhythm as well as concordance as a significant aspect of sound. Does the sound produce repeating patterns? If so, it has an element of rhythm. The human brain may be hard-wired to find rhythm,[5] since living organisms depend on it, and the human mind is particularly alert to sound patterning because it pertains to the acquisition of language. Rhythm is an element of comfort, and as such it can in some circumstances serve as an antidote to annoyance. Thus, sounds that are loud, contain discordant combinations of frequencies, and lack predictable patterns of repetition are most likely to be perceived as annoying,[6] at least by listeners accustomed by their culture to listening for harmonic concords.

The vocalizations of cats and dogs can demonstrate the characteristics of noise annoyance if we subject them to a process of musical analysis. To a musician's ear, the dog is classified as a percussive instrument while the cat is a melodic one. The cat's melody may not be pleasant – feline voices range from the sweet trilled mew of the Maine Coon breed to the guttural shriek of the excited Siamese – but it has the musical qualities of pitch (a note), attack (onset), and decay (offset), and can therefore be interpreted as resembling music or speech. To the human ear, accustomed to some degree of melodic inflection in language, the cat may sound contented or annoyed, inquisitive or demanding. Cats are often credited with being talkative, and their tones are easily imitated by children, adults, and trained singers alike.[7]

In their size and weight as well as their vocabularies and vocal ranges, cats resemble newborn humans, and may benefit from our species' innate response of affection for and attention to infants. Cats' most annoying sounds, their notorious blood-curdling yowls, are produced in states of emotional agitation: warning calls to other cats perceived as rivals, mating calls, or protests. Such calls resemble the distress cries of infants and can easily alarm the human nervous system. The cat's range of vocal expression includes a counterweight to the yowl in the form of purring. The purr is a sustained vibration of contentment that relaxes the cat as well as giving voice to its achieved state of relaxation. For this reason, cats may purr when under mild stress and even while injured or giving birth.

Unlike the cat's relatively sustained and melodic tones, a dog's bark is abrupt. The bark results from the sudden expulsion of air through the gap between vocal cords rather than sustained vibration of the cords. It is analogous to a cough, which is also likely to startle the human hearing apparatus. Since a percussive sound contains a wide range of frequencies carried by pulses of short duration, it can activate all parts of the basilar membrane in the inner ear, and is therefore difficult to ignore.

The bark, again like the cough, is percussive without the organizing principle of rhythm.[8] The dog's voice is consequently less like human speech than the cat's. Dogs also sigh, snore, and yawn audibly, as well as giving soft whuffs or yips of greeting. These sounds, unlike the bark, are usually regarded as endearing. The dogs most likely to be credited with melodious voices are hound breeds that bay rather than bark – again, a more "human" sound than the bark, analogous to singing. Is our species attuned by evolution to sort the voices of animals for similarity to our own vocal characteristics? Could our perception of annoyance be linked in some way to this sorting process? Whatever your opinion of cats and dogs, you are more likely to welcome the purr or the whuff of greeting than the yowl, growl, or agitated bark.[9]

The analysis of cats and dogs can serve to illustrate the basic physical characteristics of noise annoyance. An annoying noise is likely to be loud, harsh, unpredictable, or unrelenting. It may be dissimilar to human vocal sounds or similar to human vocalizations under extreme stress. The latter case is best demonstrated by the sounds of small children. A toddler shrieking in frustration or distress while playing outdoors is more likely to alarm listeners than one shrieking with laughter. A dog barking in aggression will be more disturbing to a human listener than one barking in greeting. This, too, might be a matter of evolutionary patterning. The recognition of emotional arousal and aggression in animals and fellow humans was crucial to the survival of our species in its origins, and is still an important signal. Other natural sounds will provoke responses ranging from alarm to vague unease, as well: thunder, rising wind, hail storms, heavy rain, blizzards; in mountainous regions, avalanches or the sound of rocks falling. All are related to awareness of potential danger.

Particular characteristics of natural soundscapes that may activate annoyance or stress responses are present, as well, in the sounds produced by the operation of machinery. As examples, consider the musical structures inherent in the soundscapes of daily transportation. Your car has a steady hum composed of multiple layers of vibration, essentially analogous to a musical chord. It is intensified by acceleration, punctuated by turn signals, and often counterpointed by music or talk. The mechanical soundscape – independent of music or occupants of the car – is neither loud nor unpredictable, it is mostly under your control through attention to your driving habits and surrounding traffic conditions, and familiarity is so high that you can recognize a problem with the engine by a change in the car's ambient sounds. Annoyance caused by this noise, when it maintains familiar patterns, is low.

At the next level is transport under someone else's control. This machine, a bus or train or streetcar, is larger than your car and probably louder. It is less predictable: although the mechanical noises of a daily bus or subway route may become a familiar soundscape, the level and character of noise are influenced by entries and exits, conversation, and traffic and earphone spillover, which vary according to time and day. Streetcars are relatively quiet but subject to bursts of noise from adjacent traffic and from the occasional crackle of cables. Buses add considerable mechanical noise, generally of a sustained nature and composed of layers: engine, fan, wheels, brakes, doors. These soundscapes can produce annoyance when the location and route are unfamiliar, but habit will render them easy to ignore. Trains produce the added dimension of rhythm, which is likely to impose order on the chaos of mechanical noise. Of all the machines connected with transportation, only the cross-country train has a literature of praise for its soundscape, which is the most musical of the mechanical devices considered here, combining rhythmic and harmonic effects. Murray Schafer cites the variety of traditional train sounds as "rich and characteristic: the whistle, the bell, the slow chuffing of the engine … the sudden explosions of escaping steam, the squeaking of the wheels, the rattling of the coaches, the clatter of the tracks, the thwack against the window as another train passed" (Schafer 1994, 81).

A 2004 study of reactions to train and traffic noise by Chinese and Japanese graduate students found that respondents were significantly less annoyed by train noise than by traffic noise with both at 75 dBA, a high-moderate level of loudness (Ma and Yano 2004). The association of trains with musical soundscapes, ranging from country and western ostinati to movie themes, may furnish a reason for the difference.[10] The elements that make train sounds distinct from those of other transport machinery are sustained rhythm and the characteristic whistle, a loud call with multiple pitches: in effect, a musical chord. The train may have a special place in the culture of transportation because of its evocative musical qualities as well as its history of association with both adventure and comfort.[11]

Finally, for a familiar category of mechanical noise that is consistently unwelcome, consider the drill. From the dentist's drill to the jackhammer, drills are associated with discomfort and, for many, with fear. They may be loud, often to the point of causing pain, and they may be held in close proximity to the ear. Their sound can be sustained or rapidly percussive. They do not resemble human speech or singing. What they do resemble are somewhat alien life forms and dangerous extremes of weather. The dentist's drill, which produces a range of high frequencies, recalls insects like the mosquito and cicada, more likely than mammals or birds to evoke feelings of repulsion. The jackhammer is a close acoustic cousin of the machine gun: not a component of the prehistoric soundscapes in which our species evolved, but acoustically related to hailstorms and rock slides, which were similarly – if not so immediately – threatening.[12]

Just as with natural sounds, the likelihood of an annoyance or stress response to mechanical noise will rise with sounds associated with discomfort or danger. This association might hold true even when the noise is not fully audible.

FEELING THE NOISE: LOW FREQUENCIES

The evidence on low-frequency noise is sufficiently strong to warrant immediate concern. Various industrial sources emit continuous low-frequency noise (compressors, pumps, diesel engines, fans, public works); and large aircraft, heavy-duty vehicles and railway traffic produce intermittent low-frequency noise. Low-frequency noise may also produce vibrations and rattles as secondary effects. Health effects due to low-frequency components in noise are estimated to be more severe than for community noises in general.

Berglund, Lindvall, and Schwela 1999, 55

As suggested by this quote from the World Health Organization, of particular concern as a cause of annoyance is infrasound or low-frequency noise (LFN), components of which fall below the threshold of human hearing. LFN includes sound at frequencies ranging from 20 to 200 Hz, as well as components below 20 Hz, which are largely inaudible to humans but may be felt as vibration or pressure. It is usually a product of industrial soundscapes, once limited to the developed world but now a global concern. LFN is a component of noise from major highways, railway lines, airport runways, excavation and construction sites, oil drilling rigs, compressor and pumping stations, and industrial ventilation systems. People exposed to it may report pressure on the eardrums, headaches, nausea, fatigue, insomnia, and/or chronic anxiety even when repeated testing suggests that the complaint is out of proportion to the measurable degree of noise (Persson Waye, Björkman, and Rylander 1985; Pawlaczyk-Łuszczyńska et al. 2005). Extreme infrasound is known to produce pain in the ear canals and even impairment of breathing (Berglund, Hassmén, and Soames Job 1996). Given that the low frequency components of an annoying environmental or mechanical noise may be felt rather than heard, if they are consciously perceived at all, what is driving the reactions?

Natural causes of LFN might provide an answer. Sensitivity to subsonic noise may have an evolutionary aspect, since its most dramatic natural sources are volcanoes, hurricanes, tidal waves, and earthquakes. Like other mammals that panic and flee at the early signs of seismic disturbance, our hominid ancestors may have responded with reactions of physical and emotional unease that are now routinely ignored amid the sensory overload of urban and technological life. Low-frequency noise was not a component of a safe and stable environmental soundscape, and the healthy response to it was to get far away as quickly as possible. When a homeowner today finds a mechanical source of LFN moving into the neighbourhood, the flight response is countered by the desire to stay put and get rid of the noise. Anxiety may be aggravated by the conflict of responses, and the consequent chronic stress could account for some of the reported physical symptoms that are not directly traceable – at least with current systems of measurement – to the noise itself.[13]

Low-frequency noise presents a set of particularly well-documented problems in workplaces that require a high degree of concentration on hazardous physical tasks. A team of researchers in Poland pursuing studies of LFN in workplaces including factories, chemical plants, aircraft cabins, and military shooting ranges since 1999 have found that – while sensitivity to LFN was not necessarily correlated with sensitivity to noise in general – "LFN at moderate levels might adversely affect visual functions, concentration, [and] continuous and selective attention, especially in the high-sensitive to LFN subjects" (Pawlaczyk-Łuszczyńska et al. 2005). The team defined "moderate levels" of LFN at 50 dBA, approximately equivalent to ambient office noise or moderately loud conversation. Indoor sources monitored included industrial sewing machines, grinders, welders, washers, saws, and planers (Pawlaczyk-Łuszczynska, Dudarewicz, and Śliwińska-Kowalska 2007). In all of the workplaces, LFN was shown to produce varying degrees of interference with workers' tasks, a situation that increases the likelihood of accidents with potentially dangerous machinery.

Another team in Sweden found that workers exposed to LFN reported increased annoyance and showed interference with cognitive tasks as well as measurable increases in the production of the stress hormone cortisol (Persson Waye et al. 1985, 2001, 2002). A preliminary investigation of six industries in India reports that workers associated symptoms including neck pain, headache, backache, chronic fatigue, irritability, and pressure on the eyes with chronic exposure to LFN and "mid-frequency" noise – in the low audible-tactile range of frequencies – at work (Prashanth and Sridhar 2008).[14]

Because accurate measurement of LFN requires specific expertise, it is often overlooked as a source of annoyance. Even noise abatement programs may disregard it because of the expense involved in controlling it: sources and transmission paths usually need to be retrofitted or heavily insulated, which are exacting and expensive processes. Sensitivity to LFN does not always correlate with high hearing acuity or sensitivity to audible noise, and sensitive individuals often report emotional as well as purely physical symptoms as effects. Such reports may be dismissed by occupational health and safety investigators because their understanding of LFN is incomplete, because the number of workers affected is small, or because cause-and-effect relationships between LFN and symptoms are difficult to establish. Moreover, insurance and compensation plans do not always regard mental and emotional symptoms as evidence of harm to health, and most complaints about noise – regardless of frequency spectrum – involve descriptions of anxiety, depression, fatigue, or inability to concentrate. Without sufficient knowledge of the scope of noise annoyance and its effects on the nervous system, the complainants may be regarded by themselves, their families, and even by medical professionals as emotionally fragile or even mentally ill.

THAT NOISE IS DRIVING ME *CRAZY*!

Is noise annoyance a direct cause of mental or emotional illness, or simply one of a cluster of interrelated factors? In the case of noise-induced hearing loss, direct causative links between noise and illness are clear because measurements are conclusive and the physical mechanism by which harm occurs is known. In conditions related to stress and anxiety, the evidence suggests connections but not necessarily direct causation. While it is tempting to regard the impairment of hearing as purely physical and stress reactions as purely emotional, in fact the two realms are not separable. The biochemistry of stress produces somatic changes. Changes in physical function provoke emotional responses, as well: if you are subjected to frequent interruption of daily activities and sleep, you will experience changes in your ability to concentrate and to fall asleep, which in turn are likely to affect your mood and productivity.

If a noise is "driving you crazy," you have plenty of company. An Internet search of the phrase in 2019 yielded over 43 million results, a majority connected with car engines and lesser numbers with apartment and dorm neighbours, household appliances, computer fans, ambient mechanical roars, and even nocturnally active pets. The sensations involved – feelings of anxiety, irritability, faintness, and/or fatigue – are produced by the biochemistry of stress: the production and circulation of the hormones cortisol and norepinephrine, which activate the body's fight-or-flight response to danger – increasing heart rate, circulation, and respiration to fuel sudden movement. When the stress becomes prolonged or chronic, these responses – intended to motivate short bursts of intense activity – lead to states of anxiety and exhaustion.

The condition called *misophonia* (hatred of sound), a chronic state of severe sensitivity to sound, is a case in point: for people with this condition, even ordinary sounds like gum chewing, pen clicking, finger drumming, and typing can produce annoyance leading to panic or rage, although producing the same sounds themselves can calm some sufferers. Because ambient noise can be torture for misophonics, their ability to function in many work and social situations may be compromised. Misophonia is not a mental illness but a disorder involving hyper-connectivity between the auditory and limbic (emotional and behavioural) systems of the brain, a condition that may have hereditary components (Edelstein et al. 2013). There is no cure, but some individuals are helped by anti-anxiety medications.

Many sufferers from misophonia and hyperacusis are reluctant to reveal their conditions because long-held stereotypes associating noise annoyance with mental illness are still common. Especially prominent are stereotypes around the phenomena of "hearing things" and "hearing voices." Although sound is defined as the intersection between physical vibration of air or fluid and its perception by an auditory system, it is not strictly limited to physical sources. The hair cells of the inner ear can produce sound internally when irritated or damaged. In addition, the brain maintains its own internalized sub-vocal chatter of thinking in words, the bane of meditation practices and the nutrient of neuroses as well as the process of writing. In normal brain function as well as some varieties of mental illness, internal speech in the absence of stimuli can even become physically audible through the firing of auditory nerves during silence.

While the phenomenon of "hearing voices" is associated in both medical literature and popular culture with schizophrenia, it can also happen on an occasional and fleeting basis for many people in a normal pre-sleep or deeply relaxed state (Lawrence, Jones, and Cooper 2010). Such auditory hallucinations may even provide a physical component to the voices of gods, saints, and angels, heard in dreams and waking states, reported so consistently in ancient and medieval literature (Thraenhardt 2006–7). One study on the mechanism for the phenomenon suggests a possible genetic basis: one gene that increases susceptibility to the firing of auditory neurons in silence, and another that enhances susceptibility to the interpretation of such firings as meaningful speech (Sanjuan et al. 2006). A key difference between normal and abnormal responses is whether the "voices" are dismissed as half-asleep hallucinations or interpreted as commands and acted upon. Such responses may depend on measurable differences in brain anatomy (Gaser et al. 2004).

Despite the stereotypes, noise-related stress is not a direct cause of psychiatric disorders. It may be an aggravating factor in existing conditions (Lercher et al. 2002; Haines and Stansfeld 2003), but few studies have been done and nearly all were limited to children and controversial in design, and so the question is still open. The World Health Organization *Guidelines for Community Noise*, which categorized the effects of noise on mental health as inconclusive because of insufficient study at the time, mentioned that populations living in noisy urban areas show increased frequency of psychiatric symptoms and mental hospital admissions, and implicated noise above 80 dBA as potentially causative of aggressive behaviour (Berglund, Lindvall, and Schwela 1999, section 3.6).

There is still uncertainty about whether any suspected correlation of urban noise with mental illness is due to the increased sensitivity to annoyance often displayed by people with existing psychiatric disorders, or whether the noisy conditions are causative. This in turn raises the question of whether noise sensitivity and susceptibility to mental and emotional stress are both simply parts of a generalized condition of sensitivity also characterized by high startle response and perhaps associated with a pattern of hypervigilance. Early research literature raised the question about whether people who are physically reactive even to accustomed noise have a vulnerability to mental disorders, but conclusive evidence was not available (Stansfield 1992). However, a more recent investigation focusing on reactions to airport and traffic noise by residents living near Frankfurt Airport in Germany found no correlation between noise annoyance and mental illness, and concluded firmly that reported noise annoyance was a more reliable indicator of genuinely excessive noise than of individual attitudes or states of mental health (Schreckenberg, Griefahn, and Meis 2010).

What *is* known is that sustained periods of stress directly affect emotional health, which affects mood, attitude, and quality of life. When noise annoyance leads to chronic states of irritability and fatigue, the reason is often quite simple: frequent disruption of sleep.

THINGS THAT GO
BUMP/CRASH/ROAR IN THE NIGHT

While common sense alone suggests that noise will disturb sleep, either by waking the sleeper or by changing the level of sleep, actual measurement of the problem is not a simple matter. Most studies of noise-induced sleep disturbance have been done in laboratory conditions with small populations, usually healthy young adult students at universities. One researcher found no difference between laboratory and residential conditions in measuring sleep disturbance but cautioned that individuals accustomed to a noisy overnight environment may complicate the results of laboratory studies (Skånberg 2004). This is an important consideration: life outside the lab includes multiple sources of sleep disturbance, auditory and otherwise. Both insomnia and interrupted sleep have multiple causes, including "worries, anxiety, distress, depression, painful illnesses, bad indoor climate, low quality mattress and/or pillow, artificial light, alcoholism, taking naps during the day, psychostimulants, jet lag, exercising before bedtime, low socioeconomic status, and too long television viewing" (Stošić et al. 2009, 335). What is the actual role of noise in all this nocturnal alertness?

Sleep disturbance associated with noise annoyance – as distinct from the physical reactions discussed in Chapter 2 – is investigated in human populations largely by means of questionnaires, sometimes in combination with technical measurements of ambient noise audible in the subjects' sleeping quarters. Such questionnaires typically ask about times of falling asleep and waking, impressions of sleep quality, and use of sleeping pills. Human studies can be somewhat subjective because susceptibility to noise annoyance in daily life varies with individuals, and questionnaires involve self-reporting. To produce thoroughly valid results, human studies have been combined for decades with animal studies in laboratory conditions, which clearly show that chronic exposure to noise leads to fragmented sleep in which the brain becomes accustomed to waking several times during a period of sleep (Rabat 2007) as a result of imbalances in hormone production and even of DNA damage (Kight and Swaddle 2011). When the laboratory studies are combined with evidence from the large-scale population surveys conducted in Europe by the World Health Organization in the 1990s, definitive answers arise. The guidelines of the WHO "recognize sleep disturbance, which can take place at noise levels above 30 dBA, less than the level of traffic and infrastructure noise in many cities, as a significant health risk" (Berglund et al. 1999, section 4.2.3.).

As with so many aspects of noise annoyance, precise measurement and the drawing of lines between acceptable and unacceptable conditions can be a complex matter, not only because individual reactions vary, but also because social customs around sleep are not universal. Research into the history of sleeping customs suggests that, for centuries, people routinely broke their sleep cycles for an hour or two to engage in quiet activities, talk, or sex, and then returned to sleep (Ekirch 2006). Many equatorial cultures with the custom of sleeping during the hottest part of the day return to work at night, then take a short sleep until dawn. While medical research concludes that a minimum of five hours of uninterrupted sleep is essential to the maintenance of optimal physical and mental health, it is likely that the "normal" eight-hour nocturnal sleep cycle is an invention of the North European industrial culture of the nineteenth century and a result of artificial lighting. In any case, consistent interruption of any habitual sleep pattern is both annoying and potentially detrimental to health and productivity. Frequent disturbance of sleep can lead to levels of stress that themselves become habitual, eroding the restorative benefits of the time spent in sleep. Noise, particularly night-time noise above 50 dBA, is conclusively recognized as a major risk factor for sleep disturbance and, therefore, for general health.

As mentioned in Chapter 4, urban residents are especially prone to sleep disturbance: traffic, transportation infrastructure, and housing density all contribute to the problem. Traffic noise at typical intensity for urban neighbourhoods, usually averaging 45–55 dBA at night, is sufficient by itself to produce disturbances in sleep, with disturbance rising as the noise gets louder (Bluhm, Nordling, and Berglind 2004; Stosić, Belojević, and Milutinović 2009). Railway noise has long been considered less disturbing than traffic or aircraft noise at equivalent loudness, an assumption that has influenced noise control legislation in several European countries. This conclusion has been questioned, however, by researchers in Switzerland who surveyed 1,643 people about their use of sleeping medication to counteract the effects of noise at night. They found no significant difference for levels of disturbance between railway and traffic noise, at least for railway noise above 55 dBA (Lercher et al. 2010), which is at the louder end of the averaged scale for nocturnal ambient noise. Since traffic and trains produce quite different frequency patterns, the discrepancy in results suggests that both conclusions may be correct: train noise might be less disturbing than traffic until it reaches a particular level of loudness, but disturbance is disturbance. Whether the sound that wakes you is a fire engine or a partner's snoring, you are experiencing noise annoyance.

JUST BE *QUIET!* SENSITIVITY AND MOOD

Noise annoyance affects mood as well as sleep. Individual sensitivity to sonic irritation can be determined by hearing acuity and by the presence of other stressors such as pain, overwork, or insomnia. Additional influences come from living and working conditions, physical and social environment, side effects of medications, individual brain chemistry, and the deeply complex factor known as "personality." Attempts to study the influence of personality on noise sensitivity and distraction have led to confusing results, in part because existing studies raise some perplexing questions about the design and ethics of such research. Experiments involve asking volunteers to complete verbal and mathematical tasks with backgrounds of silence, music, or noise. Results are then analyzed with reference to standard scales that classify traits like extroversion and introversion as well as neuroticism, which is highly correlated with startle reflex and anxiety. Determining degrees of these traits, as well as their significance, can be a complex task because of varying cultural influences as well as differences in the types of music and noise used. One British study, comparing equal numbers of introverts and extroverts in laboratory conditions, found that both groups were equally efficient at cognitive tasks in silence, while the introverts were more likely to be distracted by both noise and music (Furnham and Strbac 2002), a finding common to subsequent studies of this type (Reynolds, McClelland, and Furnham 2013). A group of Serbian researchers, also finding that introverted personalities were more intensely affected by noise than extroverted ones, suggested that sensitivity to noise may be correlated with anxiety-prone personalities (Belojevic, Jakovljevic, and Slepcevic 2003). The team's concluding recommendation that employers screen job applicants on the basis of their reactivity to noise in the workplace makes their research ethically questionable, however, especially in light of findings that an individual's mood at the time of encountering annoying noise, as well as a wide variety of personality traits, may have some bearing on the degree of annoyance (Västfjäll 2002). The relation of hearing acuity to noise sensitivity raises another question: do introverts spend less time in noisy recreational venues, and have a lower incidence of noise-induced hearing loss as a result?

The association of noise with emotional stress is particularly evident in situations where noise triggers a traumatic stress response. Combat veterans are susceptible to this phenomenon, as are people who have lived in war zones. Studies of post-traumatic stress disorder (PTSD) in American veterans of the Vietnam War found that an exaggerated acoustic startle response accompanied by elevated heart rate is a characteristic symptom of the condition (Orr et al. 1995; Morgan et al. 1996). A later study involving Vietnam veterans who had non-combatant identical twin brothers proved that PTSD was in fact a condition acquired in combat and not a genetic vulnerability (Orr et al. 2003). Children exposed at an early age to the sounds of combat may be especially susceptible, and may react with fear to any sudden loud sound. For the same reason – the evolutionary triggering of fight-or-flight reactions by sudden noise, exacerbated by combat experience or traumatic exposure to its soundscape – members of the military returning to civilian life are at increased risk for aggressive reactions to noise annoyance. This is an area in need of more investigation, not just from the medical research community but from the perspectives of social work and counselling psychology.

I CAN'T HEAR IT, BUT IT *HURTS*: ELECTROMAGNETIC HYPERSENSITIVITY

In 2004 the World Health Organization held an international conference in Prague on a new and growing phenomenon: people reporting discomfort and illness associated with a collection of symptoms that included skin irritations, muscle pain, physical and mental fatigue, dizziness, nausea, "brain fog," heart palpitations, headaches, blurred vision, and digestive irritations. Most reported an association between the symptoms and prolonged exposure to electronic devices: computers and mobile phones, and later cellphones and cell towers. The syndrome, first reported in 1976 by a computer engineer in Sweden (Hviid 2010), came to be known as electromagnetic hypersensitivity syndrome (EHS) or electromagnetic frequency (EMF) sensitivity. Since the range of electromagnetic frequencies is far beyond the hearing threshold of humans, they are not considered to be within the traditional definition of noise as an auditory phenomenon, but the syndrome fits the acoustic definition of noise as interference: some individuals – estimated to be between 0.5 and 3 per cent of the population in industrialized countries – can apparently experience activation of the auditory nerve by ultrasonic frequencies, which may be experienced as tinnitus or as soundless physical symptoms.

EMF sensitivity is currently regarded as controversial in medical literature because no direct cause for it has yet been found. One controlled experiment carried out in laboratory conditions found that there was no difference between sufferers and the control group in ability to detect the presence of EMF (Eltiti et al. 2007),[15] another reportedly did detect differences.[16] The possibilities for research and alleviation of symptoms are complicated by the fact that complainants often refuse to have testing or treatment because the electromagnetic fields given off by medical equipment in hospital emergency rooms and surgical suites, as well as doctors' offices, intensify their symptoms (Granlund-Lind and Lind, 2004). This is especially problematic if surgery is required for an injury or pre-existing medical condition, since those affected by EHS are concerned that their recovery will be impaired.

Medical explanations for the condition are scarce at this time. Studies done in the 1990s usually dismissed the idea of such sensitivity being possible, but questions are now being raised in the wake of other conditions – like AIDS, fibromyalgia, chronic fatigue syndrome, and Lyme disease – that were initially dismissed and later validated by more carefully designed medical studies. Consideration is now being given to more precise investigation (Baliatsas et al. 2012) and to re-examining theories of causation. Some theories hold that the symptoms are caused by environmental allergies and stress (Gangi and Johansson 2000; Baliatsas et al. 2014) or by poor air quality, lighting, and ergonomics in the workplace (Granlund-Lind and Lind 2004). Others suggest that chemical compounds in the brominated flame retardants that permeate bedding and furniture in industrialized countries, which are now routinely found in human tissue, might also have some connection with the development of reactivity.[17] Dental amalgams that combine gold with other metals in the mouth are being examined as well, because they produce weak but significant electrical conductivity (Bergdahl, Anneroth, and Stenman 1994; Walker et al. 2003; Sutow et al. 2004), which can have impacts on the auditory nerves and the auditory processing centres of the brain.[18]

Another view of the condition relates it to mast cells, produced in allergic reactions by an excess of histamine flooding the mucous membranes of the body and setting off sneezing, wheezing, eye irritations, itching, swelling, headaches, and a host of other symptoms that range from annoying to life threatening. Since many individuals with EHS also report environmental allergies (Gangi and Johansson 2000; Hviid 2010) – sensitivity to mould, dust, chemicals, and changes in barometric pressure or weather – and experience EHS symptoms that mimic allergic reactions, this line of investigation may be promising; however, attention is also circling back to ultrasound itself as a direct cause.

Dr Timothy Leighton of Southampton University in England, an award-winning biomedical acoustician, points out that the general urban public in industrialized nations is now constantly exposed to airborne ultrasound from cellphone towers and personal devices in the absence of valid scientific knowledge about their effects on health. Since current guidelines for safety pertain only to occupational exposure, and measurement of airborne high frequencies is extremely difficult, he believes that current guidelines for safe exposure are "based on an insufficient evidence base, most of which was collected over 40 years ago by researchers who themselves considered it insufficient to finalize guidelines" and that reports of "a range of subjective effects (nausea, dizziness, migraine, fatigue, tinnitus, and 'pressure in the ears')" resulting from occupational exposures should motivate more and better-designed investigations into the risks of continuous public exposure (Leighton 2016, 1–2; 2017).

Leighton critiques the state of medical literature on the subject as unreliable, with most of it composed of manufacturers' claims (Leighton 2016, 2) pitted against each other as a substitute for actual evidence. After taking extensive measurements in public venues including rail stations, schools, and sports stadiums, he recommends that new guidelines for safe and unsafe exposure – independent of those used to this point – are urgently needed and that "research must be undertaken in recognition that the fundamental science upon which all existing guidelines have been based is not sufficiently substantial" (Leighton 2016, 3).

In view of the current uncertainty about causes, treatment of EHS is challenging. Patients are advised to avoid using computers and/or cellphones if symptoms are mild, but this can affect employment options as well as social networks. In severe cases, they may have to retrofit their homes with shielding or relocate to areas not covered by cellphone networks – which is increasingly difficult. While awareness of the condition is relatively poor in North America, some areas of Europe are already providing assistance. Sweden, which has the longest history of familiarity with EHS, now provides some hospitals with EMF-shielded emergency and treatment rooms and offers subsidies to cover the costs of shielding homes or relocating. Some as-yet-unproven methods for neurological reprogramming claim success at diminishing symptoms, attributing the sensitivity to past emotional trauma and employing the theory of neuroplasticity to condition the brain into a less reactive state through specific sets of exercises.[19] Research might eventually validate some of these techniques, but for now their claims are based entirely on anecdotal evidence. There is also a vast unregulated market of devices – to wear, install on computers, plug into sockets, or attach to walls – that are advertised as cures. Until more information is validated and shared, this leaves many people vulnerable to fraud. As with so many other aspects of noise sensitivity, far more research is needed. So is more attention to the needs of noise-sensitive and EMF-sensitive people.

6

The Hearing Body

Hypertension is an important independent risk factor for myocardial infarction and stroke, and the increased risk of hypertension in relation to aircraft and road traffic noise near airports demonstrated in our study may therefore contribute to the burden of cardiovascular disease.

Jarup et al. 2008, 333

It is estimated that DALYs [disability adjusted life years] lost from environmental noise are 61 000 years for ischaemic heart disease, 45 000 years for cognitive impairment of children, 903 000 years for sleep disturbance, 22 000 years for tinnitus and 654 000 years for annoyance in the European Union Member States and other western European countries. These results indicate that at least one million healthy life years are lost every year from traffic-related noise in the western part of Europe.

World Health Organization, Regional Office for Europe 2011, xvii

We recognize, evaluate, and react to environmental sounds even while asleep.

Münzel et al. 2014, 831

Most of our current and growing knowledge of the effects of noise on health comes from decades of experiments that measure physical, cognitive, and emotional reactions to sound. Experimental evidence gives us information on the process of hearing and on how sound is interpreted by the brain, as well as what can go wrong with the systems involved. Industrial work and participation in the military have long been identified as causes of hearing damage; they have recently been joined by music. If played too loudly for too long, it will damage hearing as thoroughly as will the sounds of industrial machinery or artillery. While tinnitus and noise-induced hearing loss are the most common effects of noise on health, the most obviously severe impact is destruction of the eardrum through extreme sound pressure. A soundwave of sufficient amplitude – for example, one resulting from an explosion – becomes a shock wave, and if it pushes air against the sensitive eardrum with enough force, the membrane ruptures and the inner ear is injured, producing total and incurable deafness. But the ears are not the only potential site of injury.

Subtle effects on the entire body are now being recognized as significant both for their immediate impact and because they can increase risk for serious illness: the circulatory and nervous systems are both affected by common levels of ambient urban noise. One of the most extensive studies on noise as a factor in population health, the 2002–3 Pan-European Large Analysis and Review of European Housing and Health Status (LARES) project, measured traffic and ambient neighbourhood noise occurring within and around apartment buildings in several cities. It then correlated the measurements with assessments of residents' health. The research teams found that ambient urban noise was linked in adults with symptoms related to the heart and nervous system as well as with depression and migraine.

One of the central challenges in medical research on the effects of noise is dealing with the theoretical division between "damaging" and "non-damaging" categories of exposure. While it is clear to many people that noise can be damaging to the auditory system, there is less awareness of what are called the "non-damaging" aspects of noise on the entire body. The term "non-damaging" is somewhat misleading. It means that the effects of long-term moderate noise exposure are not – strictly speaking – *immediately* damaging. Instead, symptoms caused by this form of exposure accumulate over time, and repeated exposures add up to heightened risks for chronic illness.

Many "non-damaging" effects of sound include effects from noise that falls below the threshold of attention: this exposure happens routinely at home, at work, or during sleep. The potential for harm in such cases is not necessarily to the ears but to the cardiovascular system. It is here that noise moves to, quite literally, the heart of the matter: long-term exposure to "non-damaging" levels of noise can gradually disturb the rhythm that governs your life.

THE HEARING HEART:
CARDIOLOGY, HYPERTENSION, AND AMBIENT NOISE

While attempts to assess the effects of noise on the risk of heart attack were controversial within the medical community for some time (Babisch et al. 2005; Willich et al. 2006), noise from road traffic is now strongly implicated as an aggravating factor, along with aircraft noise and even the general ambient noise of urban neighbourhoods (Sobotova et al. 2010; Tomei et al. 2010; Obelenis and Malinauskienė 2007; Ndrepepa and Twardella 2011). The risk of hypertension – high blood pressure – is also a major concern, and chronic long-term exposure to occupational noise has the potential to cause elevated blood pressure as well as abnormalities in heart rhythm (Tomei et al. 2010).

Although some early studies on blood pressure were too small in scope and short in duration to give clear results, more recent research has expanded to include measurements taken with increasingly accurate technology, and the studies are more conclusive: loud noise is now firmly associated with increased heart rate, increased arterial blood pressure, and risk of heart attack (Obelenis and Malinauskienė 2007). Factory workers are at particular risk.[1] One early investigation, a large-scale Israeli study of factory workers consistently exposed to industrial noise, measured levels of cholesterol and triglycerides in the workers' blood and found a correlation between noise-induced stress and risk factors for heart disease. The most marked effects were seen in males under the age of forty-five, possibly because risks multiply and are more difficult to isolate after that age, while inconclusive results were reported for females of all ages (Melamed et al. 1997).

A preliminary study conducted in an Italian metalworking factory investigated blood pressure in factory floor workers whose hearing was already damaged by workplace noise, conducting electrocardiograms as well as multiple measurements of blood pressure and comparing the group with two control groups who were not exposed to significant workplace noise. Researchers concluded that noise exceeding 90 dBA was a risk for raising blood pressure as well as damaging hearing, and that auditory damage was linked to cardiovascular damage, although they cautioned that individual responses vary (F. Tomei et al. 2000). Their finding was later confirmed at a metalworking factory in India, where exposure to continuous machine noise at or over 90 dBA was associated with increases in arterial blood pressure (Singhal et al. 2009). At a metalworking factory in Korea, 530 workers were tracked annually from 2000 to 2009, with comparison among four groups of employees: office workers exposed to less than 60 dBA of noise on the job, work site inspectors intermittently exposed to levels of 50–85 dBA, factory floor workers steadily exposed to more than 85 dBA and using one type of hearing protection, and factory

floor workers using two types of hearing protection for the same exposure. The research team found small but statistically significant rises in
systolic blood pressure in all groups except the office workers, with the
largest rise shown in the group of factory workers with one source of hearing protection (J. Lee et al. 2009).

Industrial workplace noise may not be the only sonic hazard to heart
health: the soundscapes of large clerical workplaces, urban traffic and commuter trains, and urban neighbourhoods are also under scrutiny. A meticulously documented small-population study in Japan compared blood
pressure and pulse responses in three groups of participants: one group listened to three types of music, another to steady noise, and the third to fluctuating noise samples contoured to resemble the samples of music
(Sakamoto et al. 2002). While listening to music produced only minor
changes in blood pressure, probably explained by emotional arousal, researchers found a consistent and medically significant rise in blood pressure with exposure to steady noise at a loudness level of 54 dBA and above.
Since 54 dBA does not represent an unusual level of loudness – it is easily
reached in an office, for example – this suggests that some degree of physical harm related to noise might occur well below the levels of loudness
usually considered hazardous.

Do the findings of the studies of noise and blood pressure indicate a
clear danger for everyone? Reported differences in pressure amounted on
average to only 2–5 mmHg,[2] not a dangerous rise for healthy individuals.
Since occupational health studies are done on subjects screened for reasonable health in order to avoid putting anyone in danger, the reported
risks are considered statistically significant but not conclusive. For individuals already diagnosed with hypertension, rises in blood pressure
equivalent to those found in the factory studies could represent a hazard.
The risk is small on average, but highly significant to the individuals
affected and to their families. Knowledge of risks is particularly important
to workers who are exposed to more than one factor: specifically, males in
middle age with a diagnosis of hypertension along with smoking, excessive weight, *and* employment in high-noise conditions.

Since sound impacts the entire body and not just the ears, hearing protection may not provide a successful barrier against the effects of noise on
the cardiovascular and nervous systems. A recent analysis of medical literature on the subject recommends that industrial workers routinely
exposed to high levels of noise be given electrocardiograms on a regular
basis (G. Tomei et al. 2010).

Airports, Roads, and HYENAS:
Is Your Blood Pressure Rising?

The low roar or whoosh that is always audible in the background of city soundscapes consists largely of transportation noise: freeways and major streets, bus routes, aircraft flight paths, and subway and elevated trains. Its components, ubiquitous at levels well below the decibel danger zone for hearing damage, are the sounds of commerce and convenience, essential to city life. Researchers have been exploring its effects on urban residents for the past three decades, and the results are clear: whether the ambient roar excites or annoys you, it is affecting your health.

Air transportation is a major cause of noise-induced physical stress. Jet engines routinely produce levels of sound pressure far above the threshold for hearing damage and well into the range of physical pain. Airport runway workers are required by law and by union regulations to wear hearing protection, and airports – at least those in parts of the world where population health is a matter of concern – are monitored for noise levels and held responsible for strict noise abatement programs. Far lower levels of sound pressure affect the general public in the proximity of airports and flight paths, but the potential for annoyance, aggravation of stress, and even harm to health is real. Aircraft noise at 55 dBA or above was estimated to affect 44 per cent of the North American population in 2005, compared to 18 per cent for both Europe and Asia (He 2010). A controlled study of the health and stress levels of two thousand residents of US neighbourhoods adjacent to an airport in the Minneapolis–St Paul area found that "all health measures were significantly worse in the neighborhoods exposed to commercial-aircraft noise. Respondents with the worst health status tended to be among those experiencing commercial-aircraft noise of the greatest severity" (Meister and Donatelle 2000).

The risk of hypertension is now known to increase with regular exposure to aircraft noise, at least for males. Indications for females are less clear. An early Japanese study conducted only on women living near airports and comparing them to suburban women for blood pressure levels did not find any differences, although it was found that the airport group had greater likelihood of smoking and daily drinking, possibly as a response to stress (Goto and Kaneko 2002). However, a later large-scale study of 29,000 people living near US military airfields in the Ryukyu Islands of Japan found a significant correlation between aircraft noise levels and blood pressure regardless of sex. The risk of hypertension was reported to increase with the loudness of the noise and was inversely proportional to how far test subjects lived from the airport (Matsui et al. 2004).

The most comprehensive study of aircraft noise exposure to date has been the Hypertension and Exposure to Noise Near Airports (HYENA) study (Jarup et al. 2008). This wide-ranging European study found that exposure to aircraft noise indicated a risk for hypertension, especially over long periods. Over 4,000 test subjects aged forty-five to seventy who had lived for at least five years near major airports in London, Berlin, Amsterdam, Stockholm, Milan, and Athens were studied. The data – even when multiple socio-economic, diet, and general health factors (country, age, alcohol consumption, education, and exercise) were taken into account – demonstrated the subjects' marked risk for hazards to cardio-vascular health. All it took was a 10-dBA increase in exposure to night-time aircraft noise above general noise volumes to produce changes in blood pressure.

People living under airport approach routes may report that they are accustomed to the noise and that it does not particularly bother them. Why, then, can it produce measurable challenges to their health? The reasons lie in the science of psychoneuroimmunology, whose convoluted name illustrates its focus: interactive connections among the brain, nervous system, and immune system. Mind and body, far from being separate entities, are in fact aspects of a unified system. The brain is a physical organ as well as the matrix of thought, and its intricate chemistry is duplicated in other sites in the body: the heart and large intestine produce the same hormones that regulate and communicate stress reactions and emotional responses in the brain. When illness upsets the balance of health, both physical and emotional systems are affected. A familiar example is the common cold, which is more likely to overwhelm your immune system when you are tired or stressed. The difference between resisting the virus and succumbing to it is often to be found in your patterns of sleep, a critical factor in the healing process and also in the maintenance of health. Stress is detrimental to both. Since noise affects the quality of sleep, it affects immune-system responses. Compromised immunity results in compromised health.

It doesn't take living near an airport to cause cardiovascular changes, however. The HYENA research team also found that increases in average daytime road traffic noise of 10 dBA over normal exposure yielded similar results to the airport studies in hypertension risk. A branch of the team in Greece, investigating night-time blood pressure and heart rate through measurements taken every fifteen minutes on 140 subjects during sleep, found significant rises in blood pressure along with small changes in heart rate (Haralabidis et al. 2008). A cross-sectional public health survey conducted in Sweden on more than 24,000 adults aged eighteen to eighty concluded that round-the-clock exposure to road traffic noise was a significant risk for high blood pressure, especially in people aged forty to sixty (Bodin et al. 2009).

Since the HYENA studies, other lines of investigation are showing that living near major transportation routes can affect the body in several ways. Analysis of hospital admissions and death records in the city of London suggests that long-term exposure to traffic noise at and above 55 dBA, considered a moderate level in large cities, is affecting the health of 1.6 million residents – approximately 20 per cent of the city's population – by increasing their risk for cardiovascular incidents, especially stroke in those aged sixty-five and over (Halonen et al. 2015).

The intricate feedback loops between body and brain are amply illustrated by the effects of both sudden and chronic noise exposure on the cardiovascular system. Noise can affect heart rhythm as well as blood pressure. When a sudden loud noise wakes you at night, parts of your brain are activated by auditory signalling to produce hormones that start the behavioural fight-or-flight response. As your body subconsciously prepares to run or fight, your heart rate increases in order to provide more oxygen to your muscles.[3] When you realize that the sound was harmless, your rhythm returns to normal. This example is a familiar one, but researchers are also finding subtler effects: chronic exposure to moderate noise can disrupt the variability of people's heart rate.

Although heart rhythm seems to be perfectly regular, it actually varies by microseconds of time between beats, even at rest. This tiny variability, resulting from the communication of heart anatomy with the nervous system and measurable only with diagnostic equipment, is now known to be a sign of heart health. When variability is consistently decreased, it can serve as a warning of high risk for heart attack (Stauss 2003).[4] Since research on ambient traffic noise shows that even moderate levels, below 65 dBA, have the potential to alter micro-rhythms, increasing attention should be paid to the need for aligning urban policy planning with medical science: keeping urban traffic noise controlled and setting building codes for efficient noise insulation will save both lives and health care expenditures.

Night Noise

The HYENA study noted that noise was potentially more harmful to the heart at night than in the daytime (Jarup et al. 2008, Münzel et al. 2014). Why are aircraft and traffic noise of particular concern at night, if residents can sleep through them? One reason is that the nocturnal soundscape is more quiet, and noise produced by aircraft or heavy transport trucks stands out against the background more distinctly than it does in the daytime, when continual traffic masks some of the impact, putting the nervous system on alert to some degree even in sleep. Another is that the nervous system reacts to sound in sleep in a variety of ways. Sudden loud noises at night wake you as a response to potential danger, while gradual increases in loudness – like the accustomed approach and passing of aircraft and traffic – may simply change the level of sleep or produce somatic effects without your being aware of them. The changes in blood pressure documented by studies mentioned above were not enough to cause harm to healthy individuals, but could be of concern to people with chronic hypertension, especially if the condition is unrecognized and untreated.

Shifts in the level of sleep are not necessarily trivial.[5] Since the purpose of sleep is to allow the body's systems to recover from the stresses of daily activity and release toxins, frequent interruption of the process is detrimental to general health as well as heart health. Synthesis of nutrients, appetite regulation, insulin secretion, hormonal activities, and memory formation happen during sleep. Blood pressure normally dips slightly in deep sleep, as part of a complex chemical reaction that allows the cardiovascular system to regenerate from the stresses of the day. Repeated waking at night prevents all these processes from functioning optimally (Dettoni et al. 2012; Münzel et al. 2014). Interruption of sleep is also a concern because of increases in heart rate that accompany partial waking (Griefahn et al. 2008) as well as the general effect of disrupted sleep on health. An analysis of the impacts of nocturnal transportation noise in Europe cited it as a major cause of sleep interruption, and stated that "habitual short sleep ([less than] 6 hours per night) is associated with obesity, diabetes, hypertension, cardiovascular disease, and all-cause mortality" (Münzel et al. 2014, section 4).

Because proximity to a sound source intensifies the reception of sound, even the way that residences are arranged on city streets is an important variable in determining whether ambient noise will interrupt sleep. A study in Sweden found that residents in apartments are more prone to noise-relat-ed sleep disturbance than those in detached houses, and recommends that bedrooms should not be built facing the street (Bluhm, Nordling, and Berglind 2004). Sleeping with closed windows will reduce the likelihood of disturbance from street noise, but in hot seasons or climates it may not be feasible. These factors expand the subject of noise-induced sleep distur-bance from the well-defined medical realm into the vast reaches of sociolo-gy. Access to a quiet environment, or at least to one in which noise is muted and constant rather than loud and intermittent, appears to be central to the quality and consistency of sleep. Such access is readily available in rural communities and in many suburban ones, but low-noise environments within cities are generally limited to high-income residential districts. Low-cost housing is usually situated close to major traffic routes, rail lines, or air-ports, and built with inadequate insulation against noise. Poverty is there-fore (because of noise and so many other reasons, including high stress and poor nutrition) a major cause of chronic illness.

Growing knowledge of the effects of noise on health leads to an ethical imperative to take action: reduction of urban noise through regulation, abatement, and social consensus has the potential to produce improve-ments in health and quality of life for people of all ages in all social and economic conditions. Keeping cities as quiet as possible, especially at night, has the potential to save lives and to reduce the life-changing effects of heart disease and stroke, thus producing savings in the huge expendi-tures for health care systems around the world. Thanks to the LARES and HYENA studies conducted under the auspices of the WHO, the European Union (EU) currently has the strongest and best regulations for noise con-trol in the world. Canadian federal policy follows the regulations of the Canadian Centre for Occupational Health and Safety (CCOHS) for work-place noise, but regulation of night-time levels and controls on noise from utilities are under provincial jurisdiction. China is beginning to take the matter seriously and is currently conducting studies in its most populous cities, with India following closely. In the United States, the Environmen-tal Protection Agency estimated in 1981 that 100 million people were routinely exposed to noise at levels that would harm their health, but Congress has not revised regulation of noise since 1972, despite rising numbers of complaints in major cities (Hammer, Swinburn, and Neitzel 2014). Along with legal prohibitions against pollution of air, water, and farmland, policies for controlling noise in the interest of population health should have international scope and must include legal limits on pollution of the soundscape.[6]

The Mind's Ear

Reading, as soon as we attend to its sensorial texture, discloses itself as a profoundly synaesthetic encounter. Our eyes converge upon a visible mark, or a series of marks, yet what they find there is a sequence not of images but of sounds, something heard; the visible letters ... trade our eyes for our ears.

Abram 1996, 124

The inherent promiscuity of the senses recurs in everyday as well as literary or artistic language. We report "chilling sights" and "hard looks"; voices variously grate or caress, while the eyes of lovers conventionally reach out to touch or devour the adored other. The visible becomes audible in dramatic ways ...
Colours are represented across all the senses, but register most forcibly in terms of noise, as in "a riot of colour" or "screeching yellow."

Bailey [1998] 2004, 27

Reading is normally superimposed on a foundation of listening. The ability to listen seems to set limits on the ability to read.

Lundsteen 1971, 3

Along with receptors for the physical process of perceiving sound, described in chapter 2, the human auditory system includes components for interpreting what is heard: the auditory-processing regions of the brain. Different sites on the primary auditory cortex, located on the temporal lobe of the cerebrum, are responsible for sorting sound frequencies, determining the sound's direction of origin, equalizing the time-lag that results from sounds reaching one ear before the other, distinguishing whether the sound is random or patterned, reacting to levels of loudness, and interpreting different types of sound. Other sites on the human auditory cortex are dedicated to recognizing patterns of rhythm and frequency, enabling the development of language. The organ of Corti, along with the cochlear nerve that carries its signals to the brain, facilitates balance as well as hearing and thus plays a role in all rhythmic motion: crawling, walking, skipping, jumping, running.

Auditory sites are also found on the brainstem, which relays and coordinates signals between the peripheral (sensory and motor) nerves and spinal cord and the upper regions of the brain, where signals are interpreted and voluntary responses are initiated or inhibited. The brainstem controls autonomic (involuntary) functions including alertness, arousal, breathing, blood pressure, and heart rate. All of these functions are shared to varying degrees with other mammals and with all vertebrates: they are integral to the ability to wake from sleep, move in rhythmic gaits, react to danger, and locate food. Our identities as mammals, as humans, and as individuals all incorporate the ways in which our auditory systems respond to the world.

Although hearing is the "second sense" in our species, taking up a smaller proportion of cerebral real estate than vision, such territorial concepts are not entirely accurate. Sensory perception may be less easily divided and categorized than we have been led to believe (Seitz et al. 2007; Shimojo and Shams 2001). Rather than five discrete senses, we experience a spectrum of sensory cues to which our brains and nervous systems respond with mixed signals as well as distinct ones.

The mixing of sensory cues, called *synesthesia*, is extreme in some individuals, who may experience music as a flow of colours or taste spoken words. In its full manifestations, synesthesia is associated with measurable differences in brain function and may have a genetic basis. While technical synesthesia is rare, some interweaving of sensory cues in everyday experience is quite common. When you see a plate of fresh muffins, just out of the oven in the kitchen or bakery, you smell them at the same time and can easily anticipate their taste.[1] Now look at a picture of muffins in an ad: can you virtually smell and taste them?[2] Does a picture of a lemon, the word *lemon*, or even seeing the colour of one cause you to salivate? If so, you are experiencing an everyday form of synesthesia. In fact, the integration of sensory perceptions into an overlapping spectrum is not even limited to physical experience. We constantly summon sensory memories,[3] and use them in the service of the imagination, constructing or reconstructing scenes and conversations, journeys, emotions, events, meals. Memory feeds imagination as familiar experiences are moved into the service of new contexts to provide the basis for creativity, humour, and even science fiction movies that echo with the calls of alien monsters. We are able to remember sounds as well as visual images, tactile and kinesthetic sensations, odours and tastes.[4] Is the voice of a former friend still familiar in your memory after many years' absence? Can you recall the sounds of the street you lived on when you were a child, as well as the way it looked?

Visual imagination – the stuff of dreams, fantasies, films, art, animation, and music videos – is deeply familiar territory to most of us and also to neuroscientists. Its intersection with auditory imagination is perhaps best exemplified by radio drama. Since visual details – the facial and physical characteristics of the characters, their surroundings, and costumes, all clues to social status, culture, age, ethnicity, and personality – cannot be conveyed directly, they must be implied by a combination of dialogue, vocal tone, accent, and sound effects. In the absence of a set, a scene played outdoors might require recordings of bird song or traffic in the background, and its historical veracity can depend on whether the traffic noise involves motors or hooves. In the absence of cinematic close-ups, every scene depends on the actors' ability to project emotion by means of vocal tone quality, timing and inflection of phrases, and breath control. The listener is led to form images of the scenes and characters, effectively creating a personal film that is cast, costumed, lit, and decorated by the operation of auditory-visual synesthesia.

Less attention has been given to auditory imagination, although it is equally familiar: the silent rehearsal of anticipated wording for a request, the delayed "staircase wit" retort, and the recalling of dialogue from a film are some of its most common manifestations. Auditory imagination has rarely been studied, although it is starting to attract the attention of neuroscientists investigating the improvisation of music. Chilean researchers scanned brain activity in musicians and in a control group as the subjects imagined and recalled musical phrases and then imagined modifications to them. All of the musicians could evoke and modify imagined musical phrases. In contrast, half of the control group could evoke, while only 17 per cent were able to modify, and their imagined phrases were less elaborate than those of the musicians (Goycoolea et al. 2007). This suggests that auditory memory and imagination, along with actual listening skills, can be strengthened by training and practice. The importance of doing so? Learning processes, literacy, and language skills all depend on auditory processing, which is not strictly limited to input from the sense of hearing. It is intimately connected with the senses of touch, motion, and vision. Such intersections make possible some of our species' most remarkable skills: language, music, dancing, reading, and writing.

HEARING WITH THE BODY,
SEEING WITH THE EARS

The reception of sound waves is not limited to the ears: it is tactile as well as auditory. The vibrating sound wave strikes skin and then bone and the watery substance of individual cells with approximately equal force, although only the ear has evolved to amplify the wave and transmit it to the brain. In fact, the tactility of sound contributes to its auditory qualities, since the physical process of hearing is initiated by an act of percussion – compressed air striking eardrum – and sustained by vibration within the middle and inner ears. The intersection of sound with touch is most easily demonstrated by listening to live music with low frequency components. If you visit a church or concert hall with a traditional pipe organ and hear it in full play, you will experience the low bass notes through the floor as well as the air; they will vibrate your feet, legs, and spine as you hear them. If rock concerts are more to your taste, you already know that the bass will drive you to move, that your dancing is incomplete without it.

Drums alone will drive dancing because rhythm activates the basal ganglia, a part of the brain that governs voluntary motor control and the selection of action (Grahn and Brett 2007; Grahn and Rowe 2013). Much of the traditional music of Africa, which overlaps individual drum patterns into an intricate tapestry of communal polyrhythm that provides the impetus for spontaneous movement, is based on that connection; so are jazz, rock, and hip-hop, its American descendants. The most compelling drum of all, of course, is the most subtle: the heart. Its *short-long, short-long* pattern is the basis for speech, poetry, and dance in a variety of cultures. Ancient songs and dances pair it with lively melodies, Shakespeare's iambic rhythms mimic it, the rapturous waltz of nineteenth-century Vienna swept Europe into a frenzy of (*ONE-two-three, ONE-two-three*) spinning embraces.

Musicians draw on the intersections of hearing and touch with heightened attention to the feedback loop between the interaction of body and instrument – arms and hands to keys or strings, breath to length and speed of phrase, entire body to rhythm – and the sound produced by that interaction. Talented dancers inhabit the feedback loop, perceiving and responding to music with movement in the moment rather than simply counting beats. The best actors and orators ride the interplays of speech, breath, and meaning like surfers on waves. Mothers in all cultures rock their infants while crooning. Even before birth, our nervous systems are

formed in darkness but not in silence: the fetus floats in sound-conduct-
ing fluid, moving to the muffled whoosh and boom of the mother's
heart, breath, and voice. Regardless of race or culture, gender or language,
we all begin with bodies that vibrate and move to sound.

Even our perceptions of space and orientation within it are influenced
by hearing. In *Spaces Speak, Are You Listening?*, their ground-breaking
study of the aural properties of landscape and architecture, interdisciplinary researchers Barry Blesser and Linda-Ruth Salter (2006) point out that
the ability to perceive space in considerable detail by means of sound
exists in our species, although it is unlikely to be developed or even recognized except by individuals with significant visual impairment and
their teachers. It is a "hidden" skill, developed by evolution but overshadowed by the visual perception of space (309–16). Their examination of the
auditory-spatial sense delves into history and neurology, but also into the
emotional and aesthetic effects of sound as it is shaped and amplified by
architectural space and ambience, concluding that architects and communities need to balance the visual qualities of design with far more
attention to its qualities of amplification and reverberation of sound.[5]

The integration of hearing with spatial orientation and motion enables
fine motor skills as well as whole-body movements. For example, taking
notes during a meeting or lecture, whether by handwriting or typing,
depends on the intersection of auditory processing with haptic memory,
the ability to recall motion. Also called "muscle memory," haptic memory
is present in all of us but most strongly developed in the bodies of athletes and dancers and the hands of musicians. Experienced pianists can
benefit from virtual practice on tabletops because their brains have
learned to map hand motion to key position and key position to musical
pitch. They can even recognize familiar pieces of music by watching
another player's hands on a video screen without any sound cues (Hasegawa et al. 2004). Haptic memory plays a role in the learning of language
through speech and writing as the muscles of mouth, tongue, hands, and
fingers become accustomed to particular patterns. Like the clarinet or sax-
ophone student practising scales to develop an effective *embouchure*, children learning to speak begin to link facial muscles to sounds. When they
start to read, links are built between sounds and visual symbols; writing
adds hand motions to the network of audio-visual connections.[6] Research
comparing the effectiveness of handwriting and typing now suggests that
longhand – which uses individualized movements for each letter – fixes
written content in the memory more effectively than typing (Mueller and
Oppenheimer 2014).[7]

Our brains connect sound directly to vision as well as motion. Audio-visual connections are easily demonstrated by the skill of lip-reading, or even more universally by the layers of information added to speech by facial expression, posture, and gesture. Sounds stimulate parts of the visual cortex in certain circumstances (Molholm et al. 2002; Romei et al. 2007; Vidal et al. 2008), most strongly in the blind but also in sighted individuals (Cate et al. 2009). The interrelation of these sensory modes is not surprising, although its manifestations are sometimes unexpected.[8] A recent discovery in brain neuroscience confirms that auditory signals can activate peripheral vision: sounds that are perceived as significant signal the areas of the brain that process vision to *prepare* to see the source of the sound as you turn your head to find it, even when it is initially outside the range of vision (Cate et al. 2009). This neural connection had obvious survival value when the source of the sound was a leopard that almost – but not quite – managed to sneak up on you. The leopard sometimes benefited, though, from another phenomenon discovered in the same experiment: sounds that do not attract attention do not activate peripheral vision. Evolution favours the wary but protects against sensory overload: if every sound in the environment required us to have a look at its source, not much else would get accomplished. Our nervous systems are finely tuned to balance attention to the environment with mental focus. When an environment is polluted with excessive noise, the balance is destroyed. Important signals are masked and distraction overcomes focus. When the complex process of learning new skills is accompanied by high ambient noise, learning can suffer.

LISTENING TO LEARNING

Among the early predictors of educational success for children – including social class, parental culture and income, and nutrition – the most directly significant are listening skills and literacy. Listening to spoken information and instruction is central to education in any culture, whether it involves the socialization of children or the introduction of older students and workers to new skills. Reading is the key to future employment and economic success in a technological culture, as well as social proficiency and the navigation of complex issues in a modern democratic society. It might even be said that both functional democracy and the post-industrial economy depend on a literate population, and people with challenged literacy skills face considerable disadvantages.

Learning disabilities affect the majority of high-school and college dropouts in North America and Europe, and undoubtedly contribute to their reasons for leaving school (Bruck 1987; Murray et al. 2000). Functional illiteracy is common among prison inmates, both youth offenders and adults. While marginal literacy is certainly not a cause of criminal behaviour, people with poor or non-existent reading skills are typically denied access to educational and employment opportunities open to those for whom reading comes easily. For this reason, literacy is often seen as a key to economic development for entire nations as well as for individuals. The ability to read well opens doors to prosperity and contentment for large numbers of people. It is a skill valued and promoted by benevolent governments because it affects national productivity and therefore prosperity.

The causes of substandard literacy and learning skills are many: malnutrition, poverty, inadequate instruction, cultural pressures and stereotypes, disabilities. Another, less obvious, cause is the distraction and confusion caused by noisy conditions in the home or school. Neuroscientists are now aware that auditory memory, the ability to store and recall sounds, is a crucial component in the process of developing vocabulary and reading skills (Čeponienė et al. 1999; Glass, Sachse, and von Suchodoletz 2008). The development of auditory memory is enhanced by auditory activities: speaking to and with the child, reading aloud, singing, and training in music (Trainor, Shahin, and Roberts 2003; Shahin, Roberts, and Trainor 2004). This is because language is, at least initially, a largely auditory activity. Its progress depends upon several levels of interaction that involve the auditory system (hearing and listening), vocal apparatus (formation of speech sounds with the larynx, lips, tongue, teeth, and palate), respiratory system (regulation of air flow using the diaphragm and lungs to control the speed and loudness of speech), and parts of the brain that govern the comprehension of speech as well as its rhythms and inflections. The pre-verbal child can produce an enormous range of abstract vocal sounds – babbling, crooning, and squealing – that are gradually edited, through listening to surrounding adults, down to the vocabulary of a native language. Language learning in its earliest stage is in part a matter of listening for which sounds elicit responses from adult caregivers and therefore carry meaning. Our speech follows a feedback loop: we listen as we speak, learning as children to make subtle adjustments in phrasing, pacing, tone, and loudness in order to optimize meaning and response from others. As children we also learn to imitate familiar speakers, picking up the accents of our parents unless they are modified by the surrounding community or the school. The speech of children with congenital deafness is hesitant and "out of tune" because such feedback is lacking, and diligent practice through speech therapy is necessary to compensate for the lack of internal auditory reinforcement.

Even reading has an auditory foundation. The process of learning to read involves sub-vocalizing, sounding out phonemes and words internally by means of the auditory imagination, in a reprocessing of the auditory and phoneme-formation skills practised by the very young child learning to speak. The practice of silent reading is a fairly recent development in historical terms. In traditional oral cultures, information is conveyed by speech and preserved in memory. The growth of literacy in late medieval Europe – when hand-copied books were rare and readers were few – was marked by the *cognoscenti* reading aloud to others, and even to themselves.[9]

Because reading results from so many complex interactions, impairment at any point – in hearing, vision, perception and processing of visual and auditory patterns, distinguishing of speech from background sounds and ability to form accurate phonemes, as well as ability to sustain attention on an abstract task – can produce challenges beyond the usual ones. The specific neurological processes involved in both normal and impaired reading are not yet fully understood. Research on learning disabilities is controversial and sometimes inconclusive, with differing schools of interpretation vying for recognition. It is now evident, however, that auditory perception plays an important, though not exclusive, role in the process of learning to read and the lifelong enjoyment of literacy. When this process takes place in the presence of excessive noise, distraction leads to confusion, and crucial neuronal connections can be delayed or poorly developed.[10] This is true of very young children learning to speak, older ones learning to read, and anyone learning a second language (Nelson et al. 2005).

The Role of Auditory Processing in Learning Disabilities

Among the causes of poor reading skills in adults – which include ineffective preparation in childhood and undiagnosed visual and hearing disabilities – are problems with the processing of auditory signals. Children with weak reading skills have difficulty in discriminating between sounds in speech, a skill that depends on detection of sound frequencies and intervals of time. The difficulty persists into adulthood: while adults who received additional training were able to improve their scores on skill tests, they were still not fully fluent (Ahissar et al. 2000), although training with computer programs has been found to be helpful for children (Tallal et al. 1996).

While ambient noise in classrooms is widely recognized as a potential distractor from learning processes in all children, the effect is particularly marked in children with attention deficit disorders (ADD). One US study that examined a wide range of aggravating effects concluded that "the classroom situation with its high stimulation level (noise, visual distractors, large class size) is likely to reveal or accentuate instability, impulsivity and inattention" in children with the ADD diagnosis (Purper-Ouakil et al. 2004). However, one group has been shown to benefit from selective exposure to a particular category of noise, at least in laboratory conditions: children with diagnosed attention deficit hyperactivity disorder (ADHD) appear to be beneficially stimulated in conditions that are detrimental for children without the disorder (Uno et al. 2006; Söderlund, Sikström, and Smart 2007).[11] Experimenters used electronically generated white noise in a controlled laboratory setting as the stimulant,[12] so the effect may well not apply to other types of noise, but the authors of one study found that such "noise exerted a positive effect on cognitive performance for the ADHD group and deteriorated performance for the control group, indicating that ADHD subjects need more noise than controls for optimal cognitive performance" (Söderlund et al. 2007, 840). This result may occur because the noise stimulates cognitive activities in the brain that are otherwise impeded by the particular chemistry of ADHD, and – in the case of white noise – does so without the distraction of specific auditory content. Thus, while ordinary ambient noise during learning activities is likely to be especially harmful to children with ADHD, controlled use of soft electronic noise through earphones may be helpful.

Further evidence for the significance of auditory processing in ADHD is suggested by the finding that children with the condition have specific difficulties with speech: "Children with ADHD were perceived to have significantly more hoarseness, breathiness, and straining in their voice[s]. They were also louder compared to controls" (Hamdan et al. 2009). The authors of this study conclude that children with ADHD show vocal behaviour that differs from normal development. Recognition of the interlocking processes that drive development of hearing and speech production in children may well produce new insights into the diagnosis and treatment of ADHD as well as related conditions.[13]

Impaired auditory processing, which can occur with normal hearing because it affects brain function and not the auditory system itself, is also implicated in dyslexia (Shaywitz et al. 2002), which is now known to be a result of difficulty with auditory processing of phonemes, the sounds that make up language (Schulte-Körne et al. 1998; Boets 2007).[14] Dyslexic children have trouble recognizing rhymes and distinguishing syllables within a word, skills that help build understanding of the connection of symbolic visual patterns to speech sounds (Stefanics et al. 2011). Dyslexic adults have greater difficulty than normal readers in distinguishing speech elements from background noise, even when their hearing is normal (Chait et al. 2007). An Israeli study of adults classified as "disabled readers" found that approximately a third of them had trouble detecting differences in sound frequency and loudness, direction of sound sources, and distinguishing tones from background noise: "Taken together, these results suggest that a large portion of disabled readers suffer from diverse difficulties in auditory processing" (Amitay, Ahissar, and Nelken 2002). A study that looked into the high rate of failure in US literacy education in the 1990s concluded that phonological processing – the ability to divide spoken words into their component sounds – was predictive of future literacy skills, and defects in the process caused a high risk for dyslexia in pre-school children (Tallal 2000).

Phoneme awareness is influenced by *sonority* – the relative loudness of a spoken sound in relation to others of the same length, stress, and pitch (Yavas and Gogate 1999) and by rhythm. Each language, and each regional accent or dialect within it, has its characteristic speed of delivery, length given to syllables, and accentuation patterns. These, even more than phonemes, are what distinguish human speech from computerized replicas: the computer can reproduce phonemes and frequencies, but cannot yet imitate fully realistic rhythm. Rhythm and accentuation patterns may also distinguish native speakers of a language from reasonably fluent foreign speakers, and expert fluency involves mastery of them. So do effective public speaking and acting. Talented actors, comedians, and individuals famed for their skills at oratory all demonstrate finely tuned awareness of timing, pacing, phrasing and accentuation in speech. All depend on auditory attention. In fact, some modes of learning depend on listening as their main foundation.

Auditory Learners

When learning new information, do you feel a need to repeat or summarize it vocally in order to remember it? Do you verbalize phone numbers before entering them, translating the written number symbols in your phone or phonebook into the names of the numbers? Does comprehension "click" and memory "ring a bell"? Did you talk, chant, or hum to yourself as a child for reassurance or for confidence in unfamiliar situations, and do you still want to do so? You may feel a need for constant musical background to your actions, or for complete silence to avoid distraction. In either case, you may be what is known to theorists of learning styles as an *auditory learner*.

Theories of learning styles recognize a large spectrum of individual and categorical differences in modes of learning and retaining information.[15] Some are based on preferences for abstract or concrete reasoning and for linear or random patterning of cues, others on sensory modes of processing information, including visual patterning and hands-on experience. While the concept of learning styles is controversial in the literature of cognitive science because quantitative studies are scarce and have not been conclusive (Pashler et al. 2008), it is sometimes used in primary and secondary education, and can be helpful for students, teachers, and parents if regarded as a menu rather than a dogma.

The *auditory learning mode*, based on speaking and listening as the preferred means of transmitting information and recognizing significance, is part of the VARK (visual, aural, read-write, kinesthetic) system developed in the 1980s as an aid to students and teachers in primary and secondary schools (Fleming and Baume 2006). Students who identify with the auditory mode – "auditories" – are especially responsive to information gained through lectures and podcasts, and get particular benefit from discussion, verbal question-and-answer reviews, and storytelling exercises. Auditories are said to be particularly sensitive to nuances of inflection and timing in speech and music, and may have talent as public speakers, actors, teachers, writers, storytellers, and/or musicians. They are also potentially more alert to soundscapes and more easily distracted by noise, but in fact *all* learners – especially children and anyone learning a new language or a new skill – are susceptible to the effects of masking and distraction caused by ambient noise. This is especially true of noise within and around classrooms.

THE CLASSROOM AS A SOUNDSCAPE

Vast amounts of research have been conducted since the 1970s on classroom noise: an Internet search reveals more than 63 million results. Much of it has been driven by fear of lost productivity due to distraction, but there is also a larger issue to consider: neurological reactions to noise, and how they affect cognitive development in children and ability to concentrate in adults. Noise is a problem for the process of learning because learning depends in large part on two foundational skills, listening and reading. In the former, physical sound in the form of instruction or verbal interaction is central. In the latter, cognitive connections between vision and speech are established and reinforced. If learning to read involves auditory processes, learning in general also involves auditory awareness and attention. Whether situated in a formal classroom, a kitchen, or a tent, learning involves verbal interaction as well as reading. When the student cannot hear or pay attention to the instructions or dialogue, much of the content in both verbal and mathematical instruction is lost (Pimperton and Nation 2010). Intrusive noise from outside the place of learning can be a cause, but noise inherent in classrooms – including the mechanical sounds of ventilation systems as well as the noise made by active children – can also be a significant problem. Noise produces distraction, luring students' attention away from the often repetitive tasks that constitute early stages of learning. Its most direct detriment, however, may result from masking, in which one acoustic signal effectively hides or scrambles another, as when the roar of a ventilation fan causes students at the back of the room to miss what the teacher is saying about preparing for the math quiz.

Such auditory clashes, sound against sound, can be minimized by the acoustic design of classrooms, but the design of school buildings rarely takes acoustics into account except to insulate gyms and music rooms. A pamphlet produced by the Technical Committee on Architectural Acoustics of the Acoustical Society of America in 2000 warned that many US classrooms had a speech-intelligibility rating of only 75 per cent. This meant that a child in the middle grades with average, unimpaired hearing and full fluency in English was missing every fourth word spoken in the room (Seep et al. 2000).[16] Children with English as a second or subsequent language and those with any degree of hearing impairment – whether genetic, congenital, or resulting from temporary ear infections – could be expected to miss even more of the content of spoken instructions. So could those with attention deficit disorders. Small children in the earliest grades, unable to rely on known context to keep them alert to what was actually being said, were found to be at particular risk.

Open-plan classrooms, popular in the 1960s and '70s for their symbolic levelling of traditional teacher-student hierarchies as well as for their flexibility and the relatively low cost of construction, were considered by the Acoustical Society researchers to be especially problematic. Even when they are modified with moveable partitions that reduce visual distraction, the overlapping sounds from several classes or study groups that are active simultaneously can produce distraction and confusion. Open-plan construction declined toward the end of the twentieth century with demands for more traditional styles of education, but it is now being revived as a method for decreasing construction costs (Shield, Greenland, and Dockrell 2010). Continued use of older open-plan buildings, as well as their ongoing construction, requires careful planning of schedules to avoid placing noisy activities next to tasks best conducted in quiet, as well as effective use of space so that students are close enough to their own teacher to hear instructions. Even the competing voices of teachers in open-plan schools can distract students from following their own teacher: in one study, the "irrelevant speech" of instructions overheard from adjacent teaching areas was reported by the children themselves as an annoyance (ibid.). Whether the noise is recognized as a problem may depend on whether the children are accustomed to noise or quiet in the home as well as on cultural expectations. But lack of recognition is not the same as lack of harm: another investigation of speech perception and listening comprehension found that children were not necessarily aware of classroom noise as a detrimental environment even when it was having significant negative effects on their performance (Klatte, Lachmann, and Meis 2010).

Noise is of particular concern where children are involved because the neural circuitry of their brains is still developing and their attention can be easily fragmented. Since critical attitudes to learning are also in formative stages, any interference with success can set up expectations that inhibit confidence.[17] Children are not usually able to diagnose the causes of their fear or frustration, and parents may be at a loss to do so. A complaint like "I can't understand the teacher" – when student and teacher are fluent in the same language – might lead parents to schedule a hearing test and a meeting with the teacher, but testing the acoustics of the classroom is unlikely to occur to them as a potential solution to their child's increasing frustration with school.

Classroom noise is also a problem for teachers, especially those working with young children. Instructions have to be repeated frequently and loudly over the ambient noise of building systems, mechanical toys, noisy activities, and chatter. Unless teachers are provided with the techniques that enable them to project their voices without strain, vocal fatigue and sore throats result (Södersten et al. 2002). These can be overcome with rest and with appropriate vocal exercises, but prevention is the best policy. The design of classrooms and awareness of ambient noise can contribute to a remedy, even when activities within the classroom continue to generate noise. Adequate insulation of walls and the use of sound-absorbing materials for ceilings and floors is especially important in classrooms for the primary grades. Additional solutions are now being provided by free noise-monitoring apps for smartphones that will give a decibel reading and present interactive graphics to show students how much noise they're making, enabling teachers to motivate awareness of noise levels and how to reduce them (Carmichael 2017).

LEARNING AND COMMUNITY SOUNDSCAPES

Just as classroom acoustics influence what is learned, the soundscape that surrounds a place of learning can predispose students to success or failure. If ambient noise in or near the place of instruction is consistently above 55 dBA, the upper limit recommended by the World Health Organization, masking of verbal instruction and distraction of attention can easily result. This is often the case in schools located near airports, trains, or highways (Bronzaft n.d.; Bronzaft and McCarthy 1975), in urban schools, and in buildings with open-plan classrooms or insufficient soundproofing. Since high-noise conditions are most likely to occur in densely populated urban areas with low budgets for school maintenance and improvement, support from school boards, communities, and governments is crucial.

Attention to the design and location of schools is especially important for communities with a large proportion of families who are not familiar with the dominant language of the culture, be they recent immigrants or established minority groups. Like children first learning to speak, both children and adults learning a new language must absorb an unfamiliar set of phonemes, rhythms, and intonation patterns. Comprehension and retention depend on hearing them accurately. Knowledge of the issues involved – which include student success, graduation rates, and employability – is important for governments and school boards as well as educators and parents, particularly when use of earbuds is already challenging the hearing of many children and teens.

In order to determine whether a particular type of noise affects learning or work, measurements of cognitive performance are taken for a variety of tasks that may involve proofreading, verbal reasoning, dexterity, and timed reactions. Tests account for variables such as age, physical condition, and hearing acuity. Responses are measured with and without the noise being investigated, and there is usually a comparison in the form of a common, less noticeable noise. If performance of the tasks is significantly hindered by the investigated noise, it is deemed to be harmful to cognitive performance.

Studies of this type have firmly established the detrimental effects of noise from aircraft flight paths and urban traffic on learning. Research conducted in England (Matsui et al. 2004) and across Europe (C. Clark et al. 2006; Stansfeld et al. 2005) confirms earlier findings that aircraft noise can interfere with auditory perception, aggravate stress, and disrupt learning environments. Aircraft noise was also found to have stronger effects than traffic noise at the same levels of loudness, possibly because it contains components of low-frequency noise (Clark et al. 2006). The pan-European study, which included eighty-nine schools located near airports in England, the Netherlands, and Spain, tracked the test scores of 2,844 children aged nine to ten years. Statistics were adjusted for such variables as language spoken at home and education level of parents, as well as diagnosed dyslexia and hearing impairment (ibid.). After all adjustments, a correlation remained between aircraft noise and delays in learning, whether the noise was experienced at school or at home. The authors concluded that a chronic environmental stressor – aircraft noise – could impair cognitive development in children, especially with regard to reading comprehension.

Schools exposed to high levels of aircraft noise are now regarded as unhealthy educational environments (Stansfield et al. 2005). Given the preference of most parents for schools located close to their homes, residents of communities near airports are likely to receive this information with dismay. The choices for improving their children's options without sending them outside the community are numerous, but not inexpensive. One is providing the best possible noise insulation for school buildings. This is best done in the planning stage for new schools, but repairs to an existing building can afford an opportunity to increase insulation by installing sound-absorbent drywall and/or flooring materials. Amplification systems can improve speech-comprehension rates for lecture rooms in secondary schools and universities (Arnold and Canning 1999; Crandell and Smaldino 1999), although they may not compensate for the effects of masking and poor listening skills (Smaldino and Crandell 2000).

Because staff and students in schools with high noise levels may be so accustomed to the soundscape that they do not notice it, objective measurements of loudness levels are necessary. Simply asking people –

especially children – whether they are bothered by the noise will not provide useful information about whether they are affected by it, but a noise audit will show which areas of the school receive the most impact.[18] Insulation can then be increased where needed, and activities within the building organized to minimize effects. Restructuring of schedules with attention to noisy and quiet activities, and to peak periods of traffic and aircraft activity, can help to reduce the distraction.

Case Studies: Classrooms across Cultures

Along with aircraft noise, consistent effects on learning result from the soundscape of traffic in major cities and from the activities of children within classrooms. Poor insulation of walls and windows contributes to the effect, but so do the location of schools and the social customs that govern tolerance for noise and appropriate behaviour for children in school. Three studies from Europe and Canada published in the same year, among many investigating the detrimental influence of noise on learning, provide examples of the questions that have shaped research:

- How much noise is there? (UK)
- Does the noise have impacts on learning? (Canada)
- Does the noise affect emotions and behaviour as well as learning? (Macedonia)

Taken together, the three studies suggest a constellation of factors that, if accounted for and reported, might guide the design of future research and result in even better comprehension of the problem.

The British study took measurements of ambient noise outside 142 elementary (primary) schools in London, situated away from major airports and flight paths. The major source of external noise was road traffic, measured at an average of 57 dBA. Measurements were then taken inside 140 classrooms during a variety of activities. The majority exceeded the WHO threshold, and – not surprisingly – levels depended on the number of children in the room. The external noise was not considered by the researchers to be a major problem except when children were involved in quiet activities. No specific conclusions were drawn about indoor noise except that it was over recommended levels (Shield and Dockrell 2004). This represents a basic level of investigation: quantifying the levels of noise by means of measurements taken with sound-level meters and reporting them as decibel levels. The study is typical in finding that noise levels were consistently above the thresholds recommended by the WHO.

The Canadian study went a step further, examining forty children, ages five to eight, for the effects of noise on a specific skill: speech comprehension while indoor ambient noise averaged at 65 dBA, which was taken to be typical of classroom ambient levels. The researchers found that all children had some difficulty understanding what they heard. At lower levels of noise, the youngest children (Kindergarten and Grade 1) had considerably more difficulty than the older ones. The authors conclude: "These results suggest that the youngest children in the school system, whose classrooms also tend to be among the noisiest, are the most susceptible to the effects of noise" (Jamieson et al. 2004, 508). This type of study requires the researchers to administer cognitive tests to students as well as taking quantitative measurements of noise levels, and then to interpret the results of the tests to determine whether there was a cause-and-effect relationship between noise and cognitive effects. Whether the results hold true for all communities or just the one studied in a particular case depends on the size of the population studied (small in this case, but convincing when combined with many similar studies) and on whether the results are similar to other studies of comparable groups.

The third investigation involved yet another set of measurements. In a study of urban and suburban schools in Skopje, Macedonia, researchers took neighbourhood ambient noise readings with sound-level meters, averaged them for eight- and sixteen-hour periods, and found the urban district sound levels to be above 55 dBA and the suburban levels below it. They then administered several psychological tests to a group of approximately 260 school children, aged ten to eleven, in each district. Samples of the children's saliva were also taken in order to measure stress hormones. Socio-economic differences among the children's families were accounted for, and did not prove to be significant. The researchers found that children exposed to noise above the WHO threshold during school hours, a situation that is common in urban schools everywhere, had "significantly decreased attention and social adaptability, and increased opposing behavior in comparison with school children who were not exposed to elevated noise levels" (Ristovska, Gjorgjev, and Jordanova 2004, 473). Although anxiety levels were not found to be significantly affected by noise, other psychological and emotional consequences – self-reported

Table 7.1
Suggested parameters for further research on classroom noise

Conditions/environment	Students	Effects to be studied
Outdoor soundscape (dBA measurements)	Age	Sound masking (e.g., audibility of teacher's speech)
Indoor soundscape (dBA measurements)	Gender	Distractions
Location (urban, suburban, rural)	Number studied	Emotional states
Structure and condition of school buildings	Tests administered, including hearing acuity	Behaviour
Degree of sound insulation in walls and ceilings	Social customs, cultural norms influencing children's behaviour	Cognitive development (measured)
Number of individuals per classroom	Language(s) spoken in the classroom	Acquisition of skills (measured)

stress, attention deficits, and increased levels of the stress hormone corti-sol – were found. The authors concluded that attention should be paid to chronic noise exposure in urban schools when assessing the psychological health and learning skills of children. They also reported that the coping strategies used by children to screen out distracting noise might themselves cause deficits in attention to spoken instructions.

When taken together, the research results from each of the studies show that ambient noise in urban schools is often excessive and that speech comprehension is affected. The question about emotional and behavioural reactions is still open, however, because cultural expectations of children's behaviour differ, complicating interpretation of results. Did the Canadian or English children show any behavioural effects as a result of noise exposure, like the Macedonians, or could the degree of aggression shown by the Macedonian children under the influence of noise have been at a level considered normal in the UK and Canadian cultures? Such questions might be answered more definitively by accounting for a greater variety of parameters in future research. Additional information to consider would include the gender of children, the teachers' expectations of noise and quiet, the children's activities and hearing acuity, and even the amount and quality of soundproofing materials in classrooms.[19] Even the language spoken in the room might prove significant, since languages vary in tone qualities and vocal production (focused, breathy, nasal, guttural, tightly or loosely articulated, and so on). Consistent recording and comparison of such details would lead to a better understanding of how best to allot resources to solving the problem. Conclusive research that can be used to guide policies about noise abatement in schools should ultimately account for a range of environmental, architectural, developmental, social, and cultural variables, as summarized in table 7.1.

Follow-up studies and longitudinal tracking of test subjects over a decade or more might theoretically bring to light any links between noise-induced learning problems in childhood and long-term effects in adults, an area that has apparently not been studied. However, such projects would probably be ruled out by the difficulties of getting long-term compliance from parents – and later, the children themselves – as well as by the complexity of consistently tracking noise levels. Nevertheless, such investigations might eventually provide critical information about the importance of soundscapes in processes of learning.

8

Born Loud

Evidence has been accruing to indicate that young children are vulnerable to noise in their physical environment. A literature review identified that, in addition to hearing loss, noise exposure is associated with negative birth outcomes, reduced cognitive function, inability to concentrate, increased psychosocial activation, nervousness, feeling of helplessness, and increased blood pressure in children.

Viet et al. 2014, 105

[I]nfants are at a greater disadvantage than adults when processing speech in noise and … concern over the effects of a noisy environment on the acquisition of language is justified.

Nozza et al. 1990, 339

Just as noise can disrupt and delay the development of connections in the brain that underlie learning, it can affect the physical and mental health of children. As awareness grows regarding the importance of home and school environments on the development of babies and young children, positive and negative influences are becoming clear. Nutrition, family income and social status, parenting styles, and media are all coming under current scrutiny for their effects on the physical, mental, and emotional growth of children – a recognition that environmental factors contribute to shaping both physique and personality.

In 2009 a team of scientists from two prominent US universities and the National Institute of Child Health began an investigation of feasibility and costs for a planned National Children's Study, with the intention to track five thousand children from before birth to age twenty-one. Their plan, which will span two decades, includes the monitoring of noise levels in the homes of participating families as well as questionnaires related to noise annoyance. The research team's preliminary estimate, based on data published in 2008, is that almost thirty million adults and more than five million children in the United States already "suffer from irreversible noise-induced hearing impairment, and more than 20 million are exposed to dangerous levels of noise each day" (Viet et al. 2014).

The National Children's Study should, in time, produce important information that will help to fill gaps in knowledge about the specific effects of noise on children. While ambient noise in the classroom has been studied extensively, ambient noise in the home and even in the expectant mother's workplace have not; nor is it known whether children are more sensitive to noise annoyance than adults, or whether they are affected by electromagnetic frequencies to a greater extent. Noise begins to affect the health and hearing of children well before they are ready for school, and continues to do so as they interact with their environments and develop preferences in recreation and music. In fact, the acoustic ambience of our lives starts before we are born. Along with an expectant mother's heartbeat, breath, voice, and digestion, her baby hears quite a lot in the womb, but does not consciously process the soundscape that surrounds them both. If the mother is consistently subjected to noise annoyance, does the chemistry of stress hormones in her nervous system incline her baby to heightened sensitivity? If her surroundings make her calm and contented, will her baby be that way? What exactly is known about the effects of sound on expectant mothers and fetal development?

MUSIC AND MATERNITY

In the absence of extensive knowledge about prenatal noise reactions, we can start with what is suspected about fetal responses to music. Because the ability to hear develops *in utero*, many parents believe that playing music during pregnancy will enhance a child's intelligence. Research suggests that newborns have a memory of melodic lines played to them in the last trimester of pregnancy (Granier-Deferre et al. 2011; Partanen et al. 2013) for about four months, and that hearing the mother's voice and heartbeat may enhance brain development in extremely premature babies (Webb et al. 2015). Animal studies suggest that prenatal music listening advances the development of spatial learning in day-old chickens (Sanyal et al. 2013) and rat pups (Rauscher, Robinson, and Jens 1998), but there is more to intellect than spatial orientation, and far more investigation on humans needs to be done. Nevertheless, an extensive industry of advice and products – from "Baby Mozart" recordings to speakers and audio belts meant to be worn over the womb by pregnant women – has been launched, with major claims but little validity about outcomes for the baby (Swaminathan 2007).[1]

The consensus at this point is that hearing music during pregnancy will probably not affect the child's intellect or personality in measurable ways, but that it will relax and/or cheer the expectant mother and consequently flood the fetal nervous system with beneficial neurotransmitters, neurohormones, and other important biochemicals that induce calm and contentment. Listening to music while pregnant is worth doing, with or without special equipment, whatever the eventual effect on the child turns out to be.

HEARING BEFORE BIRTH

The human auditory system develops at twenty weeks of gestation. Although it takes far longer for sounds to be interpreted by a baby's brain, the sounds themselves are heard in the fetal stage, and reactions can be observed with special monitoring equipment. Most prominent are the sounds of the mother's heartbeat and digestive processes (Parga et al. 2018). It is now known that the mother's speech is also audible. Just as newborns learn to move their jaws, tongues, and lips in patterns similar to their parents' speech well before they can produce the appropriate sounds, fetuses at thirty-two weeks begin tiny stretches of their mouths in parallel with their mothers' speech patterns (Reissland et al. 2016). The developing muscles that will eventually produce speech are gradually conditioned to move in response to sound.

Less is known about the impact of ambient noise as it is perceived within the womb. What *is* known indicates that the human fetus startles at loud noise. Fetal startle responses are measured by applying a device called the artificial larynx, which generates a brief 100-dBA burst of sound to the skin surrounding the pregnant womb. This intervention, which is presumed to be harmless to the fetus because part of the sound pressure is absorbed by the mother's body and the amniotic fluid, produces both sound and vibration, causing the fetus to react with a quick burst of movement (Divon et al. 1985). The startle response develops at twenty-seven to twenty-eight weeks of gestation, and appears to be sex-specific by thirty-one weeks: males show a stronger response and a longer developmental curve, suggesting that their nervous systems may mature more slowly than those of females (Buss et al. 2009).

What all of this tells us is that the body is affected by sound even before the mind develops a state of awareness, and the instinctive startle response is only one example of how we begin the life-long process of hearing, absorbing, and listening to noise. Can the soundscapes of pregnancy and infancy account for sensitivity to sounds? Do these sensitivities then lead to the ways in which our children react and learn how to deal with the environmental sounds that parents and the environment place in their lives?

In the absence of extensive knowledge, there is a temptation to attribute small children's reactions to noise simply to personality, raw intelligence, or social milieu. No doubt these are all important in determining a child's ability to learn and develop, but could there also be contributing factors that come from the earliest exposure to ambient noise? Does the pregnant mother's response to noise affect her baby? Would high exposure accustom the baby to noise and reduce reactivity, or increase the startle reaction and raise sensitivity?

NATAL NOISE

Recently developed technology makes it possible to record sounds heard inside the pregnant womb, enabling speculation about what a developing fetus can hear. Intrauterine sound is quite loud. Sound outside the womb itself is also perceived. Although it is modified by amniotic fluid and masked by internal sounds, noise can affect fetal development.

Since exposing pregnant women and newborn babies to potentially harmful noise is not an option for researchers because of ethical concerns, laboratory experiments to determine risks for noise-related harm are done on birds and animals.[2] From these investigations, scientists are able to estimate – by comparing anatomical similarities in the auditory and nervous systems of the animals with those of humans – what risks might be faced by human newborns. Evidence of harm can be a subtle matter: does it take the infant laboratory rat exposed to noise longer to develop its instinctive skills than a rat of the same age gestated in a quiet environment? Is there any reduction of its hearing acuity or ability to navigate using sound cues? Is its behaviour the same as others, or different?

Animal studies confirm that exposure to loud noise before birth – unlike exposure to music – delays the development of spatial orientation and memory in day-old chickens (Sanyal et al. 2013; Kumar et al. 2014) and spatial learning and neural development in baby rats (Kim et al.

2006). Low-frequency noise is implicated specifically in the underdevelopment of the respiratory system in newborn rats, as well (Oliveira et al. 2001). Proof that exposure of pregnant mammals to noise can be perceived by their fetuses – and that it can alter the development of the newborn's auditory system – has come from experiments on sheep (Griffiths et al. 1994; Huang et al. 1997).

Evidence regarding human pregnancies and newborns also comes from situations where mothers-to-be have been exposed to excessive ambient or occupational noise, not for the purpose of research but because of where they live or how they make a living. When their babies are brought in for routine medical check-ups, evidence of low birth weight, delayed development, or health problems may be recorded by their clinics and later analyzed by researchers. The science of noise-related health risks to babies has been built by collecting small fragments of evidence over a long period of time and then assembling them into a picture, like the placing of mosaic tiles to make up a scene.

It is now believed that newborn humans are not endangered by ambient neighbourhood noise. Prematurity, birth weight, and immune-system response do not seem to be affected by it, although there are still questions about blood pressure (Hohmann et al. 2013).[3] However, there is reason for concern with noise in the expectant mother's workplace that stays consistently at or above 85 dBA: it is associated with low birth weight, premature birth, and abnormalities in the development of the immune system (Nurminen 1995; American Academy of Pediatrics 1997; Sobrian et al. 1997). Since factory work employs hundreds of millions of women worldwide, the issue is not a small one. In 2010, more than 50 million women in the United States alone were employed in the manufacturing sector (US Department of Labor, 2011). In the ongoing exodus of manufacturing jobs from countries where wages are high and protective laws available, to those where there are few laws and no unions, young women are often the cheapest labour available and therefore desirable as factory workers. They are enticed, or forced by poverty, to take jobs that are stressful or even unsafe just at the time when they are starting families. Worldwide, 85 per cent of garment factory workers are female. The potential degree of risk to their hearing and their babies' health was suggested by sound-intensity measurements taken in textile factories in Vietnam (averaged levels above 90 dBA; Nguyen et al. 1998) and Thailand (101.3 dBA; Chavalitsakulchai et al. 1989). Both sets were well above the threshold for hearing damage to the mother – and distress to the fetus

– during an eight-hour workday. A more recent study confirms the presence of oxidative stress, a biochemical cause of damage to DNA, as well as hearing loss in Turkish textile factory workers exposed to 105 dBA (Yildirim et al. 2007).

Early investigations suggested that high noise conditions during pregnancy could produce strong reactions to noise in newborns whose mothers were reactive (Ando and Hattori 1970) and might be related to low birth weight (Ando and Hattori 1973). The risk of early spontaneous abortion was also found in one study to be slightly elevated in factory workers on shift work routinely exposed to workplace noise levels consistently above 80 dBA (Nurminen and Kurppa 1989). Because most research studies were too small to produce fully conclusive results (American Academy of Pediatrics 1997), questions persisted about the difference between slightly raised risks and definitive cause-and-effect relationships between prenatal noise exposure and low birth weight. Clear correlation was found, however, with prenatal hearing impairment in a meticulous large-scale Swedish study, especially if noise over 75 dBA at the expectant mother's workplace contained low-frequency components (Selander et al. 2016). The expectant mother's use of hearing protection at work is important for her own well-being, but it will not protect her fetus: because sound waves strike the entire body, the mother's abdomen cannot be a sound-proof chamber. Quite the contrary: although most sound waves, including those produced by high-frequency ultrasound scanning devices, are somewhat muted by uterine fluid (Glover 1995), waves of low-frequency infrasound are amplified, and mechanical noise in factories usually includes infrasound components (Richards et al. 1992).

When a baby is born and begins the process of learning how the world operates, whatever conditions are present in the home become the standard for what is familiar and therefore "safe." If a soundscape of loud or layered noise is constantly present, children come to regard it as normal. It does not attract their attention: without much experience of quiet for comparison, the child assumes that the world is usually loud, and develops the ability to tune out noise at hazardous volumes rather than being alarmed by it.

Nor do parents necessarily recognize where limits should be set. Loudness has become a potent symbol of the youth culture, designed for teens and young adults, that dominates the music industry with a vast machine for generating profits. The ability of a home entertainment device, a car stereo system, or a popular music event to generate high decibel levels is a selling point: the attraction is freedom to create your own world, to make yourself (represented by your music of choice) heard, to drown the pressures of looming responsibility in a sea of music. Aging out of the "youth" category does not change the attraction of loud entertainment, nor does the next cultural category – and marketing demographic – for many: parents of young children.

Is your radio turned up, or your music device docked into speakers that blast at maximum volume, while the baby is sleeping and you have time to vacuum? Is the big-screen television on loud enough for you to hear it in the kitchen, over the sounds of the refrigerator and blender, while your toddler sits in front of it? What effect is this soundscape having on the child's hearing acuity and attention span?

In the course of the twentieth century, ambient broadcast noise in the home became normalized; since the 1940s, recorded conversation, music, and the urgency of advertising have been a constant presence in the lives of children from birth. So little attention has been given to the phenomenon of domestic mechanical sound that even the "sleep machines" now marketed to promote sleep for babies have been found to exceed sound pressure levels hazardous to hearing if left on for eight hours. In the absence of regulation for their manufacture, no limits have been set (Hugh 2014). Loud noise is even presented as fun, an idea that has driven the growth of industries for children and youth since the 1960s, when it became a selling point for toys advertised directly to children and particularly to boys.[4] Toys today – firecrackers, cap pistols, pellet guns, and other toy weapons, as well as mechanical and pull-toys, to say nothing of players used by parents as educational devices – can be detrimental sources of sound pressure levels sufficient to cause high frequency hearing loss, tinnitus, and even perforated eardrums (Yaremchuk 1997; Fleischer 1999; Segal 2003).

PLAYING WITH HEARING LOSS

Canada sets limits on noise emissions from toys by law rather than recommendation, which makes it a rarity, but the maximum decibel level, established in 1970, is considerably higher than the limit favoured by audiologists today. Health Canada caps sound emissions from toys for children of all ages at 100 dBA, a level the World Health Organization considers safe for only fifteen minutes per day. A report published in 2004 with the sponsorship of Industry Canada tested forty toys, finding that two exceeded the limit, two more were measured at 90–99.9 dBA, and the remaining thirty-six were at or below 90 dBA (Charbonneau and Gold-schmidt 2004, 20). The authors pointed out that little was known about how the auditory systems of children from newborn to age three reacted to noise, and that adult tolerance of 85 dBA for an eight-hour workday might not apply to children, particularly very young ones who were like-ly to hold toys close to their ears. When measurements of the toys' noise emissions were taken at distances that imitated the habits of children under three observed in two daycare settings, most "considerably exceed-ed the WHO 7.5 minutes safe exposure limit" and were likely to cause hear-ing loss in less than ten minutes (ibid., v).

In the United States, limits are by recommendation, so a national con-sumer organization has stepped up to the task of investigating toys. The Sight and Hearing Association (SHA) based in St. Paul, Minnesota, pub-lishes an annual list of toys that produce hazardous levels of noise, timed to appear on its website in anticipation of the Christmas buying rush.[5] It regularly includes products sponsored or manufactured by known and trusted companies like Disney, Fisher-Price, Hasbro, Mattel, and Sesame Street. SHA researchers measure the sound pressure of each toy at two dis-tances: right next to the ear and at arm's length, averaged at ten inches for children. In 2013 the website warned that seven of the eighteen toys test-ed were measured at levels over 100 dBA, which can potentially impair a child's hearing in fifteen minutes or less: whether the impairment is per-manent depends on how close the toy is held to the ear. On the 2019 list are twenty-four toys, nineteen of which are louder than the 85-dBA for eight hours limit for adult occupational safety (an eight-hour time span is unlikely for toys, but the threshold for harm to children is suspected to be much lower). One surveyed toy, the Fisher-Price "Smart Learning Home" equipped with a solar panel to teach about energy conservation, sounds off at 104.6 dBA.[6]

Standards for the manufacture of sound-producing toys were pub-lished by the American Society for Testing and Materials (ASTM) in 2003, but they are regarded by audiologists as excessively lenient. ASTM Regulation 963 limits toys held close to the ear to 70 dBA for averaged continuous sound at a distance of ten inches (twenty-five cm), which is considered to be arm's length for a child – but small children are more likely to hold sound-producing toys right next to their ears. Hand-held, table-top, and crib toys are limited to 90 dBA of averaged continuous sound. That would damage hearing after several hours' continuous exposure: this is more than the limits of a child's attention span, at least for the auditory system, but we should also consider the additional risk to the child's feeling of safety and comfort. The limit for impulse-sounding toys, including toy guns other than cap pistols, is set at 120 dBA for peak levels; the total peak limit for all toys is 138 dBA (American Society for Testing Materials 2003). These noise levels fall within the threshold for pain and for hearing damage in a matter of seconds at close proximity.

Despite growing awareness of the potential for toys to be too loud – dozens of websites produced by audiology clinics and organizations supporting children's health and safety are now available to concerned parents – the situation is not improving. The SHA regularly compares the sound pressure levels of toys to those of loud domestic machinery. In 2004, only three toys on the annual list were found to be louder indoors than a weed trimmer measured at 100 dBA. In 2008, fourteen out of nineteen toys listed were louder than a chainsaw at 100 dBA. In 2011, nineteen of twenty-four toys listed were louder than the chainsaw. The list for 2014 included twenty-one toys, of which four were louder than 100 dBA; eighteen emitted sounds above 85 dBA when held next to the ear, and one – a Disney "Sing-Along Boombox" – was measured at 84.8 dBA at ten inches' distance from the ear. The list for 2018 included sixteen toys louder than 85 dBA. Two of these, manufactured by Bright Starts and by LeapFrog, were measured at 102 dBA, "which can damage hearing in less than 15 minutes when placed at a child's ear. Both toys are engaging and educational, but do toys really need to produce deafening sounds to teach us rhythm or our ABC's?"[7]

The majority of toys listed by the SHA are recommended by the manufacturers for children aged three and above, but some are deemed appropriate for babies. In 2008 one toy listed by its manufacturer as appropriate for eighteen-month-olds produced 114.5 dBA, enough to damage hearing in less than forty seconds if held close to the ear. In unsupervised conditions with a background of high ambient noise and more than one small child in the room, exposure of half a minute or more can easily occur. Nor do risks to infants necessarily result only from accidental overexposure. The SHA report for 2013 stated: "This year's top toy offender is marketed for infants. The Baby Einstein Company's 'Take Along Tunes' exposes babies to classical music, yet the tunes sound off at a rock concert level, with numbing crescendos peaking at 114.8 dBA. According to the National Institute of Occupational Health and Safety, exposure to [those] decibel levels at a close distance would cause hearing damage almost immediately."[8]

Is It Too LOUD?

The insufficiency of ASTM regulations is compounded by the fact that compliance by manufacturers is voluntary, and many exceptions apply: music players and musical instruments, game cartridges, squeeze toys. Cap pistols are exempt because they are covered by U.S. federal regulations, but there is no regulation at all for "toys that are connected to or interfaced with external devices, such as televisions or computers, where the sound level is determined by the external device; and toys with wheels where sound is produced as a result of the wheels making contact with the play surface" (American Society for Testing Materials 2003). The SHA report recommends removing batteries to mute the excessively loud sound effects provided by the computer chips pre-loaded into these toys, not to mention refusing to buy them in the first place. Of course, a question is easily raised: who determines that the toy is too loud? A parent who may well be among the growing number of young adults with some degree of noise-induced hearing loss? A teacher or daycare worker who is distracted by keeping up with a large room full of small noisy children?

The hazard of loud toys goes beyond damage to hearing. Consider a home or daycare with more than one small child and several toys from the list available, all in use. The TV is on in the background, the children are doing enthusiastic vocal imitations of trucks, guns, or monsters, and a parent or care worker is attending to a toddler who has fallen and is screaming. As the adult in the room, would your cortisol level reflect perfect calm? Would your child feel happy and secure? Children whose hearing survives such assaults intact can still be deeply frightened by sudden noise and by pain in the ear. Once exposed, they may develop an exaggerated auditory startle response: tiny muscles in the middle ear that regulate the position of the ossicles tense in anticipation of pain or sensory overload. The result is *hyperacusis*, extreme sensitivity to sound, in which the startle response occurs even when the actual sound is not loud (Marshall, Brandt, and Marston 1975).

Because hearing loss is cumulative over a lifetime, brief exposures to extreme noise in childhood can give a significant start to the process of noise-induced hearing loss as well as hyperacusis. Researchers, pediatricians, and audiologists consistently warn that allowing children to play with loud toys will compromise their auditory health. The effects can be hard to identify without testing, because extremely high frequencies deteriorate first and the loss may not be noticed until it affects comprehension of speech. As well, children are not aware of what "normal" hearing is like, and don't complain about its loss unless it becomes severe. Unlike sudden injuries, accidents, or choking, there is no dramatic onset. The best defence is an alert parent.

The Sight and Hearing Association recommends the following interventions: listen to toys before buying them, report loud toys to the SHA for its testing list, and put masking tape over the speaker to reduce volume.[9] Parents wanting to take an even more active approach might consider not inserting batteries in the first place; but if you buy the toy you are supporting its manufacture. If a store is not willing to take the toy out of its packaging and insert batteries so that you can hear it, buy it elsewhere; better still, don't buy it and let the company that makes it know why. Companies that design toys will pay attention if customers insist that "improved" does not mean "louder."

Even if you take precautions in your own home, what about your child's friends' toys and the ones at the daycare or school? Legal controls on sound pressure levels in toys would eliminate much of the problem, but there has been relatively little attention to developing them (Sułkowski 2009). The European Union (EU) enacted legal guidelines in 2009 for the safety of all products intended for the use of children: Directive 2009/48/EC. Noise produced by toys is first mentioned in the document within a guideline for the formulation of actual laws:

> 27. In order to protect children from the risk of impairment of hearing caused by sound-emitting toys, more stringent and comprehensive standards to limit the maximum values for both impulse noise and continuous noise emitted by toys should be established. It is therefore necessary to lay down a new essential safety requirement concerning the sound from such toys.[10]

The actual directive is given as:

> 10. Toys which are designed to emit a sound shall be designed and manufactured in such a way in terms of the maximum values for impulse noise and continuous noise that the sound from them is not able to impair children's hearing.[11]

The directive does not specify what maximum values are to be used, in part because the science of measurement is subject to change as technol-

ogy improves, and the EU's member countries have made different decisions about where to set the thresholds for legal definitions of excessive
noise. As well, differing standards in different countries can make enforcement difficult in a free trade zone. In the United States, there is a 2009
requirement for toy manufacturers to follow the guidelines of the ASTM,
but the SHA lists suggest that it does not have the force of national law and
is not always followed. Most countries outside of Canada and Europe have
no regulations at all. Why, despite substantial and growing evidence
of harm, is attention to regulation lax? The answer, unfortunately, is that
loudness sells.

Loudness sells throughout the entire span of childhood, into the teen
years, and then into adulthood. From toddler to primary school student,
many young children seem oblivious to the loud world in which they are
raised. Do young parents, attracted to loud music and themselves oblivious to noise, raise children with compromised hearing? And do parents
unwittingly create attitudes in their children that promote obliviousness
to noisy environments?

Recent investigations of children's attitudes toward loud music show a
pattern. A 2012 survey on hearing-conservation programs in schools
found that a majority – approximately 60 per cent – of Canadian pre-
teens aged eight to twelve thought that extremely loud noise and music
are harmless because "it's just in the environment." They further dismissed
the idea of hearing protection at school dances because its use would be
"weird," and they believed that "music should be loud in social settings"
(Lowther 2012, 41–3). The acculturation to loud recreational soundscapes
and reported lack of concern may be consistent across cultures (Holmes
et al, 2007; Tuomi and Jelliman 2009; Danhauer et al. 2009; Muchnik et
al. 2012).

How have such attitudes come to be? The responsibility cannot solely
rest with manufacturers who have sold loudness as a lifelong habit. Nor
can the parents be fully responsible: they just buy the toys and media and
make the home environment to the best of their ability, unaware that they
are potentially contributing to a growing pediatric problem.

THE QUESTION OF DOMESTIC DIN

If toys can impair the hearing and speech development of young children, what about ambient noise in the home? Very little is known, for
example, about the use of broadcast media as an indoor soundscape,
probably because measuring such effects would require permission from

large numbers of parents to place recording equipment within their homes – an unlikely prospect. What *is* known about TV's effect on toddler learning is that pre-school children in North America easily memorize advertising jingles by the age of three (Levin, Petros, and Petrella 1982). Hundreds of studies have been done on the persuasive effects of advertising on children, but little on the auditory ambience that they experience at home as a result of it. Whether the soundscape consists of television, talk radio, podcasts, or music, when it forms a constant ambience, it is classified by the child's mind as familiar and therefore safe. Since designations of safety by means of familiarity are no guarantee of actual safety (for example, children who grow up in war and conflict zones all over the world come to consider the associated sounds as familiar), it is advisable to question the effects of noise levels as well as the content of advertising and programming on young ears and young minds. The lack of consistent research in this area should be a call to governments, research institutes, and funding organizations to approach the question more enthusiastically.

Consideration is also due to the question of layered soundscapes in the home: the television on in one room, a computer game in another, and children plugged into personal audio devices at the same time. How are hearing, listening skills, and learning being affected? A clear answer to this question is not yet available, but a study on rats published in 2003 suggests that attention should be paid to finding one. Infant rats raised in noisy conditions showed developmental delays in the neural connections and responses of the primary auditory cortex in the brain; mature development of the cortex did not occur even in adulthood. Because development of sensory perception follows a common pattern in all mammals, the authors concluded that noise was also implicated as "a risk factor for abnormal child development" (Chang and Merzenich 2003). Even the title of their article leaves no ambiguity about the seriousness of the implication: "Environmental Noise Retards Auditory Cortical Development."

NOISE COURTESY (FOR CHILDREN AND ALL OF US)

Children are noisy; not just when they play chasing games on the playground or video games indoors, but when their emotions overflow in sheer excitement. School grounds get loud simply because excited children yell and giggle and shriek, and cheer fiercely for their friends at sports. Children in classrooms shout because they *know* the answer!!! As one elementary school teacher said to me, "just try to keep them quiet when there's several dozen of them excited about just about anything."

Like any form of noise, children's exuberance is regulated by their culture and/or social class: whether their community favours spontaneity or control, whether their parents are sensitive to noise, whether they are taught early to be aware of the needs of others in the home (*sssh, the baby's asleep*) or the community (*remember, we don't shout in the library*). As a small child you may have been told – repeatedly – about the difference between "indoor" and "outdoor" voices, and when to use them. You may have passed on the same advice to your own children, or be planning to do so when you have them. These are examples of *noise courtesy*, which also includes how loudly you speak on your phone in public and where you set the volume on media players outdoors and in public venues.

While no one should advocate total suppression of a child's expressive nature, learning noise courtesy will give school-age children valuable skills for knowing when being loud is appropriate and when it isn't. Lessons in science class about sound waves and the structure of the human ear (as well as those of dogs, cats, hamsters, cows, sheep, or whatever pet or farm animals are familiar) can turn abstract rules into something meaningful. Exercises in noticing and describing sounds will turn children's attention toward what they hear. Discussion of what makes a sound attractive or repellent will teach discrimination; it will also demonstrate diversity of perception, as likes and dislikes are shown to depend on the listener (Schafer 1992).[12] As children taught to understand soundscapes grow into their teen years, learning about sound as environmental ambience and communicative signalling can provide them with the awareness to question the commercial soundscapes that surround them, accepting or rejecting them consciously.

Growing Up Loud

One finds excessively loud music at elegant nightclubs, sporting events, wedding parties, cinema theaters, popular concerts, customized automobiles, and in the ear buds of kids walking around in a daze. Loud music is a world phenomenon unrelated to social class or cultural status. What function does loud music serve?

Blesser 2007, 2

From a systems theoretical point of view, it is not an easy task to change health risk behaviors since it is not just individual attitudes that have to be changed. It is most likely the whole system of interrelated levels such as laws and regulations, peer norms and family values etc., that has to be changed ... Health promotive strategies should focus on changing not merely individual attitudes, but also societal norms and regulations in order to decrease noise induced auditory symptoms among adolescents.

Landälv, Malmström, and Widén 2013, 353

Adolescents and young adults hear the world around them, but they sometimes attempt to shut it out in order to seize control of their own experiences. To some, the expectations presented by their parents and their culture are overwhelming, and the only safety they can find is in creating imaginary alternatives. The virtual realities offered by their electronic devices tempt them with rabbit-hole entrances into a magical world in which they are able to connect with others at a safe distance, to internalize and rehearse the emotions found in pop songs, to take trial runs at careers and love, to be winners of every game. The rabbit hole is often one of sonic dimensions. When earbuds stay firmly snugged against their ear canals, no competing voices – of parents, teachers, cultural norms, their own anxieties – can intrude. They are safe from having to make life-changing decisions for a little longer, and insulated from the doubts and critiques that will eventually result from their choices, from the process of becoming adult in a society that expects hard work and material success as the benchmarks for self-respect. They are safe, temporarily, from their own futures, but not from harm: noise-induced hearing damage among teens and young adults has been rising for decades.

Because of "recreational noise," a category that includes live and recorded music, noise-induced hearing loss (NIHL) is no longer a condition necessarily associated with advanced age and industrial work. People in their early twenties who had never worked in factories were assessed in the early 1990s with degrees of hearing loss comparable to industrial workers in their fifties (Spaeth et al. 1993). Researchers in Nottingham, UK, surveyed 356 subjects aged eighteen to twenty-five about their noise exposure in all aspects of life and compared the results to findings from earlier studies. Their conclusion was that "social noise exposure" had tripled during the 1990s in the United Kingdom (P. Smith et al. 2000, 55). Of subjects who attended rock concerts or night clubs, 66 per cent reported tinnitus or temporary effects on their hearing (ibid.), a finding also reported in a similar German study (Spaeth et al. 1993) and a French one (Meyer-Bisch 1996).

In 2002, the US Centers for Disease Control reported that 12.5 per cent of American children under the age of eighteen had some degree of noise-induced hearing loss caused by recreational devices and activities: personal music players (PMPs), mechanical toys, snowmobiles, concerts. That's one out of every eight children. That number was based on a study completed in 1994, well before the widespread use of PMPs and recent changes in the technology of amplifiers for music.[1] More recent statistics for the United States are hard to come by, since a comprehensive national survey has not yet been done in the twenty-first century, and estimates from the Centers for Disease Control include congenital and disease-related deafness along with NIHL in the category of hearing loss. Canadian estimates, examined in an article assessing information presented by a Canadian Health Measures Survey taken of 2,434 children and youth aged six to nineteen in 2012–13, suggested that the national rate for any type of hearing loss in young people was at least 7.7 per cent at that time. The authors commented, however, that much of the information leading to the estimate was based on self-reported surveys, which have questionable accuracy, and that listening to personal devices with ambient background noise over 65 dBA – the average level of city traffic – would increase risk, leading to the conclusion that the national result may have been underestimated (Feder et al. 2016).

According to Israeli research published in 2012, the risk to young teens may be particularly high. Comparing self-reported listening levels with actual measurements from ear simulators and microphones suggested that teens aged thirteen to sixteen had a 27 per cent risk of NIHL, while risk to university students was estimated at 23–4 per cent (Muchnik et al. 2012). A more recent international survey of research articles reporting on the listening levels of teens and young adults found that 58.2 per cent of youth "exceed the current recommended 100% daily noise dose" with their personal audio devices (PADs) and have "significantly worse hearing thresholds," even when they don't notice any deficits (Jiang et al. 2016, 197).

While NIHL is more prevalent among male youth, females may have closed the gap. A US research team compared results from audiometric testing done on more than 4,000 teens aged twelve to nineteen during National Health and Nutrition Examination Surveys (NHANES). The researchers found "a significant increase" in NIHL among females between surveys taken in 1988–94 (11.6 per cent) and in 2005–6 (16.7 per cent), which they attributed to "amplified music at concerts and clubs" (Henderson, Testa, and Hartnick 2011, e44). The authors point out that the NHANES researchers at the time of the second survey did not account for this factor in their questionnaires, which casts doubt on their assumption that the risk of NIHL was relatively low.

The process of arriving at estimates for both prevalence and risk of NIHL is particularly complicated because of the number of variables involved. Self-reporting surveys can produce quite different results from clinical measurements, and funds for using sophisticated equipment to test the hearing of thousands of students are hard to come by. As well, since NIHL is an erosion of ability to hear certain frequency ranges rather than a blackout of the entire sense of hearing, it can be difficult to measure precisely. Researchers arrive at probable numbers for risk by means of statistical adjustments to their data, but results can also vary because of differences in culture and access to technology, as well as methods for data collection and reporting. Inconclusive results are common (Keppler, Dhooge, and Vinck 2015), but this does not mean that there is no problem: in medical research, "inconclusive" means that more, and more precise, investigation is needed. Bearing in mind that levels for hazardous listening pertain to both loudness and length of exposure, as well as the type of listening (ears in a live venue? earbuds? earmuff-type headphones?), one European research team concluded that, although the majority of youth are not likely to be at high risk, a significant minority are. Since the incidence of tinnitus in surveyed northern European and American youth populations has apparently tripled since the 1980s (Śliwińska-Kowalska and Davis 2012), it is easy to suspect that the incidence of hearing loss has also risen significantly.

Thanks to the technology of amplification, live concert audiences as well as PMP users can easily sustain noise-induced damage to their hearing. Concern in medical literature about hearing damage to the audiences at rock concerts has been growing for decades. A German study in 1999 found high incidence of temporary hearing loss as well as tinnitus: "In the majority of examined patients (67%) the hearing loss developed on the basis of one-time exposure at a rock concert or pop concert" (Metternich and Brusis 1999, abstract). While the cases of hearing loss were temporary and responded to treatment, 33 per cent of those experiencing tinnitus did not recover, and the authors caution that repeated bouts of post-concert hearing loss can signal permanent damage, although they assessed risk as low compared to the risk of tinnitus. A 2005 Canadian survey of 204 people with an average age of twenty attending four "large rock concerts" in Toronto found that 85 per cent of those responding to the investigators' questionnaire reported experiencing tinnitus after the concert.[2] Only 3 per cent of the sampled population wore earplugs; another 42 per cent said that they would be willing to wear them if they were provided free at the door (Bogoch, House, and Kudla 2005).

MY EARS ARE RINGING!

Tinnitus, usually the first sign of injury to cochlear hair cells, is temporary in its early stages and therefore easy to dismiss. What is its prevalence? In 2002, Swiss researchers surveyed 700 youths aged sixteen to twenty-five. Over 70 per cent of them reported experiencing tinnitus after attendance at a live concert (Mercier and Hohmann 2002). More than 70 per cent of 1,500 Dutch teens surveyed in 2009 reported that they "attended discotheques" with live music at least once per year; 24.6 per cent of them were regarded by researchers as being at high risk for tinnitus and later stages of NIHL because of attending performances of music over 100 dBA for more than an hour per week without the use of hearing protection. The difficulty for youth of considering long-term future consequences was cited by the study's authors as a primary reason for the risk; another was the formation of listening habits through repetition and peer-group influence. In other words, *if everyone's doing it, it must be safe – and who cares what's going to happen to my hearing in thirty years anyway?* (Vogel et al. 2010). One research group surveying university students found that approximately 90 per cent had experienced transient tinnitus after attending live concerts, and that tinnitus became permanent in 15 per cent. Only those who developed permanent tinnitus were likely to use hearing protection at concerts (Gilles et al. 2012).[3]

Since youth world-wide are so thoroughly accustomed to the culture of loud music and its potent attractions of emotional thrill and shared experience, trying to turn this particular tide is widely reported by all recent researchers as ineffective. Earplugs are not well accepted by the majority of people attending loud concerts. A Swedish study of 1,547 students aged thirteen to nineteen who attended "pop concerts and discos" in Sweden found only a 21 per cent rate of earplug use (Widén and Erlandsson 2004).

THE ACOUSTICS OF ADOLESCENCE

Internet websites referring to the problem of NIHL in youth are now proliferating, but nearly all repeat the 2001 statistic of 12.5 per cent and fail to connect the problem with anything more than personal habits. What drives these habits? Neighbourhood noise and parental choices of soundscape in the home contribute to the normalization of loudness, but most influential are the social norms dictated by commercialized popular culture, which begin their influence in childhood. Smoking was normalized and associated with high social status in the mid-twentieth century: the first cigarette was a rite of passage into adulthood for decades. More recently, processed food devoid of nutritional benefits has been marketed to parents as convenient and modern, and directly to children as colourful and fun. Critiques of both have taken decades to raise public awareness and change the norms established by advertising. Loud toys, loud cars, and loud music are now being marketed (loudly!) as commodities of choice.

Most NIHL in teens and young adults is classified as *socioacusis*, hearing loss that results from social habits. Its leading cause is recreational activity. Family fun on a speedboat, or attending a major playoff game, contributes to risk. Holiday fireworks, popular with children and teens all over the world, can produce an average of 160 dBA at a distance of two metres (6.5 feet), enough to cause temporary hearing loss (Smoorenberg 1993). But occasional games or holiday celebrations are not the most significant hazards. The *main* cause of NIHL in teens and young adults is excessively loud music.

Too Much of a Good Thing: Music as Noise

Music is an integral part of human society, and one of the glories of our species. It is easy to imagine a prehistoric tribe celebrating a successful hunt by singing with the percussive accompaniment of clapping, feet stomping in a dance, hands striking thighs, and bones from the catch lifted and clashed together. Plan a celebration of anything, and food, drink, and music are likely to be involved. Relaxing after work? Shaking off stress? Bonding with a social or spiritual community? Again, music is there. But even before the advent of electronic amplification, this was not always an unmitigated good.

Music can function as noise. Examples abound: the neighbours' backyard party that persists at 2 a.m., the telephone "hold" song you don't like, the movie soundtrack played too loud, something the kids are dancing to in the basement or the street. Cultural divisions can be seen rising along the boundaries that divide music from noise. They are evident in the attitudes that parents pass on to their children about appropriate indoor and outdoor activities as well as boundaries between private and public spaces. They are the subject of proclamations by religious leaders and of community bylaws. They can draw us together and drive us apart. The ability of sound to attract attention also means that music can distract, whether easing the boredom of workers or feeding the fantasies of shoppers. The spontaneous music of work songs synchronized and motivated physical labour throughout European history: long narrative songs for sedentary tasks like spinning yarn or thread, lively rhythmic ones for hoeing, harvesting or hauling fishing nets, and later for the building of railroads. Similar traditions in West Africa were carried to both North and South America in the period of slavery, giving eventual birth to a host of musical offspring: jazz, blues, rock, reggae, rap, hip-hop.

In the absence of recording technologies, music was spontaneous and functional. Historically, each song and each style had a cultural purpose: to motivate, celebrate, memorialize, protest, satirize, soothe, entertain, console, or educate. The development of recording in the late nineteenth and early twentieth centuries enabled far greater access: music was no longer available only to those with wealth or talent or deeply rooted community traditions. Recording and broadcasting brought music into every Western home, simultaneously uprooting it from its original social impacts and promoting its use as a commodity.

Throughout the twentieth century, recorded music moved into the realms of mass entertainment and mass marketing. First radio and then television brought short, snappy, unforgettable advertising jingles into homes. Theme songs became essential components of films and television series; social events increasingly relied on recordings to replace live musicians, whether a hired band, a friend at the piano, or a neighbour with a guitar. Recorded music was also put into service for the ambient manipulation of mood in workplaces, brought first to industrial assembly lines during the 1930s and 40s, a practice designed to lift workers' spirits and increase wartime production.

By the 1960s, ambient music specifically engineered for mood induction, known by the trade name "Muzak," was available in corporate offices. From there it spread to commercial venues, where it was designed to lull shoppers into lingering in stores and malls (S. Jones and Schumacher 1992). Corporate chain stores now produce their own endlessly repeating soundtracks, creating a carefully engineered ambience in every store that overrides local and regional cultures in favour of a "brand identity," complete with auditory components. Music in the service of advertising now bombards us from radio, television, computers, and commercial venues, each song or sound clip aimed at a particular fragmentary cultural subset and designed to repel others.

Walk through a mall today with your PAD on and you may be hearing as many as five layers of music simultaneously: the song on your own player in the auditory foreground, the inoffensive instrumentals set to patter through the corridors of the mall and probably unnoticed, and the hot hit songs from the store in front of you, the two adjacent to it and the ones across the corridor, all perceived only subliminally until you have read this sentence and remember it on your next trip to a mall. Each of the layers you hear within the mall is designed – like the visual displays – either to lure you into a store or to discourage you from entering it. Which effect the music has will supposedly depend on your cultural background, age, income level, social identity, and taste in commodities. Its purpose is to attract the people most likely to spend money in the particular store, defining a transitory subculture – music as gatekeeper.

The commercial music industry has, in the course of the past century, fragmented potential audiences into stratified categories and created specific soundscapes designed to appeal to each of them. Beyond age, gender, culture, and income, industry profits are based on urban and rural cultures, regional affiliations, and the relentless promotion of individual artists and groups. The selling of music to youth began in the 1920s with the marketing of jazz to young working people; the growth of rock in the 1950s fuelled an industry directed specifically to teens and based on compelling rhythms and loud, fast blasts of sound. Today, tens of millions of young buyers support the production of rock, pop, and hip-hop recordings.

By the 1980s, exposure to loud music was no longer limited to social events or to vehicles. Thanks to the invention of Walkman portable cassette and then CD players, music could be played at hazardous levels directly into the ears. Although some early European studies found that the majority of teens and young adults listening to personal cassette players in the 1990s kept volume levels moderate, others reported serious concern about hearing loss in teens (Ising et al. 1997; Jokitulppo, Björki, and Akaan-Penttiä 1997). The discrepancy in these results was caused by a number of factors. One was variation in the design of portable cassette and CD players, another was variation in the design of the studies. While some took account of ambient noise levels while subjects were listening, others did not. Some relied on questionnaires, others on unstructured self-reporting, still others on sound-level measurements as well as detailed questionnaires. Self-reporting can be appropriate for market surveys, but it is unreliable in situations where subtle and initially imperceptible changes in physical function can signal increased risk for disability: those need to be measured.

Changes in the technology of measurement since the 1990s have led to a reconsideration of the question of risk; so have changes in the technology and habits of listening. In the 1990s, portable music devices were not yet miniaturized to the extent that they are now and were not as affordable; nor were in-ear devices as widely used as they are today. Earbud enthusiasts of all ages are now a far more significant sector of the population, and ongoing research since 2000 has yielded clear results: portable music players, especially those with in-ear transmission, damage hearing at high volume. Given the statistical trends in increased earbud use combined with a decades-long increase in aggressive marketing to teens and young adults, it is scarcely surprising that the WHO estimates the number of youth and young adults at risk for permanent hearing loss, worldwide, at 1.1 billion (World Health Organization 2015).

What, then, is a reasonable conclusion about the prevalence of NIHL in teens and youth? When self-reported surveys are inconclusive and statistical information is hard to come by, what is actually being measured is risk. Increased risk means that more people in a given population will be affected by NIHL, but that doesn't predict the likelihood for any individual: we can't accurately tell any particular teens that they *will* lose their hearing if they keep using those earbuds, or that they will go completely deaf (they won't),[4] but we can warn of risk. If the knowledge that smoking will increase the risk of cancer motivates people to stop, the same logic can be applied to listening habits that boost the risk of NIHL.

WHO IS AT GREATEST RISK?

The study of musical taste – what people listen to, and why – includes a vast literature on the demographics of class and its interactions with ethnicity and culture. Conflicting theories abound: does wealth lead to traditional "snobbery" or to inclusive, omnivorous appreciation for a wide variety of musical genres? Does poverty align with a taste for the abrasive, or with long-held traditions? What are the roles of national and regional heritage, religion, generation, education, and commercial promotion in shaping listening habits? Do the young rebel against their parents' choices, or adopt them? Because the personal aspects of listening often resist classification, aligning taste with cultural subgroups isn't an exact science, but two general patterns are valid: first, exposure to a variety of musical styles leads to eclectic taste, while limited exposure fixes preference within a narrow spectrum of possibilities; and, second, friends influence an individual's choice of music, at least for youth. Given these patterns, it's possible for researchers to survey subgroups in a population and predict their risk of noise-induced hearing loss from recreational listening. The risk is related more to the loudness of the music than to its content – classical music played at 90 dBA will damage hearing faster than heavy metal at 70 dBA – but loudness is an inherent component of some styles and of some performance venues: we don't choose metal, grunge, or punk for their soothing properties.

On average, the individual most likely to be at risk for NIHL is a young male industrial worker who smokes and who frequently attends rock concerts and/or social events with loud background music. A major longitudinal study at Manchester University in England investigating the question of whether social class affects the risk of hearing loss followed a cohort of 9,023 men and women born during one week in 1958, measuring their hearing periodically over fifty years. They reported that "the magnitude of social class effect is comparable to that of occupational noise. Susceptibility to hearing impairment is likely to be appreciably determined in early childhood" (Ecob et al. 2008, 100).

The direct association of hearing loss with social class is problematic because it takes for granted a large number of correspondences between class and social habits that may not apply outside of specific cultures. Is class causative in the sense that it exerts cultural pressures and limits opportunities, or do the British researchers associate risk with social class because they and their subjects are members of a culture that pays considerable attention to class distinctions? Class itself would not be predictive of risk for the individual whose habits are atypical. Recreational preferences are not entirely determined by class or gender, and listening to loud music is not characteristic only of working-class males. In the case of the British study, social habits typically associated with working-class status – smoking and listening to loud music – combined with exposure to noisy industrial workplaces and reluctance to wear hearing protection on the job unless it is required, might be the actual causes for the connection.

Further studies of secondary school students in Germany (Maassen et al. 2001) and the Netherlands (Vogel et al. 2008, 2009, 2010) confirm that pre-vocational students are more likely to listen to music at hazardous volume than pre-university students and to be less receptive to warnings about the hazard. An American study of the PAD listening habits of university students suggests that their average level of risk for NIHL is considerably lower than that of secondary school students (Danhauer et al. 2009). Another American study of twelve- to nineteen-year-olds associated increased risk with low income: "individuals from families below the federal poverty threshold had significantly higher odds of hearing loss than those above the threshold" (Shargorodsky et al. 2010, 772).

Gender also bears further investigation. Males were consistently found to be more likely than females to turn volume dangerously high, and to listen to personal music devices for longer periods of time (Vogel et al. 2008, 2009, 2010). In places where factory work in noisy conditions is typically done by women as well as men – for example in China, Southeast Asia, and Mexico – the risk might be less heavily weighted toward males. In any case, is attraction to loud music as well as loud noise actually characteristic of males, or have cultural traditions played a role in creating a truism that boys are noisier – and consequently more attracted to loud surroundings and more prone on average to damaged hearing – than girls? Like so many other aspects of NIHL, this is still an open question.

Along with the hazard posed by live music and thundering speaker systems, more recent technologies are adding to the level of risk. Portable personal music players, which can place extreme decibel levels very close to the eardrum, have normalized the creeping advancement of NIHL.

DOES PAD = NIHL?

The personal audio device is the latest in a series of music carriers that began in the 1950s with the radio-equipped car (Bull 2007): the category has now expanded to phones that play radio stations, podcasts, movies, and games. The technologies that make music and soundtracked entertainments portable and instantly controllable satisfy a desire to infuse ordinary events with a measure of magic by enhancing their emotional significance. Any listener can summon memories, evoke moods, and ease boredom simply by pressing a button. The process can be as intimate and private as we choose: no consultation with others, no summoning of fellow humans with live voices and instruments and the talent to perform with them, or even of physical objects that record their sound, is necessary. So easy, so convenient is this option that it quickly becomes normative. The music is always *on*, and the sonic environment of the non-electronic world is dismissed. Ambient sounds, whether abrasive, informative, or potentially relaxing, are submerged beneath the carefully chosen soundtrack. If they intrude, we turn up the volume.

Surveys of the listening habits of high school and college students in the United States suggest that the majority keep volume levels moderate, but that a significant minority of both groups – various estimates place it at 25 to 40 per cent – are at high risk for NIHL through listening at excessive volume, especially if earbuds are used. Even the cautious majority may be at risk from combined sources of music and background noise (Danhauer et al. 2009 and 2012; Muchnik et al. 2012). Studies done during the era of the Walkman and other pre-digital portable music devices found that maximum output levels from headphones ranged from 90 to 120 dBA, with inset earphones adding another 7 to 9 dBA to those totals. One research team estimated that one hour of listening at 70 per cent of total volume on conventional headphones would equal the maximum permissible noise exposure for one day by occupational health standards; using inset earphones would decrease the safe time span considerably and unpredictably, since manufacturers' standards were variable. A limit of one hour per day at 60 per cent volume was recommended to "protect the hearing of a majority of consumers" (Fligor and Cox 2004). Research in

China and Italy in subsequent years confirmed the hazard (Peng, Tao, and Huang 2007; Cassano et al. 2008).

European researchers analyzing data available in 2012 from earlier academic studies found a watershed in 2008, three years after the entry of the iPod into the PAD market.[5] They concluded that, before that time, the likelihood of high risk for hearing loss within a five-year period was between 5 and 10 per cent, which represented approximately 2.5 million European youths. After 2008, published studies on student populations in Europe and North America indicated that a majority of youth were listening for longer periods of time at greater levels of loudness and were consequently at greater risk, some as high as 30 per cent (Śliwińska-Kowalska and Davis 2012).

A current posting on the Health Canada website, still referring to portable CD players as well as digital audio players (DAPs), gives even more sobering information: damage could be done in as little as ten minutes at 102 dBA, the maximum level for some of the players tested (Government of Canada 2019). Most of the DAPs tested at maximum volume would produce serious hearing loss in seven to twelve minutes; some – with sound levels at 125 dBA through earbuds – would do so in a few seconds. In 2007 a study on 1,512 Dutch secondary school students, which correlated questionnaires with measurements on the sound pressure levels at concerts and the output levels of listening devices, concluded that 50 per cent of the population surveyed was at high risk for NIHL from a combination of concerts and portable players. One-third of the total were at high risk from PADs alone (Vogel et al. 2010). A similar team, investigating another 1,657 Dutch secondary school students specifically for PAD habits, found that 48 per cent habitually used high-volume settings (Vogel et al. 2009).

The rate of occurrence for NIHL associated with PADs may be under-reported in popular media because medical studies – which show consistent alarm at the degree of risk to the young – are not well understood by journalists, but also because the erosion of hearing acuity can be so gradual that in most cases it is not perceived in its earliest stages even by conventional audiology tests (Okamoto et al. 2011). This makes the rise in NIHL associated with PADs, ironically, into a quiet epidemic.

Most research articles on PAD use by children and teens recommend that information about auditory health and hearing protection be included in health classes at schools. However, warnings about hazards to hearing are unlikely to have much impact on dedicated participants in PAD culture, any more than on nightclub or rock concert enthusiasts. With the intimacy of personalized listening comes an insistence

on freedom of choice. According to one research team, "most adolescents – especially male students and students from pre-vocational
schools – indicated that they often played their mp3 players at maximum volume. Although they appeared to be generally aware of the
risks of exposure to loud music, they expressed low personal vulnerability to music-induced hearing loss. Most adolescents said that they
would not accept any interference with their music-exposure habits"
(Vogel et al. 2008; see also Borchgrevink 2003).

Adults, most teens, and some pre-teens are capable of exercising judgment about how and how often to use PADs and other sources of entertainment. Younger children cannot be expected to understand issues of
control and protection, and yet their hearing is even more sensitive. As
discussed in chapter 8, they are particularly vulnerable to the effects of
excessive noise generated from a variety of sources in the home, in
school, and throughout the environment. As a result, prolonged use of
PADs with earbuds is not advisable for them. As well, because the technology for in-ear devices is relatively new, reliable information on long-
term auditory and cognitive effects on developing brains is not yet available. It may take several more years for patterns to be clear. For these
reasons, it is important for parents to be cautious. Parents should set
rules early for the use of in-ear devices by their young children and
explain why the rules are a good idea. They should keep volume settings
moderate, usually defined as 60 per cent or less of maximum volume,
although that depends on what the PAD is capable of producing and on
the device – earbuds, headphones – transmitting the music. Belgian
researchers have found that earbuds add approximately 5 dBA to the level
produced by earmuff-style headphones; this more than doubles the
sound pressure (Keppler et al. 2010). Most significantly, parents need to
model safe practices for children by limiting their own use of earbuds
and explaining why it's important.

Lowering the Volume

Some websites aimed at teenagers already promote safe use of earphones
through combinations of statistics, advice, and endorsements from rock
musicians, but the only really effective strategy to reduce NIHL through
portable music players is to regulate the manufacture of the devices and
earphones. Until that happens, protecting children's hearing begins with
knowing how to protect our own. Along with awareness of the volume
settings on listening devices, we need to consider the layering of sound

sources. If you wear a PAD while running a vacuum, for example, that special song that jazzes you up for housework might be subjecting your auditory system to an assault.

So might your daily commute. Two studies conducted on transit systems in New York have recommended the use of hearing protection by frequent riders of both subways and buses, documenting loudness levels well above the limits sets by the World Health Organization and the US Environmental Protection Agency (Gershon et al. 2006; Neitzel et al. 2009). Levels on subway platforms averaged 86 dBA, with peak levels measured at 106 dBA. Levels peaked at 112 dBA inside subway cars and 96 dBA at bus stops. Both averaged and peak levels were deemed potentially causative of hearing damage, a conclusion unlikely to surprise anyone who has experienced the sustained shriek of New York subway wheels. While these readings would not necessarily produce hearing damage on brief exposure, they could easily contribute to physical stress reactions. Adding an additional sound source – for example, a PAD set to be audible above the ambient noise of the train – can produce levels hazardous to hearing, or advance the destruction of already injured cochlear hair cells.

The same principle applies if, at a busy gym, you turn up your PAD so you can hear the lyrics of a song over the treadmills, rowing machines, and ambient background music. As for the sonic ambience of the gym itself, conclusive results from research are not yet in, but questions are being raised. One South African study surveyed 236 participants in aerobics classes with loudness levels for music set at 80, 85, 89, and 97 dBA.[6] These levels were assessed by the research team as representing a range from low risk to high, and the authors conclude that aerobics becomes a risky activity for noise-induced hearing loss when high intensity music is played in classes. Attempts to reduce this risk through instruction in hearing conservation have generally failed, possibly because participants find the high intensity music enjoyable and motivating, and therefore not "too loud" (Wilson and Herbstein 2003).

A later Canadian investigation with a small sample of twenty-four participants compared individually selected listening levels of one song in conditions of rest in quiet, rest in noise, and exercise in noise. As expected, chosen levels were higher over both noise conditions than over the quiet background. Approximately one-third of participants chose levels that put their hearing at risk. The researchers conclude: "These results indicated that increased background noise causes individuals to increase the volume on their PLDs [personal listening devices] to potentially dangerous levels and that increased noise alone was not the only factor affecting the participants as the addition of exercise induced even further increases in PLLS [personal listening levels]" (Hodgetts, Szarko, and Rieger 2009). If a gym is constructed so that classes with music and independent exercisers share space in the same room, anyone frequently using a PAD to mask the music from the class could be at significant risk of hearing damage.

Why have PADs become the accessory for so many people that they are considered a cultural norm despite growing evidence of potential harm? The combination of miniaturized portable phones with cameras and media devices that can access the internet and carry out tasks like shopping and banking, as well as keeping you informed about news and the locations of your family and friends, has proven to be irresistible: convenience is the opiate of today's masses. Cost, which has plummeted since their introduction, is certainly a factor; so is mobility. The ability to fine-tune your personal emotional ambience is also a major attraction. Like the office workers that will be discussed in chapters 10 and 11, teens and adults are using personal devices to create an illusion of privacy in everyday activities. Controllable soundscapes that contain only personally selected content are seductive for their emotional comfort, but also for their ability to serve as distraction from the world outside their control. Earphones protect us from unwanted signals. They provide our missing earlids.

EARBUDS AS EARLIDS

Current personal audio devices, following their Walkman and Discman ancestors, have become a cultural phenomenon attracting the attention of academics as well as consumers. Most scholars connect their use with urban settings, youth culture, and the ability to create a separate reality that shields the user from undesired interaction with others. Two similar but subtly conflicting theories are now prominent: that people use the PAD primarily to create a personal and mobile soundscape impervious to environmental noise and signals (Bull 2007), or that its use is effectively a re-mixing of music with urban noise and signals so that the combination of music and noise creates a new experience of music (Beer 2007). Communications scholar David Beer points out that even the sound leakage from earphones, which gives passersby momentary access to your personal soundtrack, has significance as part of the re-mix (ibid., 859). Both interpretations of the PAD-mediated urban soundscape are detailed, supported with evidence from social research and interviews, and well respected in academic circles. Neither one gives any attention to the fact that music volume loud enough to create a sonic barrier – or to leak from earphones – will damage hearing. Neither questions what might motivate high-volume users to go beyond the level of sound pressure necessary to distract them from the external soundscape.

Listening to music through earphones en route imposes structure on the randomness and chaotic properties of ambient urban noise. The rattles, roars, and clatters of traffic provide a percussion track to your chosen song; the long commute home is transformed from a source of fatigue to a private concert for an audience of one. Earphones can also signal to others that you are occupied and not to be engaged in unwanted conversation. These are commonly reported motives for PAD use in public (Bull 2007), and they are usually accessible with low or moderate levels of loudness. Might there be other reasons for some listeners to play their music at extreme volume? Is it really about the music, or about the loudness? Is extreme loudness a way to avoid hearing unwelcome signals?

The assimilation model in the psychology of personality posits a community of competing internal voices – represented by the mental dialogue of thought – that reflect, advise, question, caution, motivate, encourage, and reprimand every individual, guiding behaviour and the development of morality (Elliott and Greenberg 1997; Honos-Webb and Stiles 1998; Stiles 1999; Osatuke et al. 2005 and 2007; Reid and Osatuke 2006). Recent discoveries in neuroscience suggest that the sub-vocal words that constitute some levels of thought originate in particular areas of the brain and represent an intermediate step between thought and vocalized language (J. Palmer 2012; Pasley et al. 2012). While extreme manifestations of this diverse inner community are associated with personality disorders or schizophrenia, everyone experiences normal inner vocalization of thought to some degree.[7] Such internal verbalization appears to be a universal human characteristic that transcends time, culture, and language. Ancient practices of meditation from Hindu, Buddhist, and monastic Christian traditions were designed to calm and quiet what was characterized as the "monkey chatter" of the unfocused and unproductive mind. By stilling the chatter, they sought to develop a state of serene and alert attention to what is happening in the moment and to sincere communication with others. Modern therapies in many forms – various schools of psychotherapy, spiritual counselling, addiction counselling – advocate consciously training internal voices to give more encouragement than rebuke, enhancing focus and self-esteem. This process requires recognition of internal criticism and its negativity as something other than a valid reality, something that grows from seeds of experience with critical messages given by culture, media, community, schools, and parents as well as experiences in life.

But what if you choose not to examine your own internal chatter because it carries connections to dark images and frightening memories? There is a theory in clinical psychology that self-critical internal speech is associated with some forms of depression (Firestone 1986; Osatuke et al. 2007). Could socioacusis in some teens and adults be connected with a desire to evade their own verbalized thought because it summons frightening emotions? Self-reflection requires focused awareness that is not always available or comforting; therapy can be a slow and expensive process. The quickest option, if the chatter turns disturbing, is to submerge it. For many who are desperate, immediate relief is provided by alcohol; for others, by drugs; and, for still others – less drastically and more accessibly for teens – by drowning out the sub-verbal critique with very loud music.

Can the use of PADs at high volume present a risk to life as well as hearing in some cases? Although as yet only one formal study has apparently been done on the relationship between earphone use and pedestrian or cyclist accidents, attention is being drawn to the question by police and media.[8] In 2011, a team of US researchers searched national databases for reports of accidents that involved trains or motor vehicles hitting pedestrians wearing headphones of any type. They found 116 clear cases of death or injury, more than half of which involved trains. Most incidents (89 per cent) took place in urban settings; a majority of victims were male and under the age of thirty. The team concluded that headphones can be a risk to pedestrian safety and urged further research (Lichenstein et al. 2012).

Earphone use in traffic is coming under scrutiny for two reasons: the perception of environmental sounds is reduced by the device, and the wearer's attention is distracted. Wearing just one earphone or earbud at a time, a solution used by many bicycle couriers, is an option for increasing safety, although some may find it disorienting. Accidents can be reduced by careful attention to signals, traffic, and pedestrians as well as low volume for music.

Not all of the cases studied in 2011 were clearly accidents, however. In 29 per cent of them, a warning was sounded; the victim either did not hear it or ignored it. When the earphone-wearing victim of a fatal accident is reported as not having heard a warning, we suspect that the technology may have masked the signal and that the victim could have been unaware. When the victim is suspected of having ignored the warning, a darker question occurs: how many of the deaths, especially from regularly scheduled commuter trains, were intentional? Given the ease of using extremely loud music to submerge extreme self-criticism, disturbing memories, and fears, are some cases of earphone-wearing pedestrian and cyclist deaths actually suicides?

In 1986, the American psychotherapist Robert W. Firestone published a theory of motivation for suicide. Through interviews with survivors of suicide attempts, he found evidence that the majority of people who attempt suicide are responding to a subliminal voice or thought process, resembling a parent's voice, that is obsessive and self-destructive (Firestone 1986; Firestone and Firestone 1998).[9] Firestone advocated neutralizing the self-critical internal voice with a therapeutic technique that gives it identity as a *spoken* voice, guiding the sufferer to speak the negative thoughts aloud as the therapist encourages them to question and refute the spoken thoughts' validity. Submerging the voiced destructive thoughts in even louder sound might be therapeutic as well, or it might have the opposite effect: nothing is known because it has not been studied. While there is as yet no firm evidence for the association of listening at extreme volume with depression or suicidal thoughts, this is a subject that calls out for urgent attention by researchers. The current lack of connecting evidence does not imply lack of connection; it indicates a gap in knowledge. If even a few lives can be saved by the findings of additional research, it is well worth doing.

CELEBRATING NOISE

While music played loudly through earphones can serve to isolate the listener by blocking out both internal and external signals, live music played loudly will enhance connections within a group. It is easy to find a reason: the heady combination of emotion and rhythm catalyzes excitement, and sharing it is part of the thrill. So much of commercialized youth culture is based on the experience of shared music that generations are defined by it. While styles and technologies change, there is a constant: the music is loud.

Why is extremely loud music exciting? One reason is that sound loud enough to be felt as well as heard produces a rush of adrenalin. When its source is live rock or hip-hop, the element of rhythm compounds the excitement, shaking the floor and rising through the feet to vibrate the bones of performers and audience, motivating listeners to jump and dance, obliterating speech and producing a sensation of shared excitement. The relentless assault of it, the pounding beat and discordant frequencies, signal the nervous system to enter the primordial biochemistry of extreme stress; the race to outrun the earthquake, rock slide, lion or bear, to exult with your community in the victory of escape from danger, to move suddenly from polite social conventions into the realm of spontaneous action.

In a presentation for the Ninth International Congress on Noise as a Public Health Problem, architectural soundscape researchers Barry Blesser and Linda-Ruth Salter confronted the question of benefits from extremely loud noise and music: given the hazards to hearing, why are so many people attracted to deafening sound? Along with recognizing the appeals of social distancing and ownership of personal space, they traced and summarized research on the neurological impacts of loud sound to demonstrate another conclusion: like alcohol, drugs, and sex, extremely loud music alters consciousness. It brings the entire nervous system into a state of excitement, stimulates the pleasure centres of the brain to release endorphins, and can even produce addictive responses in habitual listeners. Sound above 90 dBA, particularly when paired with rhythm and inclusive of low frequencies, sends the vestibular system of the inner ear, responsible for balance, into high alert (Blesser and Salter 2008). Thus, extreme loudness can be a strong attractant even as it erodes hearing. It is catnip especially for the young because it mimics danger, overrides reason, and sends the joyous chemistry of endorphins coursing through the dancing body. It rewires response to the moment of immediate fight-or-flight into movement spontaneously expressed in dance – no planning or memory required. It holds the illusory promise of eternal youth.

More recent examination has identified the mechanisms of the attraction. Using a research model they call Conditioning, Adaptation, and Acculturation to Loud Music (CAALM), New Zealand audiologists David Welch and Guy Fremaux explain the reactions involved: brainstem responses to urgent signalling, conditioning to associate loud music with excitement and pleasure, memories of previous pleasurable experiences summoned by the music, group cohesion induced by dancing. In live events, contagion of emotions through a crowd intensifies the thrill, with all factors leading to a conditioning process: as more loudness is required to produce the excitement, the auditory system adapts to it. Tolerance is increased precisely because hearing acuity is diminished through damage to the cochlear hair cells. The damage is both "permanent (due to pathology) and temporary (due to adaptation)" (Welch and Fremaux, 2017, section 3). In other words, death of hair cells (permanent) from prior exposures leads to more strain on the remaining ones with each new exposure, weakening them as the auditory system struggles to receive the signals and the brain struggles to interpret them. Seeking a physiological root cause for the process, the researchers identified oxidative stress to mitochondria in the hair cells (Kujawa and Liberman 2009; Welch and Fremaux, 2017).

The attractions of extremely loud music are social as well as medical phenomena. Loud music symbolizes fun, especially to extroverted personalities. Club managers turn up volume levels throughout the night because customers drink more (Guéguen et al. 2008) and the venue becomes associated with intense enjoyment that is shared with others, creating a collective identity of nightclub and mass concert enthusiasts (Welch and Fremaux 2017). This identity now defines a multinational culture sustained and manipulated by a very profitable entertainment industry. As a result, the culture of loudness follows the young as their years accumulate, integrated into the soundscapes of workplaces as well as leisure activities.

10

Working with Noise

Open-plan office noise can have a negative impact on fatigue, motivation and performance. How much performance is impaired appears to vary with hearing status and the cognitive processes required by the tasks performed. Yet, research is needed to find the best combination of reduced sound level, speech intelligibility and intermittent sounds to enhance performance. In addition, it is important to pay attention to the environments visited for restoration, as noise exposure during a break can further decrease motivation and subsequent performance.

Jahncke 2012, 39

Noise in most working environments is an unwanted by-product of the process. In most countries, noise exposure for workers has been controlled by legislation for many years. In the music industry the "noise" is actually the "desired" product.

Barlow and Castilla-Sanchez 2012, 1

Loud recreational noise is not solely for the young, as the attractions of middle-age motorcycle clubs and senior rock-band revival tours attest. The same cautionary principles apply: hearing protection now might save you from needing a hearing aid later, or at least delay the necessity. But even if your clubbing days are a memory, you're not free from noise. It hisses and roars through everyday life: the morning commute to the office, the cubicle and lunchroom, the factory floor. It's the voice of the machinery that runs economies and the signal of productivity, pervasive and largely unnoticed. If you do factory work, it might be loud enough to gradually deafen you. If you are a clerical worker, it will be far subtler and easier to ignore. *But harmless?* Not completely. While children are particularly sensitive to it, low-level ambient noise also has impacts on the cognitive processes of adults. The soundscape of clerical and cognitive work will not expose you to a risk of hearing loss through excessive loudness. It *can* expose you to fatigue, distraction, irritability, loss of privacy, and other detriments to your quality of life, most of which will slip beneath your awareness, day by working day.

Systematic interest in workplace noise effectively began with efforts to protect hearing and comprehension of commands during battles in the First World War. It intensified in the Second World War with the search for fighter plane command codes that would be audible above the sounds of engines and guns. Concern for the auditory health of soldiers was secondary; in the midst of war, what mattered most was victory, and noise reduced the efficiency of the military workforce. Post-war manufacturing industries were the next major focus of investigation, with attention given to productivity and safety in the presence of loud mechanical noise. As munitions factories were retrofitted and workers retrained to produce cars and appliances for domestic use, the soundscape of manufacturing became a symbol of growing prosperity. Workers might notice diminished hearing, but their salaries supported families, and they could take pride in their contributions to growing national wealth. As new businesses thrived, the office became an alternative workplace for the employment of women as well as men, and the preservation of information, financial records, and correspondence created a percussive soundscape for mental activity. Subsequent changes in technology throughout the later twentieth century and into the twenty-first brought new forms of noise into the clerical workplace, replacing the intermittent clatter of typewriters with the soft insistent roar of computer fans.

The aesthetics of workplace design have often sacrificed acoustic privacy to the goal of visual harmony. The use of glass for interior partitions is an example: what is gained in attractive surroundings can be lost in noise annoyance, a fact that was often overlooked by architects and their customers until recently. Compared with soundproofed drywall, traditional glass walls – even when soundproofed – transmit 50–100 per cent more noise (Shellenbarger 2012). Progress has been made recently in glass manufacturing, with marked increases in production standards leading to reductions in sound transmission. Glass products are now regarded as equivalent to walls, beams, and other structural components in the architect's toolbox of acoustically designed materials. Categories of acoustic performance such as sound transmission class (STC) and outdoor-indoor transmission class (OITC) have been introduced and have been adopted by architects when considering basic building design.

Demand has grown in the past thirty years for STC- and OITC-rated glass, whether monolithic, laminated, or insulating. In some cases, increasing glass thickness can help to reduce noise transmission. The air space, sealant materials, and even the framing system can also become vital contributors to indoor sound abatement. All of these have been found to help worker productivity. However, sound-reducing glass is expensive: it is a feasible component of high-end offices in profitable businesses but not a general solution for entire industries or for the retrofitting of older buildings. A closer look at office activities and design will provide an overview of both problems and solutions.

FOCUS INTERRUPTUS:
NOISE AT THE OFFICE

Noise does not need to be loud to be intrusive. Yet the indoor soundscape in office buildings with sealed ventilation systems, fluorescent lighting, cubicles with incomplete walls, and multiple phones, computers, and copying machines is sufficient to produce distraction and annoyance in noise-sensitive individuals. Acoustic conditions in environments where any mental activities – analysis, evaluation, and organization of information; decision-making; writing; and a host of other cognitive processes – take place will directly affect the success of those activities.

Technological advances in the sensitivity of sound-measuring equipment throughout the 1970s and 80s enabled later studies of office workplaces that focused on the phenomenon of attention itself. Noise was observed to decrease productivity, whether through saturation of attention and consequent narrowing of ability to concentrate, through interruption and the effort of refocusing, or through sensory overload and fatigue (Jahncke et al. 2011; Jahncke 2012; Jahncke, Hongisto, and Virjonen 2013). One study found that while simple tasks involving visual pattern recognition were found to be relatively unaffected or even slightly enhanced by noise, tasks involving comparison or recognition of complex patterns were negatively affected by both continuous and intermittent noise (Cohen and Spacapan 1984). The same study points out that, although simple vigilance tasks were somewhat improved by noise, complex ones were negatively affected.[1]

The results of research into noise distraction in the workplace were largely swept aside in the later twentieth century, however, by the growing trend toward reconfigurable spaces for clerical workers that would improve efficiency by eliminating dividing walls. Office cubicles, invented in the late 1960s to provide visual privacy and flexibility in open-plan offices, became popular with business owners because they reduce infrastructure costs and provide an impression – or perhaps an illusion – of cooperative culture. Like open-plan classrooms, they are an example of design decisions appropriate for specific uses but too broadly applied. Most North American office space is divided into cubicles, and the amount of space allotted to each shrank by 25 to 50 per cent in the 1980s and 90s to a 2006 average of 75 square feet (McGregor 2014).

Designed to offer visual privacy, cubicles are notoriously poor at providing acoustic privacy – wall-free systems provide none. Cubicles in some industries are now being replaced by "benching," a system for shared workspace at long tables that eliminates free-standing barriers (Steelcase WorkSpace Futures 2010). Benching, promoted for information technology, marketing, and advertising agency teams as a way to enhance collaboration, is a transfer of the traditional manufacturing workbenches used by pre-industrial artisans to information-based industries. It is a system designed to provide workspace for itinerant contract workers and to enable natural lighting conditions through direct visual access to the panoramic windows of glass-walled office buildings. Promotional literature for benching system furniture does not mention noise or auditory distraction, except to note that one variety of bench can incorporate "limited" acoustical barriers (Carroll 2019). The primary reason given to businesses to adopt the system is that it saves expensive office space and provides maximum flexibility for open-plan designs.

Research conducted from the 1970s to the present shows consistently that noise annoyance and distraction are significant aspects of open-plan culture. A 2005 overview of the findings reported by studies that interviewed workers in open-plan offices stated that a majority – between 54 and 67 per cent – of workers found noise to be a distraction that affected concentration, productivity, and mood (Banbury and Berry 2005). These perceptions of surveyed workers are supported by studies using two approaches to research in cognitive psychology. In the quantitative method, tests are set up in laboratory conditions using volunteers with normal hearing acuity who perform mental tasks involving memory or pattern recognition with words or numbers. Recorded sounds – speech, traffic noise, or white noise – are then played continuously or at specific intervals while the volunteers work. The degree of distraction is measurable by means of comparing accuracy and speed of answers in noisy and quiet conditions. Such tests consistently show disturbance by sound for a majority of volunteers, with little difference shown among the types of sound unless some are significantly louder than others. In the more qualitative type of research, questionnaires are used to determine individual responses to conditions in the workplace itself. The most useful surveys account for variables like age, gender, length of employment, nature of tasks performed on the job, and even attitudes toward the job and the workplace.

CASE STUDIES: THE OPEN-PLAN OFFICE

Since the open-plan office is unlikely to be abandoned in favour of increased privacy, its presumed advantages are worth examining. From the standpoint of management, does cost-effective use of space offset decreases in productivity caused by auditory distractions? From the perspective of workers, do physical proximity and absence of auditory barriers produce a more cooperative workplace or a more defensive one? While definitive answers to these questions may be elusive, since they depend partly on the nature of work being done and on the personalities of the workers, studies indicate both causes for concern and clear directions for developing solutions to problems.[2] Some examples demonstrate the scope of the issues involved.

Studies conducted in Europe show that typical noise resulting from work and communication in open-plan offices causes distraction and discomfort for a majority of employees. A Danish study found that recorded noise from an open-plan office, played at 55 dBA in laboratory conditions to a group of thirty test subjects for three hours, resulted in complaints of distraction, fatigue, and dissatisfaction with working conditions (Witterseh, Wyon, and Clausen 2004). A questionnaire-based survey of eighty-eight open-plan office employees in the United Kingdom determined that most workers regarded routine office noises as disturbing (Banbury and Berry 2005). Phones ringing unattended were cited as a particular annoyance; so was overheard speech, even if it was not loud enough to be heard clearly over the ambient noise of the office. Moreover, even the *thought* that other people might overhear the respondents' own conversations was cited as disturbing (ibid., 31). Computers and printers, as well as people passing through corridors between cubicles or entering cubicles to chat, were distracting as well. Nearly all (99 per cent) participants reported at least one source of disturbance by noise; a large majority (82 per cent) found three or more sources disturbing. The top-rated three were unattended phones, printers, and conversations. The questionnaire also asked respondents for suggestions about remedies to the problem. These included better sound-proofing materials, the provision of a break room where conversations unrelated to work could be held, interview or meeting rooms for work-related conversations, and designated quiet areas for those who preferred to work in them (ibid.).

A small qualitative study carried out in Finland focused on a group of thirty-one office workers moving from private enclosed offices to open-plan cubicles. The workers were surveyed by means of questionnaires about their reactions to both workplaces, their perceptions of productivity, and their preferences. Descriptions of the open-plan workplace indicated a significant reduction in speech privacy after relocation, as well as increased distraction. Self-rated loss of work performance because of noise doubled (Kaarlela-Tuomaala et al. 2009). The most significant impact was on cognitively demanding tasks and on work-related phone communication: these are not surprising, since sound from overheard conversations distracts auditory attention and masks acoustic signals in speech. The authors conclude that while open-plan offices can be advantageous for groups of individual workers focused on interactive creative work, they are detrimental to the type of cognitive tasks required of professional employees and to job satisfaction for most office workers. They recommend that employers considering moving to open-plan offices first analyze job descriptions and tasks involved in the average workday, in order to determine whether the move will enhance or decrease productivity and worker satisfaction.

Further investigation of the issues inherent in open-plan workplaces is provided by a review by a team of Dutch occupational health researchers of forty-nine articles published from 1972 to 2004 on the subject. They found "strong evidence that working in open workplaces reduces privacy and job satisfaction. There is also limited evidence that working in open workplaces intensifies cognitive workload and worsens interpersonal relations; close distance between workstations intensifies cognitive workload and reduces privacy; and desk-sharing improves communication" (De Croon et al. 2005, 119). Their recommendations include providing "acoustic and visual protection" in open-plan workplaces. Moveable barriers can be used to isolate small working groups visually; providing an enclosed and soundproofed conference room for occasional use is even more effective.

Additional evidence comes from Canada. An unpublished survey carried out in 2007 by university students who were also employed as office workers on their colleagues, all working in open-plan corporate head offices, came to similar conclusions.[3] The thirty-one voluntary participants, primarily in marketing and sales jobs, were asked questions pertaining to their preferences for levels of privacy, perceived productivity, and sources of annoyance. Nearly 80 per cent reported a negative impact on their productivity because of noise: interruptions; distractions; the temptation of personal conversations and office gossip; the difficulty of ignoring other peoples' phone conversations, whether related or unrelated to work. Several also reported a feeling of unease about using phones to contact family members or doctors during the workday because their conversations would be overheard by other workers: while this may be seen as a bonus for productivity, it actually resulted in the workers taking more frequent breaks in order to carry out personal business away from their cubicles. One respondent reported unease with the fact that the location of her cubicle next to a manager's office caused her to overhear sensitive information about fellow employees. The students examined responses for evidence of changes in personal behaviour at work, finding that a majority of their office colleagues – particularly those who had changed from private offices to cubicles – reported becoming more secretive and seeking periods of isolation. As in the Finnish study, acknowledgment was given to the advantage of open-plan offices for some forms of creative and collaborative work.

More recently, US researchers took a broad-spectrum approach in investigating whether the stated advantage of open-plan offices for collaboration could outweigh the detriment of noise distraction. They combined evidence from a literature survey of studies published from the 1980s through 2012 with comparative statistical analyses of the studies' results. The team came to the conclusion that the "benefits of enhanced 'ease of interaction' were smaller than the penalties of increased noise level and decreased privacy resulting from open-plan office configuration" (J. Kim and de Dear 2013, 18). Even the supposed advantages of open-plan spaces for younger workers in creative teams is now being questioned: millennials accustomed to multitasking may be less overtly stressed by noise and speech-related interruptions than older workers, but their ability to recover concentration when interrupted takes longer (Konnikova 2014).

Solutions to the acoustic problems of office cubicles are no longer scarce. Companies that manufacture modular office equipment are developing designs utilizing materials that absorb sound, but they are not inexpensive. While innovative design solutions and new materials are available, they are most easily accessible to companies that can afford a large expenditure for workers' comfort. The most effective solution is to plan ahead: professionals in every aspect of sound abatement now agree that the concept of acoustic protection by means of design is important. When the distractions of an open-plan office are not taken into account at the planning stage, subsequent upgrades to the acoustic insulation of the space, or its reorganization, will be costly in time and funds. Since acoustic properties and problems are rarely considered until complaints are raised by workers already using the space, foresight is unusual. For this reason, the most common "fix" for problems with acoustic privacy is the addition of a layer of content-free "white" noise.

ACOUSTIC PRIVACY: A STUDY IN WHITE?

White noise, composed of a signal spectrum with equal power in all frequencies, is a pervasive element of workplace soundscapes in the twenty-first century. So called because it is the auditory equivalent of white light, white noise and its variants are produced by mechanical devices that contain engines and fans: computers, hair dryers, and vacuum cleaners as well as air conditioners and ventilating systems.[4] It is most easily exemplified by the ubiquitous soft hissing or roar of ventilation systems in sealed-system public buildings or the interiors of commercial aircraft. White noise is now being used intentionally in the design of office environments to mask the distinctive sounds of conversation, producing an acoustic neutrality in which the content of speech cannot be easily distinguished. It is advertised as a means of promoting acoustic privacy in cubicled offices and call centres, an attempt to neutralize any sound that might draw attention by masking it with a wall of noise rather than isolating it with the barrier of a physical wall. Acoustic privacy results not from seclusion but from interference, the addition of signal upon signal. The practice is analogous to dealing with an unpleasant smell by spraying perfumed "air fresheners" at it: rather than fresh air, the result is unpleasant odour plus chemicals. The use of white noise was first suggested by research studies done in the 1990s that found noise distraction to be worse in small open-plan offices than in large ones, because a large number of people speaking at once will mask the content of conversations.[5] What results is a dull roar in which distinctive speech phrases cannot be distinguished, making them less likely to deflect attention from cognitive tasks.

Using white noise as a mask for open-plan office soundscapes can be considered a band-aid, a quick fix that is not necessarily effective in all conditions and may even be counterproductive in some. Office workers already spend the majority of their waking hours in the haze of white noise produced by ventilation systems and punctuated by phones, conversations, and office machines. When a second layer of white noise is added by design, some workers are soothed while others are further annoyed and distracted, particularly when complex cognitive tasks are required of them. For many, the immediate reaction to white noise is drowsiness. Widely used as an aid to sleep, white noise is effective for many adults and infants (López, Bracha, and Bracha 2002; Forquer and Johnson 2005) and available to the public through low-cost electronic generators for home use. When office workers accustomed to using it as a sleep aid encounter it at work, the results are likely to be counterproductive.

Very few controlled studies on the effects of white noise on occupational health and productivity have been published. In one Canadian study, student volunteers were divided into three groups and tested in conditions of quiet, recorded office noise at 55 dBA, and the same recorded office noise masked with white noise. The group working in quiet conditions performed best on cognitive tasks and felt the least stressed by their environment; the group with masked noise fared less well but better than the unmasked noise group (Loewen and Suedfeld 1992), presumably because they were less distracted: steady moderate background noise reduces the ability of intermittent foreground noise to startle.

A more recent study was carried out on laboratory rats, subjecting them for a three-week period to fifteen minutes per day of white noise at 90 dBA, loud even for humans. The rats were found to have increased heart rates, raised blood pressure, and inflammation of the intestinal lining as a result (A. Baldwin 2007). The researcher concludes that "90 db white noise reduces stimulation of the parasympathetic nervous system and also induces an inflammatory response in the intestinal mucosa, resulting in structural damage. These results are consistent with a stress response" (A. Baldwin 2007). Is this finding applicable to humans? Because the loudness of the noise as well as its frequency components are at issue in this experiment, perhaps not: the white noise used for masking conversations in open-plan offices is kept to about 45–60 dBA. Yet, in the sense that mammalian responses to stress are common across species, the findings probably are applicable. So, we might ask whether a level of noise-induced stress that causes structural damage in a rat will cause chronic mild irritation and/or inflammation in a sensitive human. There is as yet no clear evidence to answer that question, but it is certainly worth raising. So are questions about another common remedy for the lack of acoustic privacy: sonic isolation through the use of personal soundtracks.

THE DISCRETE CHARM OF THE PMP

Do you wear earphones at work because the office is too crowded and noisy? Do you keep a radio on at your desk, or stream from your smartphone, because your solitary office is too quiet? In either case, you have plenty of company.

As the use of personal music player (PMP) technology increases in shared workplaces, personalized music has taken over what used to be a contested commons. The soundscapes of twentieth-century clerical workplaces consisted of ambient noise: typewriters and conversation plus traffic sounds through open windows; then computers, printers, copiers, and HVAC (heating, ventilating, and air conditioning) in sealed environments. Information-age offices now resound with HVAC roar plus additional manufactured white noise and/or commercial background music. Workers who object to the choice of music or to the white noise increasingly mask it with their own choices on portable listening devices. Media scholar Michael Bull remarks that "the office space becomes two simultaneous privatized environments – one existing in radio sounds, the other in iPod sounds," enabling individual workers to "manage" the time and space in which they operate (Bull 2007, 116). Bull and other commentators on PMP culture frame the dispute as a contrast between the repressive dictates of management represented by radio (and presumably by other broadcast soundtracks or Muzak), and the individualistic freedom of the PMP or phone as music device, whose users overcome their distaste for mundane repetitive tasks by fine-tuning their moods with their favourite songs. PMP use is thus characterized as a progressive rebellion against the conformity of corporate office culture.

The post-industrial information-age workplace may well be developing into an Eden of contented workers whose mood and personal autonomy are "ultimately uncontrollable by others, the culture industry and society in general" (Bull 2007, 122), but that autonomy depends on commercial products and on a desire for isolation from contact with others. Is the brave new world of the PMP enthusiast an emotional utopia that will render antidepressants obsolete, or a new form of addiction that will render cooperation obsolete?[6] Neither scenario is likely, but further research is needed. Since the ambiance of an open-plan office is far less loud than a factory floor, there is considerably less risk of hearing damage for workers, but is the risk negligible? Is the productivity of workers performing cognitive tasks affected by the white noise used to mask conversation? By the music used to mask the white noise? By the personal soundtrack superimposed on the ambient music and/or noise?

The process of answering these questions will give rise to more questions about ethics as well as efficiency: Do workers in open-plan offices have a right to control their personal soundscapes, or do employers have a right to set conditions? Does the benefit of a personal soundscape on the job – the contentment of individual workers – outweigh the potential for distraction? Such questions are important because there is little information yet about the effects of either white noise or superimposed music on the performance of cognitive tasks in clerical workplaces. A certain irony is involved. If the most common remedies for auditory distraction are also auditory stimuli, can they cause additional stress? According to two studies on the reactions of computer system software developers – whose work requires extreme accuracy and involves high levels of stress – to music in the workplace, the opposite may be true, at least for most people. Both efficiency and mood tended to improve with music listening during work when participants were able to choose their individual listening menus, although workers unaccustomed to background music experienced a "learning curve" before positive effects appeared (Lesiuk 2000, 2005). Measurements in both mood and efficiency declined during a week in which music was forbidden for purposes of the study, and rebounded when music was reinstated (Lesiuk 2000). The author points out, however, that some personality types did not experience a reduction of anxiety, and that the categories of music used by participants were not monitored, so no conclusions could be drawn about the advantages or detriments of particular styles of music.

Two more recent investigations with Taiwanese university students suggest that what a person listens to while doing "desk work" *can* make a difference to productivity levels. One found that "lyrics in background music negatively affected worker attention performance," and its authors concluded that music without lyrics produces less distraction, making it a better choice for workplaces (Shih, Huang, and Chiang 2012, 577). The other compared the effects of silence, classical instrumental music, and hip-hop on a reading comprehension task. The control group, working in silence, fared best; the classical music was considered to be mildly distracting. Hip-hop, because of its heavy bass and compelling rhythms as well as rapid-fire lyrics, was considerably more distracting (Chou 2010).

Listening to lyrics engages the same areas of the brain as reading and analytical thinking; when attention is divided by multiple stimuli, efficiency drops. Add a musical style with attention-catching features, and concentration on a mental task becomes even more difficult. If you use music to mask noise and lift your mood at work, concentrating your playlist on soothing instrumentals, or listening at a volume low enough to blur lyrics, is likely to make your workday more productive and your supervisors more impressed with you.[7]

Even your attitude toward the music you hear might affect aspects of your productivity on the job. An experiment in the United Kingdom measured performance on a memory task – recalling a list of eight random words presented in a particular order – and found that music was detrimental in comparison with quiet. What made this preliminary experiment especially interesting, though, was that participants fared worse when the music was familiar and well liked than when it was unfamiliar and not particularly liked (Perham and Sykora 2012), possibly because familiar music encourages the focus of attention to settle on memories rather than tasks.

It seems, then, that whether listening to music on the job will benefit productivity and worker satisfaction – which are not necessarily equivalent goals – may depend on the type of work being done, the content of the listening, and whether the music is played on personal devices or a piped-in ambient system. If shared ambient music is appreciated by everyone within earshot, it can improve mood, but will it also distract? If the combination of music – whether shared or individual – and ambient background noise reaches levels loud enough to interrupt concentration, is it still beneficial? What if some workers enjoy the music, while others find it annoying and regard it as noise? All of these possibilities should become matters for discussion and decision making among employees, teams, and managers when considering the soundscape of an open-plan workspace.

UMM ... WHERE WAS I?

We have seen that noise causes distraction as well as detriments to health, and that it is, or should be, a major concern in office workplaces: unpleasant sound and uncomfortable temperature are believed to be the major causes of annoyance in offices, ranking above lighting and office layout (Mak and Lui 2012). Beyond annoying and distracting employees, the combination of computer fans, printers, ventilation, conversation, music, phones, and message notifications interferes with the ability to get work done.

If productivity is reduced by the soundscapes in offices, what is the cost to the industries involved? Attempts to quantify loss of productivity have estimated it at 8 per cent in the presence of mechanical background noise (Roelofson 2008) and as much as 66 per cent in the presence of irrelevant speech during activities involving memory recall for language-related and mathematical tasks (Banbury and Berry 1998). Add the hypothetical costs of increased employee stress to health care systems, and the price of intrusive workplace noise becomes incalculably large. But office work represents only a portion of the spectrum of noise as a socio-economic problem. The other components? Noise in industrial, transportation, communication, and entertainment workplaces.

WORKING LOUD

As mentioned in chapter 4, it has been known since the Industrial Revolution in nineteenth-century Europe that noise hazards abound in industrial workplaces, and that workers in many industries are prone to impaired hearing as a result. Today, hearing impairment caused by industrial noise is a worldwide problem, a price that people pay for being steadily employed in jobs that benefit economies. Since – unlike blindness, back injuries, or loss of a limb – it does not usually result in a disability that interferes with continuing to do the same job, it may be ignored by the worker, the employer, and the health care system. Worldwide, risk is higher for males than females, and higher in developing countries than in technologically advanced ones. World Health Organization (WHO) guidelines point out that industrial noise levels in developing countries are particularly high. This is because access to expensive noise-abatement technologies is limited and because legal protection of workers' health and safety is minimal or non-existent. With globalization and the trend toward outsourcing manufacturing jobs to countries where occupational health and safety standards are low, noise-induced deafness is becoming a matter of concern for societies that are new to industrialization, even as the outsourcing countries continue to improve their protection of workers' health.

A study involving 247 workers at municipal landfill work sites in four Chinese cities provides an example. Disposal of waste metals and building materials that could not be reused or recycled was carried out, exposing the workers to high levels of noise and toxic substances. Data from the workers was compared with a control group that had no exposure to workplace noise. The results were instructive: the average rate of NIHL in all exposed groups was 23.5 per cent, with one site calculated at 36 per cent – both rates significantly higher than a 2006 global average of 16 per cent. The authors estimated that failure to use hearing protection devices contributed to the results, along with the presence of ototoxic substances at the sites (Y. Liu et al. 2015).

Factories

Today's improvements in manufacturing technology, designed to maximize efficiency and profit, are not always designed to minimize noise and may present new problems that challenge even industries that are considered safe to maintain their safety. Among the numerous sources of noise in manufacturing are rotors, gears, engines and their fans, pneumatic equipment, compressors, and drilling, blasting, pumping, and crushing processes. Moreover, these sounds reflect off of floors, ceilings, walls, and equipment.

Impact processes in manufacturing and construction are especially problematic, but many industrial processes routinely produce sound pressure levels well above the threshold for hearing damage, usually set at 85 dBA over an eight-hour workday. Some industries – including foundries, shipyards, breweries, and sawmills – can produce averaged values from 92 to 96 dBA and recorded peak values between 117 and 136 dBA. At the peak levels, unprotected workers can experience hearing damage within a matter of minutes.

The risk of noise-induced hearing loss (NIHL) for factory workers, already the most vulnerable group, is increased by exposure to industrial toxins including styrene, a solvent used in the manufacture of plastics (Śliwińska-Kowalska et al. 2001; Morata et al. 2002). Some insecticides are also implicated. Organophosphates and pyrethroid products can induce damage to the central auditory system in pesticide industry and pest control workers, apparently even in the absence of exposure to damaging levels of noise, although the damage is significantly worse with exposure to both chemicals and noise (Teixeira, Augusto, and Morata 2002 and 2003). The risks are particularly great for workers in the developing world, where medical, legal, and educational systems may be inadequate to the tasks of prevention, protection, and treatment. While China, for example, has made great strides in protecting workers' hearing within factories in large cities such as Beijing and Shanghai, the measures adopted by the government are far from universal: in rural areas, there are factories that go unmonitored for extreme levels of noise production.

An epidemiological crisis is brewing in China with regard to hearing loss. Currently available published data on NIHL in Chinese research journals appear to confirm a high rate of noise-related deafness among workers "from various industries, including manufacturing, textile, metal works, railroad and packing, [that] range from 12.15 to 49.02%" (Shi and Martin 2013, section 5). Chinese research has provided an abundance of evidence demonstrating that the "incidence of NIHL increases as the duration of exposure increases," with a "prevalence of NIHL as high as 80.7% ... reported in railroad workers with a working career longer than 10 years" (ibid.). As China's population adapts to further industrial and technological growth, noise exposure and NIHL among both workers and the general population will require diligent monitoring.

As research in occupational health continues, audiologists and medical researchers are gaining knowledge that can help protect the general public as well as workers in specific industries. The breadth of focus required to produce a fully effective policy for prevention and treatment of occupational NIHL is challenging, however, because of the variables involved:

- If the number of occupational health claims increases, does that reflect better reporting or a failure of protection programs?
- Has noise just below the legal threshold of 85 or 90 dBA caused the majority of cases in the years since the enactment of legislation in North America?
- Is recreational noise responsible for some of the increase?
- How much of the increase is attributable to employers' failure to comply with legislation (Daniell et al. 2002), workers' failure to comply with rules regarding hearing protection (Daniell et al. 2006), and problems with methodology in research (Kurmis and Apps 2007)?

All of these questions require further investigation before answers can be found. As well, the methods used to gather and interpret data about NIHL need improvement: publications often fail to include information about genetic history, employment history, and recreational habits. Genetic factors are suspected to increase susceptibility to noise sensitivity or noise-induced hearing damage, so including family histories regarding NIHL would be useful. Information about whether the people involved in the study have been in the military, work out wearing earbuds at noisy gyms, or play in amateur rock bands would also be extremely significant, but it is available only if researchers ask for it and research subjects provide it.

Compliance with workplace regulations can also be problematic. Even when workers recognize the hazards of noise exposure and understand that NIHL would have a major impact on their quality of life, there is reluctance to wear hearing protection because it is uncomfortable and may interfere with warning signals or job-related communication (Svensson et al. 2004; Daniell et al. 2006). As with many solutions to workplace problems, cost is a factor in the comfort and efficacy of protective equipment as well as the design of spaces used for high-noise activities. Where governments mandate or subsidize protection, it is more likely to be effective. Where costs of prevention are borne by the private sector and compensation for injury by government plans, factory owners and managers may have less incentive to promote prevention.

Because, as discussed in chapters 5 and 6, the effect of noise on factory workers is not limited to hearing damage, both the lives of workers and the budgets of health insurance plans can be adversely affected by policies that do not emphasize prevention. For example, a Polish investigation of the effects of factory workplace noise on general health compared employees exposed to noise sufficient to cause annoyance with a control group working in quiet surroundings. The noise-exposed group had a 13.5 per cent higher rate of high blood pressure and peptic ulcers, and a slightly higher rate of stroke, diabetes, and diseases of the nervous system (Bartosińska and Ejsmont 2002). A similar comparison at three metal-working factories in India showed that a majority of workers exposed to continuous machine noise at or over 90 dBA developed significant increases in blood pressure, and specifically in arterial pressure (Singhal et al. 2009). These are not isolated findings: dozens of studies conducted since the 1970s on cardiovascular effects of noise confirm a relationship between workplace noise and risk of disease, especially for older workers who also smoke.

Call Centres

With the twenty-first-century development of a global economy has come the rapid expansion of a new type of "factory." Call centres are now among the top four employers worldwide, employing millions of young and unskilled workers, as well as those with basic computer skills, in enormous warehouses full of people talking to customers on phones. They talk and listen for eight hours a day, taking catalogue orders or travel reservations, providing technical support for software, or telemarketing (Wharton School 2002). The background soundscape of a call centre heard on the phone, the low buzz of multiple overlapping speech patterns, is now such a common marker of industriousness that recorded versions are sold to small online businesses to reassure phone customers that they are legitimate and busy.

Call centre work is not a technically difficult job, and not as dangerous as working with high-speed machinery on a factory floor, but a particular hazard has come to light with legal cases in the United Kingdom: acoustic trauma, the result of repeatedly hearing the high-pitched buzz of fax machines or modems amplified by the worker's headset, as well as feedback from the headsets themselves. Tinnitus and hypersensitivity to noise result from the condition, which can make it impossible for sufferers to continue working in call centres or offices (Revill 2006).

Acoustic trauma – also called acoustic shock – is a hazard for workers using headsets at call and dispatch centres, including emergency dispatchers, air traffic controllers, customer service representatives, switchboard operators, and collections staff. One study found that 13 per cent of call centre employees were affected after three to five years of employment (John et al. 2015). These employees are at particular risk because background noise in the centres is also considered to be hazardous: when workers turn up the volume of headsets above safe levels in order to hear information over the ambient noise, sudden spikes in the sound-pressure level can shock the inner ear, causing tiny muscles to tense. While actual loss of hearing was not among the complaints found by researchers (Pawlaczyk-Łuszczyńska et al. 2019), tinnitus, headaches, irritability, and fatigue were common. If the conditions persist, they can lead to hyperacusis (McFerran and Baguley 2007). As increasing use of new technologies for communication and marketing exposes greater numbers of employees to such conditions, better design of headsets and of workspaces is crucial. Policies allowing, and encouraging, call-centre employees to take breaks in quiet surroundings will also reduce some of the effects of auditory fatigue.

The Music and Entertainment Industries

Noise is an occupational hazard for musicians, especially if they use amplification. Dozens of medical studies have investigated hearing loss in professional rock musicians: a 2007 survey of the literature found that, on average, 20 per cent of them had some degree of permanent hearing loss, and that they were more likely to suffer from tinnitus than non-musicians (Størmer and Stenklev 2007). Music students who concentrate on popular, rather than classical, styles are at particularly high risk of hearing damage and loss through extensive rehearsal with amplified instruments as well as frequent attendance at heavily amplified concerts. One UK researcher who surveyed a hundred technical university students who were taking courses with a focus on rock and pop music found that their course-related rehearsals averaged nearly twelve hours per week at an equalized level of 98 dBA. Nearly all the students – 94 per cent – also attended nightclub performances at least once per week, with measurements at venues near the campus measured at 98 to 112 dBA over four to five hours each night. He states that "76% of subjects reported having experienced symptoms associated with hearing loss, while only 18% reported using hearing protection devices" (Barlow 2010).

The hazard of music-induced hearing damage applies to students, professionals, and amateurs. A 2006 study of forty-two amateur rock musicians active for at least five years and free of exposure to occupational noise showed a significant loss of hearing acuity when the test subjects were compared with a control group of non-musicians of the same ages (all young adults), genders, and occupations. Use of hearing protection was found to be beneficial: musicians who used it had hearing acuity close to the control group. The authors report that tinnitus, hypersensitivity to sound, and hearing loss were observed "in a significant minority" of the research subjects, while hearing loss was "significantly more pronounced, at 6.7 dB higher than the control group, in those musicians who never used ear protection" (Schmuzinger, Patscheke, and Pròbst 2006). Significant risk of hearing loss, particularly for high frequencies, is also a concern for sound technicians (El Dib et al. 2008), who set the volume and balance levels for concerts and events. Hearing loss and tinnitus may lead them to set levels that are too loud for audience safety.

Classical musicians are also at risk of tinnitus and hearing damage induced by music. A 1991 study of the Chicago Symphony Orchestra obtained sixty-eight noise exposure measurements during rehearsals and concerts, calculating daily exposures based on a fifteen-hour work week, which did not take practising into account. Measurements of sound pressure ranged from 75 to 95 dBA, with the mean calculated at 85.5 dBA. Approximately half the musicians showed minor damage to their hearing in specific frequency ranges. Violinists and violists, particularly, showed loss of hearing in their left ears, which are very close to their instruments, while players of the string bass, cello, harp, and piano were least subject to hearing damage (Royster et al. 1991; see also McBride et al. 1992 and Morais, Benito, and Almaraz 2007).[8]

Another study monitored rehearsals and performances of a professional symphony orchestra in Australia over a three-year period, concluding that first-chair players in the brass section – particularly trumpet, horn, and trombone – were exposed to the highest levels of sound pressure, along with percussionists. The authors caution, however, that the complexity of the orchestral soundscape suggests that loudness is not the only significant parameter and that more research is needed (O'Brien, Wilson, and Bradley 2008). Greater recognition of noise-induced hearing loss as an occupational disease for professional orchestral musicians is called for (Emmerich, Rudel, and Richter 2008).[9]

Finally, we need to consider the supporting employees of the music and entertainment industries: cooks, servers, bartenders, security staff, technicians, and concert tour road crews. Recognition of the risk of hearing damage for employees in popular entertainment venues began in the 1990s; one groundbreaking study surveyed staff in eight New York nightclubs that featured live bands, while taking sound-level measurements at their workplaces. The results left no ambiguity: "Employees of urban music clubs are exposed to noise levels that consistently exceed safe levels. The average performance levels ranged from 94.9 to 106.7 dBA ... In addition, ambient sound levels ranged from 83.7 to 97.1, showing that substantial noise exposure may occur even when the band is not performing" (Gunderson, Moline, and Catalano 1997, 77). If the club employees had been industrial workers, they would have been required by US law to wear hearing protection provided by their employers. Only 16 per cent of the club employees surveyed in the study voluntarily used hearing protection, and no such protection was supplied to them by employers.

Additional research has confirmed both the risk and the need for protection. In 1999, a controlled study in Singapore found that discotheque staff, exposed to noise levels of at least 89 dBA throughout their work shifts, "had higher prevalence (41.9%) of early sensorineural hearing loss compared to the control group (13.5%)," and that 21 per cent also complained of recurring tinnitus compared to only 2.7 per cent in the control group (L. Lee 1999, 571). All of the cited studies conclude with recommendation for hearing protection devices for employees.

Similar concerns were raised in a 2002 study by investigators from the Institute of Occupational Health at the University of Birmingham in England. It targeted young part-time employees – servers and security staff working up to sixteen hours per week – in entertainment venues with amplified music on a university campus. The results suggest that both groups, but especially security staff, should wear hearing protection at work. "The mean personal exposure levels for security staff were higher than those of bar staff, with both groups exceeding 90 dB(A). The maximum peak pressure reading for security staff was 124 dB. Twenty-nine per cent of subjects showed permanent hearing loss of more than 30 dB at either low or high frequencies ... Furthermore, four of the 13 subjects assessed showed a permanent hearing loss of >30 dB" (Sadhra et al. 2002, 462). This degree of impairment would diminish the ability to hear normal conversations.

More recently, another UK study focused on compliance with European Union regulations on entertainment industry noise, which were revised and standardized in 2008. The authors found that regulatory limits were rarely observed either in nightclubs or at outdoor rock concerts, which typically generated sound pressure levels between 105 and 115 dBA. Seventy percent of employees – primarily bar, catering, and technical staff in their twenties – tended to ignore rules about the use of protective headphones and earplugs despite training programs. The authors concluded that employees were at high risk for hearing damage, and that both training programs and protection policies needed improvement (Barlow and Castilla-Sanchez 2012).

Even conventional restaurants and cafés that concentrate on food rather than entertainment may put employees at risk for hearing damage. A trend that began in the 1980s toward hard, shiny surfaces in restaurant decor – metal counters and appliances, open kitchens, stone and tile floors, minimal or no upholstery, tables crowded close together, hard-surfaced ceilings – ensures that noise will be reflected rather than absorbed. Add loud background music to the sound of multiple conversations, grills, espresso machines, and kitchen activities at peak hours and you have sound pressure levels climbing to the 75–90 dBA range.

Hard surfaces are practical because they are relatively inexpensive and easy to clean. The ambience of noise and crowding is intentional: customers are drawn to the excitement implied by the crowd, and the noise ensures that turnover will be high (Krause 2002, 9).[10] Owners may refuse to turn down the volume on music because it increases their profits by stepping up the rate at which customers chew, causing them to order more food and also more alcohol in a shorter time (Prochnik 2010, 99–102).[11] Customers have to shout to converse, but their hearing is not usually at risk because they're not there for long. It is the kitchen staff and waiters who may develop tinnitus and tension headaches as well as the possibility of impaired hearing. Basic earplugs can help. So can breaks that allow staff to step outside into relative quiet, allowing the cochlear cells to recover. The most effective preventions, though, are turning the volume of music down and using some absorptive materials – which need not be obvious – in walls, ceilings, or decor to offset the sounds of machinery, music, and voices.

DEFENDING YOUR EARS

Long considered inevitable in certain industries, NIHL in high-noise workplaces is now largely preventable. Although improved insulation and design of workplaces are the best solutions, prevention can be as simple as placing a barrier – earplugs or protective headphones – over the ears so that the pressure produced by the offending noise is reduced to a tolerable level. Basic earplugs made of soft plastic or foam are inexpensive and available at any pharmacy; protective "ear-muff" headphones are available at hardware and construction-supply stores as well as on the Internet. The degree of protection varies with the intended purpose and design of the device. Its usefulness also depends on the wearer's comfort: the protection you don't wear will not protect you, so choose what is comfortable as well as effective.

If you suspect that your hearing is already compromised, see your doctor or find an audiology clinic and get tested. Testing of hearing acuity is usually done with a technique called pure-tone audiometry. The painless test is performed in an insulated booth with a machine that generates single-frequency tones. It does not measure attention or listening skills, just the physical ability to hear and whether the hearing is normal or impaired. The test will give a profile of the ranges of loudness and frequency accessible to you, enabling a health care provider to determine whether intervention in the form of a mechanical hearing aid or specialized hearing protection device would be helpful.

Quiet, Please: Noise Abatement

It is useful … in the case of noise, to enhance citizens' feelings of control …
Even after establishing a maximum noise impact for a particular area, some
choice should be left to citizens, as illustrated by the example of choosing
between many airplanes flying over in a short period of time versus a few airplanes per hour over a prolonged period of time.

Bijsterveld 2008, 258–9

Noise abatement programs have an environmental justice dimension and need
to target the at-risk population.

Moudon 2009, 167

When bylaws or regulations specify that a noise is too loud, too disruptive, or happening at the wrong time of day, and measurements verify the
offence, what is to be done? In some cases – midnight yard parties, boom
boxes in apartments, malfunctioning mufflers on cars – the offenders can
be told to go indoors, turn down the volume, or get the car fixed, on pain
of paying a fine. In others – industrial air conditioning units, drilling rigs,
rail lines – stopping is not an option, and the "fix" may be both costly and
complicated. Providing solutions for complex problems is the role of the
noise abatement industry, which is also responsible for noise impact
assessments that anticipate problems and solve them *before* the new office
tower, factory, mall, or industrial park is built.

Multiple categories of regulation are commonly used in North America and Europe to determine whether a sound needs to be controlled:

- Is the sound clearly audible, without measurement, outside the location of its source? This question applies to public property, but also to
  neighbourhood noise.

- Is there a measurable level of sound that exceeds community standards or bylaws? This question relates, for example, to noise produced by construction sites.
- Is there a measurable level of sound that causes individual complaints? At this level, bylaws about maximum allowable sound levels interact with issues of noise annoyance.

While control can include such simple measures as regulating the time at which noise occurs, abatement involves reducing the actual sound pressure of the noise or limiting its transmission. The process requires expert knowledge of the physics of sound as well as bylaws, construction materials, and industrial regulations.

NOISE ABATEMENT:
STRATEGIES AND ISSUES

While the history of noise annoyance is largely speculative, because individuals vary in sensitivity, the history of noise regulation is well documented, and it gives plenty of information about both problems and solutions in modern urban soundscapes.[1] Anti-noise leagues proliferated in UK and US cities at the end of the nineteenth century, pressuring municipal governments to regulate noise on streets and at ports. The first US municipal commission on noise abatement was formed in New York City in 1929, as a response to growing complaints about the noise of traffic, trolley cars, trains, milk delivery, garbage collection, construction, street repairs, radios, and loud parties (Thompson 2002; Van Allen 2007). The US Congress passed a national Noise Control Act in 1972; however, funding to the Office of Noise Abatement and Control – a branch of the Environmental Protection Agency – was cut in 1981, under pressure from President Ronald Reagan and a host of industries unwilling to pay for noise reduction at their factories.

Substantial growth in the past decades in passenger car, small and large truck, and tractor-trailer traffic in North America have led to steep increases in the ambient noise levels, especially in cities and in communities located close to highways (Blomberg 2014). Increases in traffic noise are also occurring in developing countries. For those that are experiencing rapid urbanization, increases in traffic can happen very quickly. The resulting noise problems grow faster than the ability of municipal governments to build or upgrade infrastructure and improve regulations. In Faisalabad City, Pakistan, with a population that grew from 2 million in 1998 to 7.5 million in 2014, a study of traffic noise at road intersections in three commercial districts found that daytime rush-hour noise over a measurement period of three weekdays reached peak levels of 91–115 dBA (Inam-ul-Haq et al. 2014), considerably higher than permissible standards for Europe and North America. Porto Velho, Brazil, is a city of half a million where rapid expansion of industrial and mining interests enticed young people from farmlands surrounding the city into the city itself in search of salaried jobs and modern lifestyles. A recent study of ambient street noise there found that the main sources – unmuffled motorbikes, airport flyovers, playgrounds, and loudspeakers placed in the entrances of shops, restaurants, and nightclubs – produced sound pressure levels averaging 74 dBA in the daytime and 78 dBA at night, with peaks of 85–98 dBA. While regulations limiting noise were in place, they were not effectively enforced (Paul et al. 2013).[2]

Where legislation and enforcement have not been adequate, science can potentially come to the rescue. Throughout the late twentieth century, advances in the design of sound barriers and in the technology of sound-absorbing materials, as well as improvements in recording and measuring technologies, boosted the efficacy of noise abatement strategies. In many places, barrier design has become more aesthetically appealing, reducing opposition to ugly concrete walls. Pedestrian access to areas surrounded by noise barriers has also improved, so that the barriers don't necessarily cut off the community from services that are often clustered around high traffic routes.

Today, noise abatement is a growing industry that serves governments, urban and regional planners, developers, architects, transportation managers, and many specific industries. It is practised by engineers with specialized education in acoustics. Their work on factories, compressor stations, water treatment plants, oil-drilling and mining sites, highways, rail lines, airports, and community infrastructure reduces both noise annoyance in communities and hazards to workers. Their work on theatres, concert halls, and recording studios ensures the accurate transmission of performances and the pleasure of audiences. It also prevents structural damage to buildings, roads, and bridges that would otherwise result from vibration. While concerned primarily with applications for transportation, industry, and built structures, noise abatement engineers also apply their expertise to small-scale problems like sound-proofing a teenager's basement bedroom or conquering the scourge of snoring (X. Zhang et al. 2014).

Noise abatement usually takes the form of interrupting the transmission of sound from its source. This is an effective intervention because it is a remedy based on the physics of vibration: when the path of a sound wave is interrupted, both transmission and reception are affected. For example, when you ring a bell, its loudness and tone quality are determined by its size and weight, the force with which you ring it, the material of which it is made, and the mechanism that produces its sound. Imagine (or experience) any kind of metal hand-bell ringing inside a room:[3] it will be fairly loud, and its sound will be amplified by the walls of the room as the waves of sound bounce off the wall surfaces to hit the opposite walls. If the floor is surfaced with a hard material like stone or tile, the sound will be louder than with a floor covered by carpeting; the presence or absence of upholstered furniture will also make a difference.

Now imagine, or have someone demonstrate, the same bell ringing in a room next to the one you are occupying, with doors and windows closed. Both distance and physical barriers will reduce the amount of sound. Finally, bring the bell back into your room but wrap its outer surface in a wool scarf, a towel, or a sheet of cotton batting. You will hear far less resonance. Both the wall between the rooms and the cloth are techniques for noise abatement.

Strategies used by the noise abatement industry follow the same principles as the bell experiment. In a factory, machines that produce noise – engines, furnaces, generators – are separated from work floors and placed on and behind surfaces that do not vibrate easily. Soft materials are wrapped around surfaces that do vibrate to prevent sound waves from being transmitted through the air and along building surfaces. Methods used in indoor noise abatement include rigid containers lined with fibrous materials, barriers that direct sound away from occupied areas, design of building components so that the shapes of rooms minimize echoes, placement of panels wrapped with sound-absorbent materials, acoustic ceiling tiles and flooring materials that absorb sound, and even the generation of signals that cancel or mask offending sound waves.

But What Does It Cost?

Noise abatement strategies require the expertise of professionals, extensive analysis and mapping of noise sources, and the purchase of effective materials. They are not inexpensive. Soundproofing a new office complex or apartment building in a large city will run well into six-figure costs if done during construction; retrofitting an existing building can more than double the estimate. This means that financial resources will determine the efficacy of plans as well as the quality of materials used, and that scarce resources will lead to a lack of protection for large numbers of people. This is often the case in residential buildings constructed before protective codes were enacted and in situations where codes and bylaws are routinely ignored.

TRANSPORTATION NOISE: ARE WE THERE YET?

Among the most carefully studied sources of noise are systems of transportation. Measurement, regulation, and abatement of noise from airports, rail lines, highways, and urban road systems generate massive amounts of data pertaining to sound pressure over distance and to definitions of nuisance noise. All systems for measurement are dependent on complex variables: What is the terrain surrounding the source? What is the planned or actual layout of runways, tracks, or roads, and how close are they to inhabited areas? How many residents and commuters are affected? Which systems of measurement are being used, and do they agree? Answers to such questions depend, as well, on what machinery and infrastructure are being evaluated. The following sections, which offer an overview of technical literature about the issues faced by noise abatement engineers, will provide further insight into their work.[4]

Traffic

Communities located next to highways and freeways can be protected from the sounds of high-speed traffic by barrier walls, densely planted green belts, or earth berms (constructed mounds or slopes of soil anchored with plants). In optimal circumstances, careful consideration is given by acoustic engineers and planners to terrain, prevailing wind direction, and speed as important influences on the success of outdoor abatement strategies. Where prevailing wind speeds are high, even the depth and condition of local soil must be taken into consideration in the design and composition of barrier walls in order to prevent their being knocked over. The process requires an approach that takes a broad focus on community needs while anticipating potential problems through close attention to detail.

Noise abatement strategies must also take existing conditions into account, particularly where infrastructure has not been optimally designed. In countries and regions that developed their street and road systems to accommodate motorized traffic, street widths, road surfaces, and the distance of major roads from residential areas have had at least some degree of planning to minimize both danger and disturbance. In older cities throughout Europe, where pre-modern streets and road systems may be preserved in historic districts, residents are subjected to the echoing clattering roar of cars and buses moving over cobblestones, amplified by sound reflection off the buildings that line very narrow streets. Traffic in central districts of older cities in Asia can contain motorized rickshaws, animal-drawn carts, motorcycles, cars, and buses all contending for space, horns ablare, on lane-free streets.

Noise reduction on roads is a matter of changing what can be changed if budgets permit: thicker windows in adjacent buildings, better sound insulation in interior walls and building façades, absorbent barriers and hedges along streets and alleys, and bylaws to set effective speed limits and reduce traffic at night. Pre-manufactured noise barriers are bringing down the costs of installation and replacement of road-side barriers. Whatever the circumstances, the most effective strategies involve residents, community officials, engineers, lawmakers, and law enforcement working together.

Road surfaces are also coming under scrutiny as potential components of a noise-reduction strategy. The use of porous asphalt, which provides better drainage in wet weather as well as reducing noise, is gaining ground in North America, China, and parts of Europe (Ressel et al. 2007), although its utility in cold climates can be compromised by the chemicals used to keep roads free of ice.

Rail Lines

The noise abatement industry measures the vibration, and consequently the sounds, produced by all types of rail transportation – urban commuter trains and subways, high-speed passenger transport lines, and freight trains. All makes and models of train, and all types of track, are studied at a great range of speeds to determine the mathematical formulas that will describe their impact on the environment and their potential to generate noise annoyance. Components of the noise are precisely described: at low speeds, engine noise dominates; at medium speeds, rolling noise – generated by wheels against track and louder if the surfaces are rough – takes over as the loudest element. At the highest speeds, the strongest noise is aerodynamic, produced by the impact of the train body against air. Adding to these sources are "brake screech," "curve squeal," and warning signals, as well as track, tunnel, and bridge conditions, number of axles and brakes on each car, rates of wear on wheels and brakes, and noise from cooling fans (de Vos 2012).

While abatement solutions vary with circumstances, standard recommendations include the use of sound-absorbing panels on the track bed, the addition of absorbent barriers set vertically along the tracks, and the installation of rail dampers to reduce vibration (Vogiatzis and Vanhonacker 2015). Since trains produce low-frequency noise as well as more conventional forms, abatement processes reduce the likelihood of long-distance transmission of sound waves through the ground as well as the air.

Airports

Since aircraft noise is a cause of so many issues related to community annoyance and public health concerns, noise abatement at and around airports is given thorough attention, regulated by governments and overseen by an international organization as well as the aviation industry itself.[5] Along with measurements and technical specifications, environmental assessments and community consultation are crucial parts of the process. The potential detriments of the airport – noise and environmental disruption – are carefully balanced against its social and economic benefits to the community.

Measurement and control of noise at airports is especially challenging because of the large areas of land that are covered by runways, the varying density of activities at different times of day, and the complexity of calculations that must account for a large number of aircraft differing in size, speed, and emissions. Placing a new engine in a given model of plane, for example, will change its sound emissions, and the acoustic engineer doing the calculations has to consult manufacturers' specifications to find the decibel output rates for both plane and engine because they will differ from measurements for the plane before the engine was changed (Bütikofer 2012).

Calculations account for several aspects of flight: location, altitude, speed, and thrust (the amount of power used by the plane's engine at any point on the flight path). Information is typically derived from a grid system that maps points on the airfield and measures noise levels specifically at take-offs and landings. National regulations may differ, however, on whether measurements are taken outside the boundaries of the airport as well as within them. In most systems, noise produced by airport infra-structure, trucks, ground transport systems, and taxiing aircraft is also monitored, but this is not always the case (Bütikofer 2012).

Standards for air traffic safety are formulated by the International Civil Aviation Organization (ICAO), a branch of the United Nations, with 191 member states. Among its mandates is environmental protection, includ-ing the reduction of impacts from noise. The initiative began in 1970 and has been revised several times to account for changes in aircraft technol-ogy from propeller planes to jets. The ICAO's current documents relating to noise, published in 2004 and revised in 2007, begin with calling atten-tion to the fact that noise is a leading reason for communities to oppose the construction and operation of airports, and with the statement that the organization's primary responsibilities include the "use of objective and measurable criteria" for its policies and the resolution of disputes (ICAO 2007). Land use management – the planning and maintenance of buffer zones around airports and runways – is mentioned prominently as well. The ICAO encourages member countries to achieve a balance among four aspects of noise control: reduction at the source through bet-ter engineering and manufacturing processes for aircraft, land use poli-cies that direct flight paths away from places where they will cause dis-turbance, direct abatement through sound-absorbing materials and engine design, and restrictions on operation when necessary. The success of noise abatement policies also depends on the expertise of air traffic controllers, who direct pilots to arrival and departure paths that, as much as possible, avoid flying over populated areas and attain specific altitudes upon departure.

Aircraft are now estimated to be 75 per cent quieter than they were in the 1960s, and new designs – both subsonic and supersonic – being reviewed for use in the next decade are expected to reduce noise even fur-ther (Dickson 2014). Design cannot entirely overcome the effects of physics, but with a combination of regulation, well-designed equipment, predictable wind patterns, expert management, and good community relations, airports can be acceptable as neighbours. Because the same assumption cannot always be made about people, indoor noise abatement is also crucial to the quality of everyday life.

INDOOR NOISE ABATEMENT

When the building you live in, the office you work in, the hotel you visit, or even the hospital where you have to stay for a while is not properly insulated against transmission of noise, your stress levels rise. Sleep, productivity, relaxation, and speed of healing consequently fall. Fortunately, both regulation and noise abatement are on the job to improve the situation. As the preceding chapter points out, bylaws and building codes are crucial to the quieting of neighbourhoods. Codes for new buildings and major remodelling projects often specify not only the type of materials to be used but the level of sound insulation required and even the companies that manufacture suitable products. Acoustical analysis of each site may be required to determine levels of noise that will impact residents from outside, including traffic, nearby businesses, neighbourhood infrastructure, and human activities.

Acoustic engineers work with architects to design effective "building skins," considering exterior and interior wall construction and insulation; drywall quality; flooring and roofing materials; caulking procedures; window glazing; location of doors, windows, and plumbing; and even the placement of mail slots and ventilation grids within units. Outside the units and inside the building is the mechanical infrastructure: generators, elevators, kitchens, washrooms, laundry and media rooms, and lighting, heating, and ventilation systems. Effective insulation placed between adjoining walls and under floors will prevent these vital services from becoming noise nuisances.

Inter-unit noise sources are also significant, particularly those coming from immediately above: the people in apartment 4A are more likely to annoy those in 3A than those in 4B, unless the walls are particularly thin. This is because the structural partitions between units convey vibration, and therefore sound, vertically. As long as floors are interrupted by solid walls, horizontal adjacencies are less problematic unless occupants are very loud.

Indoor noise abatement is accomplished primarily by two methods: blocking sound with non-resonant hard surfaces, and absorbing it with soft ones. In the former category are specially formulated wallboards and barriers made with layers of gypsum or vinyl, and window glass that encloses a layer of air or inert gas between thick panes. These are the primary means of stopping sound from being transmitted between units in a building or from streets to interiors. At their highest grade, such materials provide the insulation for concert halls and theatres and the soundproofing for recording studios. Absorbent room panels, floor linings, and wall insulation, made from high-density foam, rubber, and mineral wool respectively, muffle sound within a room or corridor and absorb echoes. Additional protection comes from encasing ventilation pipes in sound-blocking materials, and from flexible airtight sealants for wall and floor joints, gaskets, and window frames. Even the placement of small openings in a wall for electrical outlets and switches, water pipes, and heat vents must be carefully planned so that no adjacent walls share openings.

Abatement of Low-Frequency Noise

Like the legal aspects of noise control, the practice of noise abatement faces challenges and controversies in cases involving low-frequency noise (LFN). Because the frequencies of LFN do not decay with distance to the same extent as higher frequencies, they produce noticeable effects at surprising distances. In addition to the impacts on health discussed in chapters 6 and 14, LFN can cause windows and light fixtures in buildings to rattle and – in extreme cases – can gradually weaken building structures. Abatement of LFN is usually expensive. Because of the possibilities of complaints and structural damage, ignoring the problem is likely to be even more so: it's considerably less expensive to reduce mechanical noise at its source than to build a barrier to reduce it after it poses a problem. For this reason and others, including the need to avoid public annoyance, abatement of LFN resulting from mechanical sources is best done in the planning stages of a project. This means that the planning team needs to know that LFN may be a problem *before* it becomes one. Noise assessments done with both dBA and dBC scales can reveal the necessary information. A difference of more than 20 decibels between the two scales, along with frequencies below 250 Hz, can indicate the presence of LFN.[6] Thus, a relatively simple test can save a company or government considerable time and money when the problem is viewed holistically and preventive measures are considered before plans and budgets are complete. Without such foresight, residential communities at considerable distance from the offending source may experience effects on health and property that lead to complaints and to declining property values.

An additional challenge to the regulation and abatement of LFN is presented by the proliferation of new sources: venues for heavily amplified entertainment in residential districts of cities. The bone-throbbing bass that is a crucial component of electronic music is now known to be the result of LFN components available through recent technological changes in amplification (McCullough and Hetherington 2005). Particularly when powerful speakers are set up in buildings not designed to isolate the vibration they produce, LFN is carried by the building's foundations into neighbouring buildings and even down entire streets. Conventional means of measurement with the decibel A system are inadequate for determining the presence and extent of LFN, as demonstrated by a study conducted in the United Kingdom. Researchers intent on selecting a reliable method for measuring neighbourhood transmission of LFN first distinguished five levels of noise associated with entertainment venues in London, as determined by subjective assessment without measurement. Levels 3 through 5 were judged likely to provoke complaints from neighbours:

Subjective level 1 = entertainment music noise is very faint and barely audible; concentration is required to distinguish the entertainment music noise over the background noise; the absence of the entertainment music noise is more noticeable when it stops than its presence when it is on.

Subjective level 2 = entertainment music noise is distinguishable but at a low level; specific lyrics are not identifiable; entertainment music noise would be masked by normal speech or television volume.

Subjective level 3 = entertainment music noise is clear and distinct; lyrics may be identifiable; audible over normal speech or television volume; sleep would prove difficult in this climate; noise would constitute a statutory nuisance if regular and prolonged.

Subjective level 4 = entertainment music noise is dominant over all other noise; sleep would prove impossible; individual incident would constitute a statutory nuisance if prolonged.

Subjective level 5 = entertainment music noise pervades entire premises where measurement is taking place; sensation of vibration may be felt; entertainment music noise audible throughout general external area. (McCullough and Hetherington 2005, 72)

The team then applied testing methods from five established systems used in Europe, concluding that only a system used in Germany and the Netherlands was sufficiently accurate to form the basis for legal criteria to establish regulations regarding LFN. A subsequent Swedish study using the same subjective categories and measurement techniques further refined the system by adjusting mathematical calculations, leading to the drafting of EU criteria for determining the effects of LFN on neighbourhood noise annoyance from pubs, clubs, and concerts (Ryberg 2009).

Because the technologies used by entertainment industries change quickly, the process of determining standards for measurement and legal action is reactive rather than predictive. As the example above demonstrates, years often pass between recognition of a new source of annoyance and the establishment of standards and regulations to control it. As well, proposals to regulate entertainment venues can pit the desires of one subculture against those of another, producing conflict within communities. The best solution is to limit loud venues to commercial districts at considerable distance from residences and to insulate them with massively heavy construction and the most effective materials available.

12

Power and Danger

Sound is often understood as generally having a privileged role in the production and modulation of fear, activating instinctive responses, triggering an evolutionary functional nervousness.

Goodman 2009, 65

Sound enforces itself in many contemporary environments, urban or rural, central or remote, with an iron fist: the brutal percussion of digging machines, the scream of an overhead fighter jet, the banshee wail of sirens, loud music and the dull undertow of motor traffic.

Toop 2010, 53

Like many artifacts of evolution, our response to loudness transports us to another world. Advertisers who present messages before movies know that loudness sells because you cannot focus on any other sonic event. You cannot escape physically or perceptually. They own your aural space, and they know it.

Blesser 2007, 2

In the summer of 2010 a distinctive social soundscape was given international coverage by news media. During the FIFA World Cup in Johannesburg, the vuvuzela vaulted to instant worldwide fame. Thousands of the buzz-toned cylindrical plastic horns were deployed by fans in the stadium and crowds in the streets, distracting players and interfering with broadcasts (BBC World News, 14 June 2010). Complaints poured in; so did defences of the vuvuzela's exuberance and cultural roots. Social attitudes toward noise surfaced and skirmished in the ensuing commentary: was the wall of sound a support to South African players or a sonic weapon against others? Was it ethical? Was it legal? Was it ever going to *stop*?

The vuvuzela, like the kazoo, is a thoroughly democratic instrument. No training is required to play it, although practice and the use of wood or metal rather than plastic can move it from the category of noisemaker to folk instrument. Does it produce music or noise, exuberance or aggression? A lot depends on your culture of origin and, in the FIFA context, on which team you support, but also on the circumstances of its use. A few vuvuzelas in a crowd can be amusing and motivating; several thousand at a time represented a clear and present danger to auditory health. The horns, which enable an experienced player to generate between 113 and 130 dBA, caused a demand for earplugs that exhausted supplies in Johannesburg (S. Carter 2010). Their toll in annoyance and fatigue on the athletes could not be measured.

The noisemaking activities of World Cup spectators became sonic warfare. Vuvuzelas were used to encourage teams favoured by local fans and to drown out the chanting of fans for rival teams. Teams that preferred relative quiet in order to concentrate lodged protests, to no avail. FIFA blog entries were divided between South Africans defending the vuvuzela as a national tradition and Northern hemisphere fans complaining about its loudness and irritating tone. Responses from other parts of Africa tended to defend the use of instruments as a cultural tradition, while complaining about the absence of rhythm and musical coherence in vuvuzela technique. A Ghanaian commentator on the BBC advocated replacing the vuvuzelas with drums, a traditional instrument at soccer games in Ghana. A commentator from Brazil complained that the vuvuzelas made it difficult for her team to hear the traditional encouragements – chanting and samba drumming – of Brazilian fans, but defended the use of loud sound as part of the game.[1] Along with vuvuzelas, the world heard the clashing of cultural traditions.

The vuvuzela controversy demonstrated the social functions of noise as well as neglect of its physical consequences; the reactions expressed depended on team allegiances and on shared cultural assumptions about appropriate soundscapes for spectators and celebrations. In each social context, some varieties of noise are encouraged and enjoyed, some tolerated, others forbidden or complained about. Top-of-the-lungs shouting, appropriate for team sports, would be considered rude at a golf tournament. In some parts of the world, funerals are quiet and solemn; in others, animated and loud. In indigenous societies, the sounds of wind, water, and birds transmit crucial information; in agricultural ones, the calls of cattle or sheep are valued as the music of safety and prosperity. In industrial cultures, the mechanical sounds of construction – digging, pounding, hauling, hammering, welding, drilling – signal perpetual activity as the drive for financial growth overrides desires for quiet. Noise gives voice to cultural values: what is permitted or forbidden, proclaimed or kept secret, celebrated or denigrated.

Whether it is produced by voices, music, or machines, noise can echo the signalling functions of language, communicating information about what is happening, whether it is important, and who has power over the territory within earshot. Loudness and proximity are elements of its grammar; its rhetoric is formed by interlaced threads of competition, cohesion, and coercion.

THE ROAR OF THE CROWD

Team sports and races are fuelled by the cheers of the crowd. Players run, jump, throw, catch, skate, kick, pedal, bat, and dodge harder and faster within the cloud of sound than they would or could in silence, each team knowing that its supporters are fuelling the game with raucous shouted love. Along with formalized cheering and chanting, spontaneous yells, groans, and hoots – *yaaaah! aaoooooowww! wooooh!* – short-circuit language to release pure emotion, voicing both victory and defeat. Sound has emotional force: could you cheer on your team or applaud a performance without it? Would you, as athlete, actor, or musician, feel motivated to give your best in the absence of audible encouragement or response from the crowd? Would you be deterred from an action, or at least halted, if a crowd shouted *STOP!!!* Noise communicates the force of communal values, approval or disapproval, even in the transitory circumstance of traffic: *honk if you agree!*

In North America, the most deafening roar is usually contained in domed or covered stadiums. Playoff games in hockey and football usually rank the highest. However, consider 142.2 decibels' worth of sound at Arrowhead Stadium, home of the Kansas City Chiefs football team, on a balmy September Monday night in 2014. The sound level was recorded by an on-site representative from Guinness World Book of Records, and was found to be louder than a jet airplane flying a hundred feet overhead. The Kansas City franchise proudly tweeted that they were the loudest franchise in the world, and this was in an outdoor stadium in a non-playoff setting (*Kansas City Star*, 21 April 2015).

Perhaps this was a one-off statistic, designed to set a record. However, frequent peak noise values of 120 dBA and over, also enough to exceed the sound of a jet taking off nearby, may be found among one of the most sustained, noisiest games in sports – playoff hockey, simply because it takes place in a covered stadium. In one study tracking the noise made by the Edmonton Oilers' hockey franchise spectators during their Stanley Cup final appearance in game 3, noise spiked at 120 dBA several times (Hodgetts and Liu 2006). Equally disturbing is the report that there was *sustained* noise throughout the first and second periods of 105–10 dBA, easily equivalent to multiple measurements taken over many years of standard outdoor rock concerts. Even by rock concert standards, hockey games, at least in the playoffs, are consistently loud. One wonders whether intermission, with its comparatively low sound volume situated beneath the red safety hazard line of 90 dBA, isn't about resting just the athletes, but also the spectators' hearing.

The noise of crowds is known to be an important competitive factor in certain sports. The cheers, comments, and emotional ambience made evident by supporters of a home team, always more numerous than visitors, can even influence the decisions of referees. One study of European football found a 15 per cent advantage for home teams related to crowd noise as referees were pressured by local fans to question their own decisions (Nevill, Balmer, and Williams 2002).

Responses to such pressures are rooted in our hominid ancestors' need to belong to a community, supporting its members and activities with emotional fervour against all competitors. Behind the crowds of sports fans are troops of prehistoric primates and hominids making threatening shrieks to defend the territory that provides their food. It worked for them because noise is inherently competitive: the loudest wins by attracting the most immediate attention, and louder noise suggests a bigger and more powerful group.

Loudness itself can fuel competition: cheerleading is an obvious example, but there are others. In his book on the culture of noise in America, journalist George Prochnik devotes a chapter to the development of car audio systems as equipment for competitive rallies in which the object is to generate enough sound pressure to break windshields, burn out stereo systems, or even generate shock waves (Prochnik 2010, ch. 7). Motorcycle rallies and car races embrace noise as a symbol of power: to what extent does this represent an atavistic masculine urge to dominate other males – and impress females – by roaring as well as winning races? The effect of the hugely amplified mechanical roar is undeniable, since it hits nearby listeners with the force of a blow and carries for long distances, exciting or annoying people who are too far away to see the result of the contest. It signals all within earshot that something urgent and dangerous is happening nearby, something both thrilling and life-threatening, something that recalls our remote ancestors hunting wild boars or mammoths. Crowds gather in excitement to witness the contest, to cheer on the victor and celebrate the triumph, to revel in their membership in a community that produces the fastest, the strongest, the loudest – or one that intends to do so when the next contest takes place. Sports give rise to sonic rivalry, but also to alignment of loyalties and agreement to compete in contests that channel aggressive impulses. Each roaring crowd in a competition is matched in excitement by its rival; embedded in competition is social cohesion.

Vox Populi

Noise is a product of communities and a voice of culture. Traditional parades and festivals proclaim collective social identity by putting cultural beliefs and customs on public display. We use noise – produced vocally, manually, mechanically, or electronically – to enhance social bonds and confirm ownership of the space we inhabit. National anthems, regional dialects of language, styles of music, broadcasting practices, and even degrees of loudness and amplification for public events can be read as signals about community values and as displays of power. So can neighbourhood customs and community bylaws about parties, festivals, the playing of music outdoors, or the times and places that noisy social activities are permitted. Along with cultural values, communal identities are formed by our approaches to social noise.

Soundscapes produced by any event can draw you into a sense of belonging, even making you feel like a participant in a virtual scene that you are observing. Any film soundtrack is an example of this, and if the hero threw his punch at the villain without the resounding thud provided by the Foley artists who create the sound effects, the audience would lose interest.[2] Live events involving crowds rise or fall on the impact of their soundscapes. Parades and festivals depend on loud music and cheering crowds to produce the emotional arousal of celebration: a sparse and quiet group of attendees would signal failure. Everyone within earshot may be drawn into a feeling of participation even when the crowd is not visible, whether or not the event is one they actively support: the presence of a crowd of celebrants, revealed by their noise, is hard to ignore. For this reason, attracting a crowd complete with cheering and contagious excitement is crucial to political rallies. Knowing that a lot of people are excited about a candidate can persuade others to follow the enthusiasm; upbeat music and resonant speeches punctuated by frequent applause and cheering are essential to building support for the candidate and rapport among supporters. Even social media follow the pattern, striving to build virtual crowd noise in a visual medium through the silent applause of "liking" and "trending." Auditory activities – discussion, debate, applause – are translated on the Internet to visual record in the form of blog comments. The process echoes the historical transition of cultures from orality to literacy, and of democracy from public-square oration with spontaneous response to formal elections.

The social aspects and impacts of noise are now a subject of some contention between the disciplines of acoustic ecology and sound studies.[3] The former, attuned to a desire to preserve natural soundscapes against mechanical encroachment, condemns noise generally as an imposition on an essential right to quiet even in urban communities. The latter, rooted in urban sociology and communication studies, regards noise as a beneficial stimulant for social interaction among diverse urban subcultures. Recent literature has drawn some battle lines. The pro-quiet group advocates and documents quests for silence amid the American obsession with noise, condemning both ambient mechanical roars and ceaseless advertising.[4] The pro-noise group defends noise as a welcome disturber of boundaries while characterizing silence as a chilling of activity: "Silence is a slippery ideal as it gathers within the imagination a set of seemingly positive values that when overlaid onto social behavior and community lends to understandings of place and placement a contentious intolerance" (Labelle 2010, 55).[5] Acoustic sociologist and artist Brandon Labelle attributes the desire for communal quiet in part to the growth of suburbs in the mid-twentieth century and the consequent preference for self-contained, exclusive, and culturally segregated residential districts, a kind of anti-urbanism with racist undertones. To him, the city is an amplifier of social interaction among individuals and cultures with noise as its *lingua franca*, and quiet is a type of poverty. Hearing each other in close proximity forces us to communicate and to learn to recognize the value of diversity. If everyone kept quietly to themselves, there would be little progress toward intercultural understanding.

In a culture that values loudness and acoustic competition for their communicative vigour, what price in compromised health and social annoyance will be considered too high? Studies on ambient urban noise as a factor in social relations suggest that helping behaviour – willingness to assist a stranger with a simple physical or verbal task, like picking up dropped objects or giving directions – has been found to decrease in the presence of loud noise and increases with a background of relaxing music or natural sounds (Mathews and Canon 1975). Even the addition of an arm cast on the experimenter playing the role of stranger-in-need-of-assistance did not increase the help given in conditions of high noise, although it did increase help in quiet conditions.[6] This is not surprising if the experiment is conducted on the street: stopping to assist someone delays the process of getting away from the source of the noise, which is a feasible reaction to auditory stress. Some of the cited experiments, however, were conducted in laboratory settings. This suggests that noise annoyance can produce a level of irritation that affects social interactions and consequently culture itself, as well as forcing a narrowing of cognitive attention that causes people to miss signals conveyed by others. If crowding and diversity stimulate productive communication, they can also reduce it. If ambient industrial and traffic noise is loud enough to produce chronic irritation, nobody benefits.

The quest for quiet is not without its limitations, either. As urbanization in many parts of the world eclipses natural soundscapes along with wildlife habitat and agricultural land, how realistic is the desire for everyone to have consistent access to outdoor silence? For that matter, how many natural habitats are actually silent? If our species has a need for soundscapes free of mechanical noise, it is largely because we relax when able to hear natural soundscapes, the subtle rhythms of our own breathing, and the relaxed speech of companions. True silence is a sensory deprivation that is deeply unnerving even in incomplete form: in fact, it is an illusion. Even in the confines of a carefully engineered anechoic chamber designed to shut out sound, the human auditory system will perceive the internal soundscape of the body: the respiratory and circulatory systems faintly hissing, thudding, clicking. The "silence" for which activists yearn is generated by an intersection of personal autonomy, freedom from the demands of others, anxiety about diminished access to privacy, and desire for the soothing presence of natural soundscapes. Thus, the quest for silence is largely a quest for solitude in natural surroundings, for escape from mechanical sounds and urban crowding, for a way of life that merges nostalgia for simplicity with a desire for enough wealth to ensure distance from others.

Again, the idea of protective earlids comes to mind. If loud music can become a refuge for individuals seeking relief from their critical internal voices, is a culture of loudness also a protection from insistent but undesired information? Do the broadcast chatter of current events commentary, the urgency of television ads, the carefully processed music in malls and restaurants, the boom cars on streets and boom boxes on beaches, all serve as aural anesthetics? If so, what messages are we, as a society, trying to avoid? When constant sonic stimulation is normalized, how do we and our children recognize what messages are important? What information (the brutalities of warfare, the injustices of poverty, the threats of economic collapse and climate change) are we masking with all that chatter and entertainment? Does distraction delay the active solution of problems, or is it simply a way to maintain sanity in the face of them?

Given a cultural norm that is competitive and noisy, calls for quiet define a counterculture. Henry David Thoreau's advocacy of solitude and silence against the encroachments of the nineteenth-century Industrial Revolution is well recalled here;[7] his principles are being revived today by noise abatement societies and by critics of the insistent culture of advertising and consumption. As Scottish social critic Stuart Sim points out, "silence takes on a subversive quality as a result [of constant advertising] and opting for it [is] a refusal to be driven purely by profit motive, or to live a life of perpetual sensual bombardment aimed at eradicating our individuality in the name of passive consumption" (Sim 2007, 170). The irony is that, in order for calls for quiet to emerge above a background of noise, they must make even louder claims on public attention – in effect, more noise. Whether literal or symbolic, protest usually involves shouting.

SONIC SOCIAL CONTROL

In 1977 the French economist Jacques Attali traced music and noise as indicators of social power in a book titled *Noise: The Political Economy of Music*.[8] While some of his analogies now seem strained, his work is influential for its analysis of the uses of sound for social control. He drew attention to the potential for misuse of power by ideologues and governments through "eavesdropping, censorship, recording, and surveillance" as well as the dangers inherent in the broadcasting of propaganda (Attali 1985, 7). Such observations anticipated the use in 2008 in Wiltshire, UK, of "The Mosquito," an electronic emitter of high-frequency noise supposedly audible only to people under age twenty-five, placed in public spaces where teenagers congregate and designed to repel them (BBC News, 4 April 2006). As a BBC news report pointed out through interviews with teens exposed to the device, the emitted sound can cause headaches and other physical pain in addition to annoyance. It also demonizes the reasonable and law-abiding majority of young people.

Consider that the ability to hear the Mosquito and the reaction to its frequency are not necessarily limited to a certain age or socially distinguishable group, and you have an illustration of the challenges inherent in using sound as a means of social control. If you belong to a target demographic, selected by age, race, class, or any other distinguishing factor, you can be effectively reminded by an enforced unpleasant sound that suspicion is placed on you. If you are not, the sound reaches you anyway, and its presence may plant in you a resentment toward the group targeted and/or toward the authority imposing the annoyance. Since noise is no respecter of boundaries, whether physical or social, its use as an instrument of power serves to remind everyone within the affected soundscape that they are under surveillance. Compared with maintaining a consistent presence of police or military, noise is inexpensive and relatively convenient to use. Unlike the posting of regulations in public places or their broadcasting in public service announcements or advertising jingles, it cannot be ignored or relegated to an acoustic background.

The People United

Loudness is a tool for acquiring and exercising power. Children know this well before they have any grasp of relational reasoning, political propagandists know it, advertisers and television pundits know it thoroughly. Whoever makes the most noise gets the attention first; whoever repeats a message often enough gets credibility, regardless of the actual content of the message. This is an unfortunate consequence of the processing of sound by human brains, as well as by cultures and political movements. Because of the hard-wired evolutionary necessity for paying immediate attention to whatever is louder than the ambient soundscape in case it turns out to be dangerous, we become susceptible to the relentless claims of advertisers and political campaigns. We practise acoustic imperialism on each other, turning our neighbourhoods and roads into stages for the type of music we personally favour, or the ideological message we believe to be the ultimate truth. We hear the roar of a crowd and want to be part of it. Caught up in the excitement of noise as an expression of community, we yearn for power: *my voice, my cause, my culture, my team!*

In the historical measurement of social power as a function of territory controlled, being loud meant freedom to be heard as an authority. For centuries kings and other dignitaries rode in processions announced by trumpets and drums as well as bright banners, summoning and stirring crowds. Groups that were denied power – peasants and slaves, foreigners, and usually women of all classes – avoided calling attention to themselves through speech or action. For people in those powerless categories, being heard to express an opinion could lead to dangerous consequences. The development of representative governments in the Western world, a process that required centuries of social and economic change, democratized social voice as well as political processes: speaking up gradually became a source and symbol of power rather than danger, a transition now being both emulated and disputed in many parts of the world. In today's measurement of social power by means of attention paid, loudness draws the currency of publicity. "Speaking up" implies confidence and self-assertion. "Speaking out" expresses the power of a group to occupy space and proclaim an identity: the parade, the street demonstration, the group of adolescent girls conversing urgently about nearly anything. *If you can hear me I can control your attention; I can be important to you, if only for a moment.*

Thus, the association of protest with raised voices is inevitable; everyone from infants to politicians comes by it naturally. Shouting in protest is an immediate and unthinking response to a violation of rights or privileges, and protests in every culture incorporate chanting or singing as ways to draw attention to a cause. Even the lyrics of historic protest songs can evoke the passions of their causes: "The Marseillaise," "The Internationale," "We Shall Overcome." Protest chants worldwide now follow standardized rhythmic patterns, with the familiar English versions often imitated in the rhythmic phrasing of other languages: "The PEO-ple (pause) u-NI-ted (pause) will NE-ver be de-FEAT-ed!"[9] or "Hey-HEY! (pause) Ho-HO! (pause), [insert name or issue] has got to GO!" Their rhythmic phrasing, compelling and easily learned by large crowds, unifies demonstrators and draws media attention to the protest, whether its source is an irate political movement or a desperate resistance against a government bent on destroying its own citizens. As the crowd grows and more voices are added, the validity of the issue is verified in the public ear. It is no accident that demands for social change are characterized as calls and cries, or that a virtual format for the public forum, credited with assisting the election of Barack Obama as US president in 2008 and calling attention to political change in the Middle East, was named for a vocal activity: *Twitter*. With calls sufficiently loud, crowds sufficiently large, governments have changed and fallen.

As noise builds, so does competition for the limited bandwidth of public attention. The result is the pervasive ambience of emotional urgency that permeates North American culture as physical noise permeates our surroundings and our leisure activities. Alternatives to urgency get lost in the frenzy: a culture bombarded by noise cannot pause for reflection. We are too stressed to admit quiet into the dialogue, denying its role as the voice of the parasympathetic nervous system from which relaxation and healing proceed. As attention and hearing acuity are eroded, so are conscious decisions to *listen*, to pay attention to signals that are not loud. *To whom, and to what, do we most easily listen?* To whatever is most insistent. The content does not need to be significant or beneficial, or even true.

The exuberance of loudness and its power to attract attention come at a price, however. There is a dark side to the lure of acoustic social cohesion. Repetition of a message can be effective as persuasion whether or not the message has any validity, as demonstrated by any number of advertising claims for products or political candidates as well as political discourse itself.[10] Repetitive speech and song are compelling: whether they carry political, religious, nationalistic, or commercial messages, they are easily employed in propaganda. Loudness counts, and because it commands immediate attention, noise can be used as a tool of aggression.

We are more likely to listen to those who have power over us than to those who don't. The inversion of this principle also holds true: if you hold people's attention by means of the noise you are able to make, you are exercising power over them. Noise is coercive, whether it carries a message, causes physical pain, or forces a change in behaviour.

Noise as a Weapon

Because of its associations with social dominance and control of territory, as well as its acoustic properties, loud noise – even in the forms of music and speech – is easily used as a means of torture. Sustained bouts of rock music played at hazardous volume were reportedly used in the capture of Panamanian dictator Manuel Noriega by the CIA in 1989.[11] The reporting of this action in the US press at the time brought to public attention the potential for broadcast music to be used as a tool of aggression not for its content – the use of songs as propaganda has a long history, but the CIA repertoire was apparently unrelated – but for its loudness and repetitive structure. Shouting at close proximity is also recognized as a means of breaking down resistance to interrogation or subordination, particularly when the content is insulting.[12] This is well known by military intelligence units and law enforcement, and shouting accusations inches away from a suspect's face is a ubiquitous staple in crime-themed movies and television dramas. Loudness gets attention, and loudness associated with the emotional force of a human voice shouting in anger provokes deep responses of fear and reciprocal anger. In crime dramas it provokes the suspect to confess; in real-life confrontations it escalates hostility and can provoke physical violence.

The social and emotional resonances of raised voices may derive from hominid and primate roots. Dominant male chimpanzees are known to scream with differing acoustical signals at members of their own groups, members of neighbouring groups, and strangers (Herbinger et al. 2009; Slocombe et al. 2010). Their vocalizations define and defend territory and the extent of potential community: who is a member; who is a stranger and potential invader? The driver of a car with booming music and intentionally disabled muffler that roars along announcing his presence to the world holds continuity with the dominant primate as well as the small boy making motor sounds as he races through the house: *I am here, I am powerful, this territory is mine! Take notice!* When the urge to be noticed becomes an urge to take complete control, when it becomes the official response of a ruler, an empowered group, or a government to any challenge, the territorial roar is no longer spontaneous. It becomes mechanized and militarized.

In an article exploring the use of sonic torture by the US military in Iraq and Afghanistan, American musicologist Suzanne Cusick discusses methods that include acoustic bombardment of civilian areas by overhead flights breaking the sound barrier and the use of engineered acoustic generators to broadcast infrasonic signals that produce nausea and disorientation. She goes on to document the use of music as torture against detainees at Guantanamo Bay and other military prisons, citing both loudness and culturally offensive content of song lyrics as reasons for its effects (Cusick 2006). Although recent technology has widened the scope of options, the concept is not new. Many ancient indigenous cultures used ritualized screams in battle, provoking panic and disorientation in their enemies while enhancing their own aggression. The Gestapo, and undoubtedly many other agents of political oppression before and after them, placed prisoners about to be interrogated where they could hear the screams of others being tortured.

The use of sound for purposes of social control is effective on crowds as well as individuals. Research by the US military into the effects of both low and high frequencies on the human body began in the 1970s as part of a larger investigation into non-lethal weaponry designed to repel, stun, or incapacitate without killing. Infrasound or low-frequency noise, first investigated as a potential disabling force, proved to be too expensive to generate and too difficult to aim and control. Most technology for sonic weaponry now makes use of high-frequency sound, usually between 2,000 and 3,000 Hz, played at extremely high volume to repel and scatter crowds.[13] The frequencies and decibel levels used may be chosen for their ability to produce annoyance or for physical effect: dizziness, nausea, and vertigo result from challenge to the balancing function of the inner ear. Regardless of whether these devices represent a safer alternative to conventional weapons or an affront to democracy and the right of free assembly and protest, they are a part of both military and police activity in the twenty-first century.

Because of the number of variables involved – loudness, frequency spectrum, range, proximity, and interaction with ambient noise and built surfaces – the effects of sonic weapons are not always fully understood either by the public or by the institutions that use them (Altmann 2001; Vinokur 2004). Conspiracy theories about their use abound: that Russia aimed them at the United States during the Cold War, that they are aimed remotely at individuals who disagree with particular governments, that they can change the weather. While these claims appear to be exaggerated – their validation would require changing the laws of physics as well as the capabilities of known technologies – sonic weapons are real. They may be available for use by police in your own city; you may encounter them if you attend a large demonstration against an established corporate, economic, or political power. Originally developed for military uses, sonic weapons are now regarded by several governments as appropriate for use by police for the control of civilian populations. Because they can be used as communication devices –

essentially mega-loudspeakers – they are classified as such and not as weapons, a situation that permits their use in public gatherings. Today, most sonic devices employ *narrowcasting*, a technique that enables the emission of directed beams of selected high frequencies that can be aimed at specific areas. The one most commonly used is the long-range acoustic device (LRAD), promoted for crowd control but also as a non-lethal defensive weapon capable of repelling hostile crowds or individuals before they can attack with conventional weapons. Such sonic weaponry was in limited use by the US military in Iraq;[14] mounted on warships, it deterred the operators of small boats from approaching the ships. A similar system, the magnetic acoustic device (MAD), was held in readiness by law enforcement in New Orleans in the wake of Hurricane Katrina;[15] another has reportedly been in use by the Israeli military for crowd control in Palestinian areas (Block 2005; Jardin 2005; Rawnsley 2011). They are also marketed as having potential for use in hostage-taking incidents, and there were reports of their use to repel pirate ships off the coast of Somalia in 2005 (Evers 2005).

It was reported in 2008 that a US manufacturer of sonic weapons had signed a contract with the government of China before the Beijing Olympics to provide a crowd-control device (Hambling 2008), although it may not have been activated because such use is controversial in view of the potential to cause physical harm (Yang 2010). Several US police departments have not been deterred by such concerns, however. Instances of LRADs being used for crowd control in the United States have been documented since 2005, when police departments in New York City, Boston, and Los Angeles were reported to be testing systems capable of narrowcasting to a distance of 300 metres (328 yards).[16] Independent Internet stories at the time reported that such devices had already been used for crowd control of large demonstrations in New York City (Forbes 2003), although this was not reported in the mainstream press at the time. LRADs have since been used around the G20 summit in Pittsburgh (2009); at the Superbowl in the same city in 2011; against Occupy demonstrators in California and New York City (2014); and against crowds protesting police violence in Baton Rouge, Louisiana (2016).[17]

Two well-documented examples provide a warning. In August 2014, local police in Ferguson, Missouri, used an LRAD to control crowds protesting the fatal shooting of an unarmed African-American teenager by a white policeman.[18] The "sound cannon" used at the protest site was the LRAD 500X-RE, capable of projecting sound 2,000 metres (well over a mile) and of reaching a maximum loudness of 149 dBA (R. Baldwin 2014), more than enough to cause permanent damage to hearing in a matter of seconds. Anyone within a distance of fifteen metres (approximately fifty feet) from the LRAD at high volume would sustain auditory injury and possible rupture of the eardrums immediately; at greater distances, headaches and nausea as well as varying degrees of hearing damage would result. Another example occurred in New York City in December of the same year, at a Black Lives Matter protest following the killing of an unarmed African-American man by police. The soundtrack of a video taken at the scene records the brief repeated bursts of high-pitched warble from a "backpack-sized" portable device reportedly set to a moderate level that would deter crowds from gathering (Tempey 2017). The model was capable of producing 137 dBA at maximum setting. Nearby photojournalists, as well as protesters, reported reactions including migraine, tinnitus, and nerve damage to the ears. A lawsuit on behalf of protesters followed, with police arguing that the LRAD was a "communication device" rather than a weapon – a claim has been used by police departments in Australia (C. Chang 2016) and in manufacturers' descriptions. The New York suit was allowed to proceed because both the department's training of officers in the device's use and the regulations for using it were deemed inadequate by a judge (Tempey 2017).

Early reluctance to use LRADs against civilian groups on the basis of human rights was well founded. Although it is considered a "humane" weapon because it is not lethal, the LRAD has the potential to cause permanent injury to the auditory system.[19] Its sound is described as resembling the tone of a smoke detector, but at far higher sound intensity (R. Baldwin 2014). It can permanently impair the hearing of people who do not know about, and cannot afford, the high-tech hearing protection given to handlers of the devices, and it does not distinguish between genuine aggressors and bystanders caught in a crowd. This is not a trivial matter, since loss of hearing injures more than the physical body. Harvey Sapolsky, head of security studies at the Massachusetts Institute of Technology, pointed out in a 2001 interview that, in marginalized cultures where literacy is not common, deafness can destroy all hope of meaningful social interaction (Vaisman 2001–2). Even partial loss of hearing can reduce employment opportunities as well as social contact in any society. When the loss is due to aggressive action inflicted indiscriminately on bystanders as well as the targets of the action – who may themselves not be legitimate targets – questions need to be raised about the legitimacy of sonic weaponry, especially in view of ongoing development of its capabilities. A portable LRAD derivative called the Inferno is now being

marketed internationally for home and embassy security as well as industrial security and police use. One report from a brief experience with the
device describes its 123-dBA high-frequency effect as "unbearable" and
"gut-wrenching" (Weinberger 2008).

Like tear gas, narrowcasted sound is pervasive and is thoroughly effective for persuading people to move away from a designated area. It is also
difficult to control: the sound can bounce erratically off surrounding
buildings into unexpected trajectories. Its potential to cause permanent
injury to bystanders and combatants alike should serve to curtail its use.
Whether LRADs are eventually banned as an indiscriminate danger to
health, analogous to nerve gas rather than tear gas, will depend on how
seriously the threat of noise-induced deafness is taken by international
law as well as whether such laws are obeyed in situations of conflict.

The most recent technologies marketed for crowd control appear to be
working around the problem of deafening people by using sound waves
to generate a pressure field that repels or immobilizes anyone approaching the user. In 2011 the Raytheon Company of Massachusetts, a major
defence contractor for the US military, filed a patent application for the
"Man-portable Non-Lethal Pressure Shield." The shield, marketed to military and police forces, incorporates a "folded acoustic horn" with multiple
apertures into a shell to be worn by each individual in a riot control
squad. Each aperture emits a "pulsed pressure beam" that can be directed
to specific human targets, giving the operator a choice of options:
"warn/stun/incapacitate" (Bostick 2013, 0010). The patent application
specifies that the frequencies and sound pressure used are carefully
designed to avoid causing deafness:

> Although the shield will produce loud pulses, its primary effect is to
> couple a pulsed pressure wave or "barrier" to a human target's upper
> respiratory tract to warn, deter or temporarily incapacitate the target.
> This can be achieved at pressure levels that do not damage the ear.
> The terms "acoustic" and "pressure" are often used interchangeably as
> related to non-lethal weapons. We use the term "pressure" to empha-
> size that the shield is generating a pulsed pressure barrier that couples
> to the respiratory tract and not merely creating loud noise. (ibid.,
> 0025)

The pressure field, described as "having a mass equivalent to trinitro-
toluene (TNT)," has the potential to cause "disorientation and debilitation"
(ibid., 0006). Does this represent a humane technology? If used as an alter-

native to lethal weapons in war zones, perhaps it does. If used to suppress mass protests by civilian populations, the pressure shield and technologies like it have chilling implications for democratic societies as well as repressive ones.

Sonic weapons represent an extreme example of noise pollution, the intentional use of the instinct to avoid pain as a means of manipulating behaviour. Since the offending and potentially dangerous noise is carried by the open air and affects everyone within range, it demonstrates the nature of the outdoor soundscape as a *commons*, a place of potential benefit to all if its benefit is put to appropriate uses and protected by the community.[20] The determination of appropriate uses, however, can be contentious. Embedded in the process are issues of cultural and personal power, tradition and iconoclasm, aesthetics, ethics, and law. As will be explored in the next chapter, whether the issue is sonic weaponry, traffic, or your neighbour's listening habits, both social consensus and individual taste shape the reactions of communities to noise.

13

The Cacophony of the Commons

The seemingly innocent trajectory of sound as it moves from its source and toward a listener, without forgetting all the surfaces, bodies and other sounds it brushes against, is a story imparting a great deal of information fully charged with geographic, social, psychological, and emotional energy.

Labelle 2010, xvi

The soundscape of a city is generally marked as something trivial ... Sound is rarely used as a positive, informative or explorative social perception instrument within the existing urban planning and heritage management.

Leus 2011, 355–6

Like individuals, communities differ in their responses to noise. Initial choices are made between indoor and outdoor culture: do we keep noise, music, and conversation private and behind walls, or do we broadcast our presence and preferences through open space? Are public parks and beaches blank canvases waiting for sonic decoration or outer *sancta* reserved for the sounds of the natural world? Attitudes about property, privacy, and social status permeate such decisions, and levels of noise are often associated with social stratification. High-income urban and suburban communities are more likely to value greater public quiet, if only because larger and more expensively built houses allow the confinement of noise to insulated basement rooms, children are often occupied indoors, and traffic is discouraged by means of community planning. Low-income urban neighbourhoods, with higher density, smaller residences, and more traffic, may generate excessive outdoor noise – traffic itself, the machinery of public transit systems, multiple activities at street level including construction sites and maintenance, bars and businesses vying for attention by broadcasting music into the street – as well as indoor noise resulting from poor insulation, crowded conditions, and conscious efforts to mask the sound coming from outdoors with more sound. These contrasting soundscapes, and the conditions of wealth and poverty that generate them, are even more extreme in cities of the non-Western world.

Regulation of noise begins with consideration of the public sound-scape. In a study of the history of US and European noise abatement regulations in the nineteenth and early twentieth centuries, Dutch scholar Karin Bijsterveld traces the impacts of new technologies – trains, aircraft, automobiles, telephones – on anti-noise movements and the shaping of public attitudes about noise annoyance. Her examples illustrate the growth of public attention to noise as a consequence of changing social patterns: the need for wider roads to accommodate mechanized traffic, the desire to discourage the ringing of church bells at dawn because the advent of electric lights and changing urban customs had expanded nightlife, and the formulation of "noise etiquette" for public places. Anti-noise activists invoked everything from mechanical efficiency to civic morality to make their case that excessive noise should be limited in order to enhance public health and the public good (Bijsterveld 2001). The association of noise with excitement and progress, however, complicates the association of quiet with civic virtue: machines that clank and roar incessantly bring prosperity to societies, if not necessarily to the individuals living nearby. Attitudes about land use, privacy, property, and personal comfort clash along lines of class and culture in the process of regulating noise: Is the new factory or shopping mall down the road from our farm a welcome source of jobs and community income, or a noise nuisance? Does the rapid growth of our city represent success or is it going to increase density and traffic, interrupting our sleep and driving our stress out of bounds?

And what of the suburbs, the haven of peace and quiet designed as a refuge from urban din? Suburban noise in some parts of North America now approaches the loudness and annoyance levels of urban areas, thanks to the mechanization of home maintenance and leisure. A 2007 feature article in the *New York Times* described the suburban soundscape as

an assembly-line sound of lawn mowers, leaf blowers, snow blowers (each in their seasons), school buses, fuel trucks, kids with basketballs, kids with motor-powered vehicles they probably should not have, dogs, errant car alarms, errant home alarms, pool parties, intestine-rattling home entertainment centers that leak high-volume audio into

the street; and, during periodic real estate market upticks, those heavy
construction vehicles that go beep beep backing up as they tear down
the old and reinvent the newest model of the suburban home's apoth-
eosis – the dream house. (Vitello 2007)

Huge tracts of these "dream houses" now radiate from North American
cities into formerly rural areas, driving out natural and agricultural
soundscapes, overriding them with traffic and gas-powered yard-main-
tenance machinery. Remaining farmland resounds with mega-
machines: tractors, harvesters, and combines scaled for farming as industry. As residential development of rural areas intensifies, so does the
potential for noise annoyance to residents and to livestock. In addition,
the development of new technologies for capturing and processing
energy has altered the rural soundscape – as well as the landscape – with
hydroelectric dams, oil rigs, and wind turbines. All are producers of
progress, prosperity – and noise. The sections that follow illustrate the
issues involved in understanding community noise even before the
process of regulating it can begin. They demonstrate the impacts of
municipal zoning decisions, culture, and new technologies on community laws and standards.

PLANNING AND ZONING

Community planning and zoning usually take account of visual
impacts early in the process, but ambient noise is not always anticipated as a potential problem. The planning of new communities, or expan-
sions to existing ones, illustrates the challenge of managing and living
with community noise. How close we build residences and recreational facilities to each other is one important variable. Moreover, while
densely populated areas are noisier than sparsely populated ones, highways, airports, industrial zones, and commercial sites can easily raise the
ambient noise of rural communities even when buildings and their
occupants are far apart.

Developments that place detached houses close together may actually generate more noise problems in summer and in warm climates than
attached townhouse complexes, as children run shouting between
houses, and air conditioners or fans set in windows send vibrations
bouncing and echoing off close-set exterior walls. New suburbs that are
built near areas zoned for industry may include parks to shield residents from unsightly views, but open land is not necessarily effective as

insulation against noise. Given the complexity of both measurement of and legislation regarding noise, clear understanding of issues is essential. When laws are vague or become inadequate as a result of changing circumstances, community action and public will can serve to improve the situation.

The 85th Street Asphalt Plant Blues

Homeowners in a newly built suburban community on the outskirts of Calgary start their barbecues, anticipating leisurely outdoor dinners on a cloudless and hot July evening, when a sound "like a jet engine" masks their conversation. No aircraft is visible, however, and the sound does not diminish; it persists until the next morning, accompanied by an acrid odour. Construction on the next street? Malfunction of someone's air conditioner? A fire, and some noisy new machinery to combat it, in the adjacent community? Calls to the Fire Department reveal that they, too, are mystified. Measurements of air quality are taken in houses affected by the smell as far as six kilometres away when the wind is strong, but the outdoor air shows identical results and both are inconclusive. The next night, and the next, and all through the summer the noise persists: dinners in backyards are abandoned, sleep is interrupted. "For Sale" signs eventually appear on streets adjoining the picturesque boundary of the community. Property values in several upscale neighbourhoods decline.

The source of the offending noise, accompanied by smells evocative of burning rubber and tar, was a portable asphalt-manufacturing plant situated in a privately owned gravel pit at the edge of the city. Asphalt – also called bitumen or asphalt concrete – is a paving material that combines a residue from the distillation process of petroleum with gravel: once mixed, the material hardens in an hour and must be used for paving before it does. The plant was operated for the purpose of re-paving a major high-speed traffic artery around the city. Because of the volume of traffic during the day, the repaving work, and consequently the operation of the asphalt plant, took place at night through the months of July and August 2007. This period coincided with a heat wave in the area that forced residents to keep their windows open at night, compounding the problem.[1]

The portable plant, unlike the permanent facilities that preceded it at nearby gravel works in the district known as Spy Hill, was placed approximately three kilometres (almost two miles) from a residential

area. It was located at the top of a small hill, which enabled it to broadcast both noise and fumes quite effectively. The gravel pit's owner
reported that the placement of the plant had been mandated by a
provincial official against his advice, presumably as a cost-cutting measure. Residents began to complain to the district's councillor, but his
office claimed to be powerless because jurisdiction for the project lay
with the province rather than the city. The plant was located within the
municipal boundaries of Calgary but operated with an exemption from
municipal bylaws. Complaints to the member of the Legislative Assembly for the constituency did not meet with any response at the time.
Some residents became restless, and a few investigated further. Attempts
were made to contact the owners of the gravel operation and provincial
officials, to determine who was responsible for the operation of the
plant, and to protest. None of the attempts produced quick or simple
results. It was well into September before answers became clear, and
they did so largely because a dedicated group of student researchers at
the University of Calgary took on the project in order to assist the residents' group.[2]

The student group began by contacting homeowners' associations in
the areas adjoining the plant site, as well as individual residents who had
launched complaints. They found that, while some residents reported
themselves or their children as suffering reactions that included insomnia
and headaches, others were unaffected.[3] Measurements taken by the city
of the noise created by the plant were regarded by an acoustic engineer
consulted at the time as inconclusive because of traffic noise interfering
with the readings and because they were carried out only within the
boundaries of the gravel pit, not in adjacent communities. Students and
residents then attempted to contact representatives from municipal and
provincial offices, but found that they were slow or unwilling to respond
to complaints. The owners of the gravel pit were helpful but cited a tan-
gle of regulations and directives from provincial officials that thwarted
their original plans for more thorough noise abatement. An official at the
Calgary Health Region expressed concern but had to wait for responses
from the provincial government before taking any action to request further testing of noise and air quality.

Further investigation by the students revealed that provincial infrastructure projects were exempt from municipal noise bylaws. This discovery motivated several of the affected residents to contact officials at the
municipal Environmental Assessment Agency, who passed on their questions to Alberta Transport and Infrastructure, the provincial department

in charge of the project. Their representative, while sympathetic, framed the residents' reactions as matters of emotion rather than science or law. Although this was frustrating to the residents, it was a reasonable observation: individual reactions to noise and other environmental disturbances can be extremely personal, sensitivities vary, and reactions are not quantifiable – and thus functional as evidence – unless a substantial number of individuals share them.[4] Some residents were so upset by the effects of the plant that they decided to move out of the area. Others were completely unconcerned.

In such a context, what defines a public health problem? What is "average" sensitivity? Who is qualified to decide: medical officials, politicians, noise abatement engineers, lawyers, legislators? If repeated interruption of sleep causes progressively greater sensitivity through physical stress reactions, what recourse is, or should be, available? Noise bylaws tread a fine line between science – the knowledge that hearing can be physically damaged by prolonged or repeated exposure to noise over 85 dBA, and sleep interrupted over 45–50 dBA[5] – and less concrete "community standards," which include the zoning of the area for residences or industry, the hours normally kept by residents and/or workers, the density of traffic, the presence or absence of schools and children, and the presence or absence of mechanical infrastructure like subway trains and compressor stations. Because of the distinction between law, which attempts to operate with clear definitions of evidence and proof, and community advocacy, which takes subjective responses and opinions into account, answers to these questions are not easy to find. A problem with noise might be covered by existing bylaws, or it might not. It might be trivialized by officials or by less sensitive members of the community. Given the lack of communication that characterizes many urban and suburban neighbourhoods, complainants can find it difficult to determine whether others share their complaint. This was the case with the neighbourhoods adjacent to the asphalt plant until a website was set up as a means for sharing news about bylaws, symptoms, and attempts to communicate with government officials.

What frustrated the residents' group even further was being told repeatedly by the city that the asphalt plant was operating within legal limits. In fact, it was, despite violations of municipal loudness limits, because of the provincial exemption. Furthermore, the municipal law takes only the loudness of noise into account and not its frequency spectrum.[6] Even with a measurement of sufficient loudness, the law may not define a noise complaint as legitimate if it is temporary, if it occurs before or after the hours designated as night-time, or if it is insufficient to cause disturbance of sleep, according to local regulations. Much is left to the discretion of the noise bylaw officer, and community standards vary. While guidelines provided by the World Health Organization are taken as an industry standard for noise abatement, some cities allow for variation. In Calgary, for example, legal limits are set 15 dBA higher than in Toronto, a much larger city.[7]

An environmental assessment for the Spy Hill asphalt plant project had been conducted prior to its construction and was published in February 2003. Its only mention of noise states that all operations approved in the City of Calgary must comply with City of Calgary Noise Bylaw requirements. The Noise Bylaw requirements specify maximum permissible noise levels as measured in daytime and night-time periods within industrial and residential areas (Brown and Associates 2003, 21). Calgary municipal noise bylaws define continuous noise as "any Sound Level that occurs for a continuous duration of 3 minutes or sporadically for a total of more than 3 minutes, in any continuous 15-minute time period" (City of Calgary Bylaw 5M2004, Part 9, s. 28). Section 38 of the bylaw specifies that continuous sounds exceeding 65 dBA in the daytime and 50 dBA at night are not permitted in residential areas. Such regulations were of limited use, however, because the asphalt plant was located in an area zoned for industry and operating under a provincial contract.

The plan for the asphalt plant operation estimated that unmitigated maximum levels would not exceed 61 dBA at night, with occasional short bursts of noise at 65 dBA, presumably from truck traffic carrying the mixed asphalt to road construction sites (Brown and Associates 2003, 47–8).[8] Noise mitigation measures recommended in the report included insulation of equipment and placement of the plant away from residential areas. Yet not all of the report's recommendations were carried out: the major omission was the siting of the portable asphalt plant on high ground, rather than the recommended site at the bottom of the excavation, where it would be screened by piles of gravel. This raised location meant that sleep disturbance in adjoining residential areas was not only possible, but likely, and yet the record of discussion at a 2003 public information session for nearby residents shows no indication of noise being raised as a potential issue either by officials or by residents.[9] The agenda for the meeting shows that air quality issues were presented and discussed at length, but because there was so little understanding of the science and the social implications of noise among public officials, politicians, and the general public, nobody was able to draw the conclusions that later became obvious to many residents: that noise annoyance would affect their everyday lives and their health through the summer of 2007.

In the guidelines for noise mitigation along highways published by the Alberta Ministry of Transport, emphasis is given to the implied cost-benefit ratio involved: the decision to implement noise mitigation must consider whether mitigation is cost-effective, technically practical, and broadly supported by the affected residents, and whether it fits into over-all provincial priorities.[10] This is at first glance a reasonable approach, especially since technical options and their cost will determine what is possible as well as practical. Most governments are sincere in their desire to protect people from annoyances and hazards, which is why noise bylaws exist in the first place. However, bylaws are useful only to the extent that they are reasonably applied and enforced. In the case of the Calgary asphalt plant, adherence to the stated municipal and provincial regulations would have spared residents the stress that caused them to complain. The exemption of provincial projects from municipal bylaws and attempts by the provincial government to cut financial corners result-ed in noise abatement measures that fell below the recommended stan-dard and proved to be inadequate.

The portable asphalt plant at the Spy Hill site was removed in Sep-tember 2007, when the paving process was complete, and the plan com-municated to residents indicated that it, or a similar portable plant, would be put into operation in the summer of 2008 in order to pave a new set of roads on the outskirts of the city. The owners of the gravel operation reassured the residents' group that the new plant would be bet-ter insulated, oriented away from prevailing winds, and placed further away from residential areas.[11] An official with the city bylaw services was reported to be drafting a proposal for a site-specific amendment to the provincial exemption from noise bylaws. The residents' group was informed in 2008 that promised improvements with regard to noise had been made.

The Spy Hill asphalt plant controversy is an example of the spectrum of policy issues that can result from rapid urban growth. The repair, extension, and widening of roads, as well as the building of new ones, was considered a priority for a city that had doubled in size over the course of a decade.[12] When a community is in need of infrastructure changes, the public good is served by implementing them. Annoyances and hazards to the health of residents living close to the asphalt plant were overshadowed by the road project's perceived benefits to the entire city. Yet the entire problem might have been avoided by more holistic forethought about land-use planning. The gravel pits in which the asphalt operation was located pre-dated the construction of nearby residences, and were originally well away from residential areas. Permitting the nearby development of residential neighbourhoods as a response to rapid growth ran afoul of several conditions: conflicting municipal and provincial priorities, insufficient consideration of the difficulty of quantifying noise annoyance and air quality issues because of prevailing winds, and insufficient consultation between governments and residents. The resulting conflict can be characterized as occurring between public good and public health.

CHANGING CULTURES

Aggressive use of loud noise in public places is generally regarded as detrimental to the health of the community, but exactly what standards define loudness and aggression? They are hardly universal. The lines that divide communication from nuisance or celebration from intolerable din can be challenging to draw. Where public opinion is uniform, decisions about what is and is not a nuisance noise are relatively easy to make. Diversities of culture and opinion, as well as changes to the technologies of entertainment and noise abatement, complicate the process of developing both policies and informal agreements. Thus, multicultural and mixed-income communities may face particular challenges both in regulating noise and in determining the boundaries between noise and music and between indoor and outdoor activities. The interaction of cultural customs with community laws is significant in such situations, as are the attitudes of residents toward differences in parenting customs and toward sports fields, outdoor performance venues, and public celebrations in their neighbourhoods. Politics as well as science and law may be involved in a process that polarizes the community, bringing attitudes about the nature of community itself into conflict with allegiances to personal autonomy or cultural identity.

Even the implementation of beneficial community programs can have unforeseen – or unforeheard – consequences for local soundscapes. Because of its residential density and the diversity of values held by its multicultural population, New York is a model city with respect to issues about noise in urban communities. In one example from the 1990s, the provision of twenty-four-hour public basketball courts in some of that city's neighbourhoods, part of the "Midnight Basketball" strategy to discourage youth crime, resulted in sleep interruption for nearby residents. In another example, the story of the drummers in Marcus Garvey Park illustrates the complexity of coming to terms with the notion of the *commons* as expressed in the soundscapes of public urban spaces.[13]

The Drummers of Marcus Garvey Park

It is Saturday evening, the second day of summer, and the air around Marcus Garvey Park in Harlem is filled with the scent of blossoming linden trees and the sound of West African drums. Across the street from the park is 2002 Fifth Avenue, a new seven-story cream and red brick luxury co-op with a doorman, $1 million apartments and a lobby with a fireplace. The drummers in the park are African-American and from Africa and the Caribbean. They form a circle and have played in the park, in one form or another, since 1969, when the neighborhood was a more dangerous place. The musicians, who play until 10 p.m. every summer Saturday, are widely credited with helping to make the park safer over the years.

T. Williams, *New York Times*, 6 July 2008

In 2008, the *New York Times* reported on a controversy developing in the Harlem neighbourhood of Mount Morris Park over cultural values concerning noise and quiet. On one side of the dispute were the drummers of Marcus Garvey Park – located within the neighbourhood – whose music follows the traditions of their West African origins: subtle, syncopated layers of rhythm that shift and interlace like threads in a huge sonic tapestry. Its complexity surpasses the foursquare patterns of conventional pop-music drumming as a Beethoven piano sonata surpasses "Chopsticks." In keeping with West African traditions, the drums may be accompanied by trumpets, flutes, bells, gourd rattles, and/or tambourines, which provide counterpoint. The style is dynamic, exuberant, compelling, expressive, and *loud*.

The drum group began playing in Garvey Park at a time when it was littered with garbage and unsafe because of local crime. The group's presence provided an opportunity for children to play and neighbours to gather on summer days. After decades of Saturday improvisations, the group became a cultural attraction, observed by tourists and joined by city musicians and dancers. The park became their stage for performances unsuited to the acoustics or the limitations of indoor venues, and the community valued the group for its history and its ability to bring pride and a sense of heritage to the neighbourhood. Members of the group who had emigrated from West Africa and were playing Caribbean drums were carrying on traditions that rely on drumming and dance to induce states of spiritual transcendence, but they were also valued in the community for the sheer joy of the ritual of coming together with compatriots to make music.

On the other side of the dispute were the residents of new luxury condos overlooking the park, described as "young urban professionals" and mostly without connections to African culture. Although some were quoted in the *New York Times* coverage as appreciating the drumming and its significance to the community, supporters as well as detractors reported that they were unable to hear their own music or televisions – or to carry on conversations – within their apartments because of it. Some residents were openly hostile: racist e-mail messages began circulating in the building, prompting intervention by the city's Parks Department.[14] The drummers were relocated within the park, further away from residences, and provided with a sign giving their space official designation as a park project, complete with summer permits. But their new location caused some pro-drumming residents to complain that they could no longer be heard from the street and were hard to find, and the drummers found that the new location placed them too close to amplified stage performances. Elderly members of the drumming group had difficulty getting to the new location, so the drummers returned to their original location in front of the new condo building (Salvaterra 2008). At one point, a state senator became involved in mediating the dispute. It was resolved in 2009 to the satisfaction of both sides by creating a "Drummers' Circle" with benches in a designated area closer to the centre of the park, a suitable distance away from the residential buildings that housed the complainants.

As one commenter on the Garvey Park controversy mentions in a 2008 blog entry, the drummers were playing acoustic instruments and invited anyone with the desire to join in to do so. More recent musical activities in the park, unrelated to the drummers' group, have been heavily amplified and less inclusive. They have the potential to cause even more annoyance to residents, but most are located on stages placed deep within the park, out of earshot of most residences and out of reach of the noise ordinance because they are sponsored by groups that obtain permits from the city. Some, however, are described by blog respondents as transient "self-amplified" amateur performances that simply set up until police appear.

Supporters of the drummers' group perceived their presence as an injustice. An article on the drumming controversy from the website Nigerian Village Square (Dobnik 2007) pointed out that the Mount Morris Park neighbourhood attracted national attention in 2001 when former US president Bill Clinton set up an office there, initiating a wave of redevelopment projects that began to change the nature of the area. Crime and urban decay were reduced, but an influx of non-African businesses and residents turned the long-established culture into a mosaic, shifting the balance of accepted criteria: what was music and what was noise? Does cultural expression outweigh the desire for acoustic neutrality of the commons? Was the Garvey Park dispute really a clash of cultural standards, an issue of race and racism, an issue of musical taste, or simply a protest by individuals exhausted from listening to high-energy percussion for eight hours at a stretch?

Blogs on the subject of the Garvey Park drums demonstrate the range of opinion, from residents in favour of the drumming to residents who were opposed, from visitors to the park and commenters who appear to have been attracted by the controversy to those without direct experience of the situation. Most of those who posted comments were careful to avoid racial slurs, although some responders inferred the implication of racist comments from the care taken with wording. Some pro-drum enthusiasts in the blogs specified that they were "white"; some pro-quiet protestors specified that they were "black." However, the comments did not appear to regard race or age as primary concerns: what divided community members most clearly was whether or not they were accustomed to the drumming because they had lived within earshot of it for a long period of time. Long-term residents had built up associations with it as a positive element of the local culture as well as an expression of a deeply valued tradition; more recent residents had not.

Opinions may also depend on what individuals in Mount Morris Park regard as suitable activities for Saturday afternoons and evenings in the summer. If leisure time means escaping the heat of your cramped apartment and catching up with your friends outdoors in the park while watching the drummers and dancers, or joining them, the presence of this convivial and cost-free entertainment – which may provide a continuity of habit reaching back to your childhood, or an outlet that attracted you to move to the community – is cherished. If your apartment is spacious and air-conditioned, and you want to stay there and shut out the world by reading, watching a movie, or listening to your own favourite style of music, the drummers are invasive of your personal acoustic space. The income levels, cultural backgrounds, and leisure choices of residents in any neighbourhood influence their attitudes toward noise. When gentrification changes the cultural demographics and thus the expectations of a community, disagreements over ambient soundscapes can occur.

Soundscapes also change with new developments in technology. Consider, for example, the potential effect on urban traffic noise of electric cars: in time, traffic roar will join the rhythm of hooves as a soundscape of the historical past. The technologies of community infrastructure – power, water, transportation, and communication, as well as geographical features like mountains and rivers – contribute to the sound of a place and mark it, in combination with the languages spoken there, as *this* place. When these soundmarks change, a sense of mild disorientation can haunt the edges of awareness, analogous to the loss of a familiar building on your street – something is missing and it doesn't feel quite right.

CHANGING TECHNOLOGIES: URBAN AND SUBURBAN ISSUES

Everyone knows – and this book demonstrates repeatedly – that cities are noisy. Walk down a busy street, or near one, and you are enveloped in a dull pulsating roar punctuated by occasional sharper sounds: cars honking, bus doors opening, music spilling out from storefronts, bicycle bells, heels clacking on concrete. In suburbs and small communities, as well as the quieter neighbourhoods within cities, the roar is less prominent and may be absent except during rush hours. However, a new menace to peaceful existence and regular sleep is increasingly being reported worldwide: *the Hum*.

The Hum distinguishes itself from day-to-day traffic noise and from the sounds made by electric generators, idling and moving trains, water pumping stations, and any other specific components of community infrastructure. It is usually not loud – often just loud enough to disturb sleep when the residence is otherwise quiet. In some cases, it is masked by furnace and refrigerator fans, audible only beyond the kitchen and only in summer. In others, it is a pervasive nuisance, interfering year-round with concentration and relaxation as well as sleep. While urban roars are traceable to their component causes, Hums usually are not.

Do I Hear a Hum?

The standard reports of Hums describe them as sounding like a diesel-powered truck or jet engine idling in the distance, usually a faint rumble or drone. They can vary in loudness or remain barely audible, easily masked by the sounds of appliances and conversation – until the house is quiet and its occupants are trying to sleep. At this point, the Hum becomes a powerful antagonist to reason as well as rest. Once awakened, those who hear it look out their windows or leave the house to look for the truck parked and rumbling in the alley or in front of the neighbour's house, only to discover that there is no truck, and the sound has disappeared. It cannot be heard outdoors. The harried householder compares observations with family and neighbours: some have heard it; others have not, and apparently they cannot perceive it even when possessed of normal hearing and alerted to its presence.

Nuisance hums have been reported all over North America and Europe since 1991, when the community of Taos, New Mexico, first drew media attention to the phenomenon. An investigation at Taos produced recordings and identified a low-frequency tone at 41 Hz, modulating over time in both frequency and loudness. Attempts to determine a source were not successful, and the research team – which included acoustic and mechanical engineers and audiologists – concluded that "there are no known acoustic signals that might account for the hum, nor are there any seismic events that might explain it" (Mullins and Kelly 1995). Ten people who reported being able to hear the Hum were asked to identify what they heard: they consistently reported sounds that were normally below the threshold of human hearing. The investigators then surveyed approximately 8,000 residents in the community and identified 161 who were "hearers," 20 per cent of whom could detect the sound at a radius of fifty miles from the town. The majority of hearers were between the ages of thirty and sixty, with approximately equal numbers of males and females. Their reports and responses paralleled those of the initial group of ten, leading researchers to wonder whether the phenomenon had more to do with the residents' hearing sensitivity than with any actual sound signals (Mullins and Kelly 1995). Did the hearers of the Taos Hum experience an extraordinary sensitivity to low-frequency noise?

Attention was called to the phenomenon again in 2003, when a Hum was reported and investigated in Kokomo, Indiana. Again, no physical cause was detected. The lead investigator dismissed claims that people were "hearing things," stating that "what they are sensing is real and is caused by something that they are sensitive to and most others are not" (Cowan 2008, introduction). Reports of physical effects that corresponded to the onsets and cessations of Hum episodes were tracked: they included "headache, nausea, diarrhea, fatigue, and memory loss," as well as sleep deprivation (ibid., section 2). Dogs were also reported as behaving erratically during episodes of the Hum.

Some of the disturbance in Kokomo was traced to cooling fans at an industrial plant. Noise-abatement measures were implemented at the plant, producing relief for some, but not all, residents. Two other facilities initially believed to be relevant were ruled out because their times of operation did not correspond to the reported episodes of Hum disturbance.

In a review of unpublished documents relating to the Taos investigation, the Kokomo report notes that, although Hum episodes did not correspond with any measurable source of sound or seismic vibration, there was a possible relation to "an elevated electromagnetic field level that was reportedly related to the local power lines" (Cowan 2008, section 3). There had also been reports from affected people related to malfunctioning of electrical appliances in and around their homes.[15]

The same phenomenon was reported in Kokomo, so the investigators there monitored electromagnetic fields in the areas where residents experienced mechanical malfunctions, "including appliances suddenly burning out and cars having remote starters unexpectedly starting in garages" (Cowan 2008, section 4). Again, levels were found to be elevated.

The investigators concluded that, despite the Hum's being perceived as low frequency, it was more likely to be caused by electromagnetic activity: ultrasound rather than infrasound. The suspected mechanism? Thermoacoustic effects in the brain that activate the auditory nerves in the absence of sound waves.[16] Recognizing that more research was needed before a valid theory could be developed, they left the matter open: mysterious Hums remained mysterious, but attention was called to the Kokomo hearers' reports that the onset of the Hum paralleled the completion of "utility work associated with telephone, cable television, or power line maintenance, or a new cell phone tower in the neighborhood" (Cowan 2008). An association of at least a portion of Hum sensitivity with otoacoustic emissions – sounds generated by the outer layer of cochlear hair cells in the inner ear – has been introduced by recent studies (Frosch 2013; 2017) that raise the possibility of the Hum being a form of low-frequency tinnitus set off by external stimuli outside the range of human hearing. While far more research is needed to establish cause and effect, a combination of biological investigation with acoustic measurement shows promise.

Two projects carried out in Canada demonstrate the complexities of Hum research. The first, in Windsor, Ontario, investigated a persistent hum, which had the characteristics of mechanically produced infrasound, by setting up specially designed microphones to record it and notify the research team when it reached critical levels of loudness. Windsor is located on the Detroit River, which runs along the border between Canada and the United States. When the Hum was traced to infrasound emanating from an industrial island on the US side of the river, the investigation took a political turn: agencies from both governments were involved in negotiating a solution to improve the noise-abatement engineering on the island. Since such solutions require extensive funding, governments and industries can disagree on assigning liability: the discovery of the Hum's source was not revealed to the public until legal and financial decisions had been made.[17]

Another Canadian investigation, which has followed Hum reports in a suburb of Calgary since 2010,[18] demonstrates additional challenges. A team of investigators attempted to survey the community but found that their questionnaire was filled out – incompletely, in almost all cases – by just 36 per cent of households.[19] Approximately 12 per cent of the residents who returned the survey reported having heard the Hum. They were clustered mainly in three specific neighbourhoods, and the team then gained permission to take extended recordings at three houses in those areas. Consistent frequencies of 40–43.5 Hz were found at all three sites: these were apparently not produced by any appliances, utilities, or identifiable infrastructure in the area. Comparison of timing for the onset and cessation of Hum episodes with the schedules of operation for various public utilities, train routes, industrial air-conditioning units, and compressor stations did not reveal any patterns of correspondence. Neither did the calculated flow rates of the major water pumping station for the area. Substations – with rates of flow too variable to be reported – were still on the team's radar, but the absence of data halted investigation at the time.

Another set of evidence complicates the matter even further. In 2003, a team investigating Hums in Bad Waldsee and Bad Durkheim, Germany, concluded that they could not have been caused by either external sound sources or electromagnetic frequency. When hearers were tested in acoustically and magnetically shielded chambers, they still reported hearing the same frequencies as they did in their homes, showing that – at least for those hearers and those locations – otoacoustic emissions were the cause (Frosch 2013).

Are Hums caused by external sources like the one in Windsor, by otoacoustic emissions within the brain, or by some combination of the two? Half of the residents who reported hearing a Hum in Calgary also reported some degree of tinnitus. Does this finding reinforce the correlation between hearing a Hum and otoacoustic emissions? If so, how does that account for the half who did hear a Hum but do not experience tinnitus, and the possibility of a consistent audible low-frequency tone being involved? Could low-frequency noise stimulate otoacoustic emissions? Could high-frequency electromagnetic emissions do so as well?

Answers are not yet clear. What we can assume is that not all Hums are the same, that the phenomenon can have different manifestations and perhaps multiple causes, and that a combination of accurate measurements and recordings with community histories and thorough documenting of complainants' experiences and states of auditory health will be necessary to produce valid explanations.

The World Hum Project

Hums of unknown origin are now known to occur worldwide. The World Hum Map and Database (thehum.info) keeps records of reported Hum locations as well as a database of complaints itemized for location, duration, acoustic qualities, and symptoms reported by hearers whose health is affected. The map is instructive: the overwhelming density of complaints comes from regions with the greatest development of electronic infrastructure, including North America (especially the east and west coasts) and Western Europe (especially the United Kingdom and Germany). Far sparser densities appear in small areas of coastal Brazil and Chile, Central America, South Africa, New Zealand, and parts of Eastern Europe. There is at this time no reported activity in vast areas of Russia, Asia, Africa, and the Middle East – but that may be the result either of low levels of urbanization or low levels of reporting, rather than a complete absence of Hums.

Since the World Hum project depends on voluntary reporting rather than official surveys, its statistics are both intriguing and challenging to interpret. Its information depends on crowd-sourced reports that may not always distinguish between "mysterious" hums of untraceable origin and common urban infrasound from factories, rail lines, compressor stations, water treatment plants, and the growing industry that generates energy by means of wind power. "Wind farms" are themselves an increasing and controversial source of noise annoyance.

CHANGING TECHNOLOGIES: RURAL ISSUES

Despite their low density of population, rural communities are not exempt from noise annoyance and the need for regulation. The location of businesses that require large tracts of land and generate traffic – shopping malls, "big box" stores, car dealerships – can provoke noise complaints. Venues for noisy entertainment – drag racing, motocross, rock concerts – may be built adjacent to farms whose owners value quiet for their own convenience but also for the health and productivity of their livestock. Even the building of infrastructure for wind energy, an option for "green" power, is contentious. Because the science documented by the wind power industry differs from the science documented by medical researchers, legal and ethical decisions are complicated by controversy.

As in the case of the Calgary asphalt plant, industrial progress comes into conflict with public health, but in the case of wind farms the technology is so new that even the most basic information about health effects is in dispute. It is not yet clear whether health risks are significant, because most research to date has been funded by industry and by governments desperate for new sources of power as the environmental cost of oil becomes more evident. Each step in the process of research – the design and funding of studies, analysis of data, reporting of results in scientific journals, and their interpretation and dissemination by popular media – is subject to pressures by interest groups that include residents, local businesses, industries, governments, and media.

The Wind Farm Controversy

From the porch of an isolated farmhouse, you can hear the thwack-thwack-thwack of the huge blades, each as long as the freight car of a train. A stark white forest of them is visible in the distance, towering eighty metres (360 feet) above the ground. Pale sentinels nourished on air, they represent a new generation of generators planted in open land to initiate a future of clean energy.

Wind farms – sites for multiple wind turbines – offer a potential for large-scale generation of electricity without the use of fossil fuels, dams, or nuclear plants. Many existing wind farms are located offshore in coastal areas of Europe and North America in order to take advantage of strong winds and to avoid proximity to densely inhabited areas, but even their impact can be disturbing to wildlife communities as well as those of humans. Bird and bat mortality results from extreme changes in air pressure as well as collision with blades (Barclay, Baerwald, and Gruver 2007; Baerwald et al. 2008). There is also concern among wildlife biologists that the noise of the turbines, amplified by water, is disrupting migration and communication patterns for sea birds and whales (Nelsen 2012).

Noise from inland wind farms, part of a growing industry located largely in the central midwestern United States and in the Canadian provinces of Ontario and Quebec, is the subject of scientific controversy.[20] It is believed by many scientists to subject nearby residents to insomnia and headaches as well as the muscle aches, anxiety, and depression that result from sleep deprivation, from low-frequency noise, and possibly from changes in air pressure caused by operation of the turbines. Whether these symptoms are the result of actual wind turbine activity, of weather sensitivity, or of stress reactions brought on by noise annoyance is not entirely clear. Because the definition of noise annoyance encompasses emotional reactions as well as physical symptoms, studies are showing conflicting results: each side of the controversy can cite extensive evidence, but neither side is convinced by the other's interpretation of research design or findings.

Residents' physical reactions to wind farms are mixed and seem to depend on the design of the turbines, their location and distance from residential areas, averages and peaks of wind velocity, and the sensitivity of the residents. Emotional reactions also vary. Some welcome the projects as carriers of economic prosperity as well as green energy; others resent the way the turbines look as well as their noise. The science of impacts on health is debated by experts and widely regarded as inconclusive. Wind farm noise research may represent a case of science shaped by competing interests, political and economic concerns pitted against individuals who blame the projects for detrimental changes in their health and property values. As well, the conflict appears to result from differences in the focus of existing studies. The physics of wind speeds and sound transmission gives one set of facts; the reported experience of many families living near wind farms gives another. Sorting valid knowledge from a morass of conflicting claims is likely to require far more investigation of human and animal sensitivities to low-frequency noise (LFN) as well as the effects of air pressure fluctuations.

One rural Ontario physician and researcher, Dr Robert McMurtry, has called for further investigation of what is called "wind turbine syndrome" – the cluster of physical and emotional symptoms associated with noise annoyance as well as LFN – which he estimates will affect 30–50 per cent of people living near wind farms.[21] American public health epidemiologist Carl V. Phillips goes even further in condemning the industry as a source of noise pollution, accusing it of manufacturing dismissive reports to counter the thousands of noise and health complaints filed against it. He joins community activists in calling for new regulations to place turbines at a greater distance from residential areas (Phillips 2011).

Other medical studies published on government and industry websites deny the existence of wind turbine syndrome. Chief among these is a 2012 document, authored by a team of US medical researchers in prominent university and government positions, that appears to dismiss the syndrome because it can be defined within the general category of noise annoyance and can therefore be explained as the result of stress (Ellenbogen et al. 2012). A letter strongly criticizing the report's position takes the stance that there is too much evidence of symptoms that start when the turbines are operating, and stop when they are not, for the medical community to ignore them or attribute them solely to stress (Krogh 2012). Disputes of this sort can pit the technical expertise of specialists against each other as they struggle to define a new hazard arising from a technology that is not yet fully understood.

When facts derived from physics and anatomy refute reported experience, questions arise about the accuracy of standard measurements as well the effects of stress placed on complainants from other causes. Are the protesting communities simply opposed to any technological development in rural areas, or do they have legitimate complaints? The debate is ongoing, with results clouded by politics as well as medical ambiguities. A study sponsored by the provincial governments of Ontario and Prince Edward Island in Canada and released in 2014 acknowledged disagreements in prior research, but concluded that "self-reported" symptoms of wind turbine syndrome – including tinnitus, dizziness, and migraines – were not actually related to the presence of turbines. While the symptoms were described as having some connection with "annoyance towards several wind turbine features (i.e. noise, shadow flicker, blinking lights, vibrations, and visual impacts)" (Health Canada 2014), they were not considered significant enough to delay progress on wind power development. A document by the Multi-Municipal Wind Turbine Working Group, an Ontario wind-power protest group, was sent to Health Canada in response to the 2014 study. It named thirteen published experts as establishing a link between wind turbines and declining health, critiqued perceived flaws in the research design of the government study, and called for a federal study (Multi-Municipal Wind Turbine Working Group 2019). Since the development of alternative power sources in the face of climate change and declining oil prices places wind power high on the list of priorities for Canadian and many other governments, the issue has considerable importance for economic policy decisions.

Do documented cases of symptoms indicate sensitivity to low-frequency noise? Since there are as yet no medically recognized tests for such sensitivity, this is a difficult question even for experts to answer. The issue may well remain controversial in medical literature for some time because variables may be impossible to control well enough to meet the expected standard of proof, which is an unambiguous relationship between exposure and reaction. If the symptoms experienced by complainants do not start at the moment the turbines are turned on and stop at the moment they are turned off, the relationship is not considered to be clearly established. Changes in weather, the presence of allergens in the air, and even the anxiety associated with the possibility of wind turbine syndrome are all regarded as possible causes for the reported symptoms or complications in the interpretation of evidence. So is negative publicity about wind farms: residents told to expect symptoms are more likely to experience them (Chapman et al. 2013; Schmidt and Klokker 2014; Crichton et al. 2015).

A literature survey conducted in 2014 on a large number of relevant databases searched for wind turbine–related health effects, which included noise annoyance, sleep disturbance, psychological distress, tinnitus, vertigo, headache, and triggering of epileptic seizures. Of the initial 1,231 articles, only 36 were deemed acceptable for analysis. The majority were eliminated because of flaws in research design: questionable methods of measurement, selection bias, limited size of samples, poor design of questionnaires, failure to account for changes in weather and wind speed, and failure to measure for the presence of LFN. The authors concluded that the questions surrounding wind turbine syndrome are still open (Schmidt and Klokker 2014), in part because exposure to LFN is known to cause structural changes in the auditory systems of animals, and analogous structures in the human ear may well be susceptible to the same changes (Salt and Hullar 2010).

Since no human studies meeting the standard for proof have yet been completed, wind turbine syndrome cannot be either definitively proven or refuted at this time (Enbom and Enbom 2013; Schmidt and

Klokker 2014). However, researchers with the European Metrology Research Program (EMRP) have verified cases of LFN-sensitive people able to hear frequencies as low as 8 Hz, well below both the presumed limit of 20 Hz and the wind farm average of 16 Hz.[22] Their study, which was focused on investigating the use of infrasound on the auditory tract for therapeutic medical purposes, cites earlier warnings about exposure to infrasound (defined for medical purposes as 0–20 Hz) and LFN (defined as 20–500 Hz)[23] as causes of vibroacoustic disease, associated with malfunctions of cellular communication in living tissue (Duck 2007; Alves-Pereira and Castelo Branco 2007). The EMRP group, even after finding that infrasound exposure may have therapeutic uses for enhancing working memory, concludes that "the identification of potentially hazardous health effects of [infrasound] is a relevant public health issue and further research is required in order to evaluate whether protection standards against low-frequency noise need to be established" (Weichenberger et al. 2015, 88), and that inaudible frequencies need to be examined for their effects on the entire body rather than just the auditory tract. Their discovery of individuals whose hearing transcends the usual norms for humans points the way to one area for further investigation; the question of cellular malfunction in the presence of infrasound points to another. Such studies will in time bring clarity to the issue of wind farm syndrome, and clarification of medical effects will lead to valid conclusions about whether legal regulations are adequate.

Design and Regulation: What Works?

An additional challenge in the wind farm controversy is regulation. Determination of setbacks – allowable distances between massive industrial turbines and residences – is often left to local authorities after they have signed a contract with the industry. Some European countries forbid the placement of industrial turbines any closer than one kilometre (.6 miles) or two kilometres if placed near any school or residence. In the United Kingdom, where the allowable distance is 500 metres (.3 miles), complaints about the effects of noise are common (Frey and Hadden 2007). While some European municipalities have regulations based on distance or on audible noise, others do not regulate distance unless communities object to industrial applications in the planning stage – a point when measurements provided by the industry are the only ones available. In the United States and Canada, determination of setback distances is left to states, provinces, or local municipalities, when it is regulated at all (Government of Rhode Island 2016; Government of Nova Scotia 2011).

New designs are being developed by the wind power industry to reduce the noise, which has been recorded in some locations as a constant roar with regular swoops or pulses, in others as merely a soft whooshing.[24] While much depends on the equipment used for recording, the shape of the turbine's rotating blades and their orientation to prevailing winds also make a difference to how much sound is produced and whether it is carried upward to disperse or downward to annoy residents. One innovation in design from Spain even dispenses with blades entirely: a narrow conical mast is fitted with magnets that regulate the direction of its wind-driven oscillation; the gathered energy is then converted to electricity by means of an alternator. That turbine is now in use in Europe. It is reported to be effectively silent and – in the absence of blades – safe for birds (Stinson 2015).

Amateur recordings posted on the Internet provide conflicting evidence for ongoing debates: are wind turbines a symbol of progressive thinking and community pride, or an oppressive nuisance that causes illness and drives residents from their homes? Are the energy companies that build wind farms being honest with residents and municipal councils about benefits, or deceiving them about drawbacks? What recourse is – and should be – available to those residents who are susceptible to wind turbine syndrome, if in fact it is caused by the turbines and not by weather, stress, allergies, or other sensitivities?

LAWS AND MEDIATION

The issue of wind farms is also indicative of even larger concerns:

- How effectively can zoning laws and bylaws protect communities from noise annoyance?
- Are noise-related laws protective of the public or restrictive of industrial development?
- Given the differences in sensitivity to noise that will occur in any population, as well as the tendency of our species to communicate, compete, and celebrate with activities that generate loud sounds, what is a reasonable scope for individual rights?

None of these questions can be answered with certainty – the first because everything depends on the formulation of laws in accordance with actual physical conditions, the second because everything depends on whose point of view is being considered the "official" one, and the third because everything depends on local culture and legal precedents. When disputes arise, mediation may be more effective than legislation.

Mediation among individuals, cultural groups, and even levels of government is often required in order to resolve persistent noise complaints that fall outside of legal jurisdiction. According to Bill Bruce, former chief bylaw officer for the city of Calgary, whose department deals with citizen complaints about unkempt suburban yards and roaming dogs as well as noise, the job required him to be half police, half social worker. His work involved devising imaginative solutions to problems in which the conflicting interests of residents are equal according to the law. When one suburban house is home to a light sleeper who rises early and the nighthawks in the adjoining one install a home theatre with multiple speaker sub-woofers placed next to a window, the law does not specify any action.

Mediation begins with an assessment of the needs and desires of each party: what can be solved by compromise? What costs and benefits are involved in potential solutions? Who takes responsibility for modifying the causes of conflict? Mediation can result in the offending home theatre being moved to the basement so that its sound is contained by a concrete foundation, or in working out a schedule so that the speakers are turned to high volume only when neighbours are out for the evening. The process of compromise can motivate backyard party hosts to negotiate the timing of future events with complaining neighbours, dog owners to provide sufficient exercise to decrease bouts of barking, and apartment dwellers to place speakers on absorbent rather than hard surfaces.

Mediation of noise disputes represents a solution based on ethics, aesthetics, and compromise. It is a balancing act performed by individuals when community standards are not uniform and law is vague or outdated: for example, a noise bylaw guideline for western Canada specifies that "a bylaw may validly prohibit noises emanating from record players on private property that can be easily heard by a person in other premises."[25] The intent of the bylaw is clear in this case, but the validity of its enforcement in light of rapidly changing technologies is less so: does a law designed for "record players" apply to boom boxes and smartphone docks?

Disputes result as much from laws and municipal planning policies that are unclear, conflicting, or outdated as from the actions and attitudes of individuals. Legislators and municipal planners, as well as residents and their elected officials, would benefit from greater awareness of such laws and greater attention to their formulation. Rather than a legal backwater, community noise is a vital example of the interlocking priorities and consequences of community planning, sociology, law enforcement, and law itself.

While laws regarding noise have a long and effective history of protecting communities and workers from the excesses of industrial and transportation noise as well as overzealous audiophiles, the process of formulating them is not a simple one. Conflicting interests, legal ambiguities, and the complexities of measurement present challenges that call for considerable ingenuity from municipal planners and noise abatement engineers as well as legislators.

14

Hear Ye, Hear Ye! Noise and the Law

Despite the evidence about the many medical, social, and economic effects of noise, as a society, we continue to suffer from the same inertia, the same reluctance to change, and the same denial of the obvious that the anti-tobacco lobby faced a couple of decades ago. This inertia and denial are similar to those that delayed appropriate action on lead, mercury, and asbestos. Now we seem unable to make the connection between noise and disease, despite the evidence, and despite the fact, which we all recognize, that our cities are becoming increasingly more polluted with noise.

Goines and Hagler 2007, 293

As sound levels, frequency and quality vary with time, it is difficult to determine what the impact of sound will be at the planning stages of any project. As well, the cumulative impact of several noise sources may be difficult to assess. Experts are now able to project sound impacts through sophisticated computer modelling programs, but caution that ongoing noise monitoring is critical.

Siskinds 2009

When an irritant leaves traces in the air that can be measured after the irritation is perceived, as is the case with noxious smells, it is relatively easy to link cause with effect and to know whether harm has been done. Abatement methods and laws pertaining to air pollution can then be set in motion with reasonable certainty. Because sound is transient, the conditions that cause noise annoyance are far more difficult to determine. The loud party or machine that awakens you at night may well be silent by the time a noise bylaw officer arrives to take measurements and ask questions. The wind will rise to confound measurements of loudness. The malfunctioning furnace will purr with perfect docility when the repairperson shows up. Sometimes, no trace of the sound will persist to provide evidence. Even recordings may well be inconclusive: their accuracy depends on the sensitivity of equipment, but also on the amount and type of background noise in the location.

Legal regulation of noise is guided by the sciences that measure the physical properties and effects of noise, and also by knowledge of the social effects that noise can have, which comes from information about injuries and complaints. While noise is a product of the intersection between physics and physiology, noise annoyance is usually a product of social interaction. People living in close proximity cause it, machines designed to improve the conditions of living cause it, and even expectations of what a soundscape should. be contribute to it. If you move to a city from a farm, you expect the ambient sounds to be different. Some of them will annoy you, but you had some choice in the matter and may take minor annoyance as part of the package. If you stay on the farm and your municipal council decides to build an airport, shopping mall, or casino on the border of your land, your degree of annoyance is likely to be considerably greater. If you live in the city and the nightclub across the street repeatedly plays music loud enough to interrupt your sleep, or garbage-collection trucks roar past your window at 3 a.m., your annoyance level may reach the boiling point. You may be motivated to band together with your neighbours and complain – not just from the standpoint of individual annoyance and stress, but from an awareness that noise annoyance harms communities.

If the perceived harm outweighs the perceived benefit of the airport, mall, casino, nightclub, or arrangements for garbage collection, the complaint may be considered valid in economic terms, which is sometimes a persuasive argument. If there is direct proof that your health and that of your neighbours and family – and perhaps your livestock – is being harmed because of the noise, there might be a law against it.

Chapter 5 discussed the aspects of sound that can cause annoyance. Although several of them can be directly associated with levels of emotional and physical harm, they are not necessarily regulated by either legal or customary controls. Among the physical properties of noise contributing to annoyance that are not usually covered by law are the duration of the sound, its variability (whether steady or intermittent; monotonous or varied in frequency and/or loudness), its tonal qualities (rumble, warble, roar, squeak, screech, mechanical, melodic, or percussive), the masking influence of other noises like traffic and wind, and even the effects of humidity in the air. Any of these variables can complicate the process of quantifying noise annoyance for legal purposes; when more than one variable is present, the difficulty is compounded. The frustrated, sleep-deprived resident of a noisy neighbourhood may be told that nothing can be done legally, or that there is no proof that the noise is causing any physical symptoms. The acoustic engineer or law-enforcement officer taking measurements to verify a complaint about noise may arrive just at the moment when wind direction changes and the ambient decibel level rises to mask the offending noise and confound measurement processes, when the noise stops altogether, or when a thunderstorm alters readings. The officer may be using state-of-the-art equipment or an older, less expensive, and less sophisticated version that does not provide entirely valid measurements even for basic levels of loudness; and the individual may or may not be fully trained in effective use of the equipment and interpretation of results.

Just as measuring the transmission of sound may be more complicated than it appears, quantifying sound reception can be problematic. Subjective variables are more likely to affect reception than transmission, and to defy measurement. They include the complainant's taste in music, sleeping habits, expectations about the community or workplace, tolerance for cultural diversity, ability to manage stress, and schedule. As British social scientist David Halpern pointed out in a 1995 study of the influences of the built environment on mental health – a term that he uses to include emotional state and attitude as well as medical conditions – it is difficult to identify direct cause and effect in examining noise. Unlike the particulates in polluted air, which can be shown to produce measurable and consistent reactions, sounds produce a wide range of responses, depending on who hears them, when, and in what circumstances (Halpern 1995, 37–8).

Because sensitivity to noise varies by individual and by culture, the formulation of regulations and bylaws can be a difficult matter. Does a motorcycle qualify as a cause of nuisance noise? A dog? A boom box? The family down the block having a sing-along on their porch? The TV in the next apartment, where the occupants are hearing impaired? The new highway that you didn't (or did) vote for? The crows or pigeons or magpies first thing in the morning, which no law can control? Given the complexity of the process, what guides the process of regulation?

REGULATION AS PROTECTION

Prevention of noise-related interference with health and communication has been the main focus of regulation for decades. The expectation behind noise bylaws is that residents should be able to expect peaceful enjoyment of their own premises. This is not difficult so long as the expectation is considered to apply to the interior of a building, but sound waves are not respectful of territorial boundaries. You may prevent your neighbours or their dog from entering your property uninvited, but you cannot stop their voices, music, or barking at the boundary between your property and theirs. When the definition of premises includes outdoor areas – yards, gardens, balconies – and the ability to open windows when indoors, the formulation and enforcement of bylaws quickly become problematic.

Investigation of noise annoyance is a more recent procedure. Noise annoyance, however, is not necessarily the same as noise damage. In legal terms, the line is often drawn between concrete physical consequences, usually measurable hearing loss or regular interruption of sleep, and emotional reactions. In the grey area between the two parameters are such factors as intermittent sleep disturbance, stress levels, and attention deficit. Some bylaws and guidelines account for them, others do not. Since bylaws are not designed to account for every possible effect of noise, they are formulated by means of reasonable community standards. This means that if you live in a densely populated city, an expectation for quiet at all times will not be considered reasonable by the law, which is designed to balance the need for quiet with the noise-producing activities that sustain and entertain the community. While cultural standards vary greatly, legal standards are usually based on the assumption that most people desire a quiet environment at night, as well as protection from noise loud enough to interfere with privacy and with typical work and leisure activities in the daytime.

While the science of noise abatement is well advanced, the sociology of noise, and its legal implications, is not. Where, then, to begin the process of assessing and regulating noise disturbance? Two organizations with global focus have provided influential guidelines for the process. Advisory to the regulatory activities of governments are two primary institutions: the International Commission on Biological Effects of Noise (ICBEN) and the World Health Organization (WHO).

ICBEN is an independent group of scientists organized into teams according to expertise (Gjestland 2002). The organization collects information and recommendations from nine investigative teams focused on:

1 noise-induced hearing loss
2 noise and communication
3 non-auditory physiological effects induced by noise
4 influence of noise on performance and behaviour
5 effects of noise on sleep

6 community response to noise
7 noise and animals
8 combined agents
9 regulations and standards

The teams meet every five years to compare findings and develop recommendations. Recent efforts include the standardization of content for questionnaires on noise annoyance to be used worldwide, and a conceptual shift: rather than limiting regulation to the control of noise, ICBEN scientists are now considering ways to advise governments on the creation of ideal sound environments for designated zones of habitation.

The World Health Organization issued a set of guidelines for loudness limits in 1999 (Berglund, Lindvall, and Schwela 1999) that are still in effect, at least theoretically, as a standard. The WHO guidelines for community noise are summarized in table 14.1. They demonstrate concern for the broad public and occupational health issues of annoyance, sleep disturbance, and hearing impairment, but also for issues pertaining to education and communication: speech intelligibility and extraction of information from auditory sources. The guidelines also call attention to the need for regulation of entertainment devices and of toys that, as we saw in chapter 8, can produce noise in excess of safe levels of loudness. They represent an attempt to create worldwide legal standards for controlling nuisance noise in the interest of public health. However, implementation and enforcement of the resulting laws are subject to a wide range of social attitudes and local budgets.

As guidelines, the WHO recommendations are intended to present advice for the formulation of noise bylaws. Federal and municipal governments may or may not choose to follow them. Even in places where they are followed, enforcement can be problematic: police forces are often stretched beyond the ability to provide full-time noise bylaw officers, and procedures and equipment for taking objective measurements may not be accessible. Furthermore, municipal employees in large urban areas can become desensitized and may dismiss complaints as low in priority. Bylaw officers are rarely responsible just for noise infractions and may have to give priority to complaints about parking, littering, loitering, snow removal, and roaming dogs.

Consider also the cultural implications of the guidelines, which are designed primarily to protect people from the noise produced by machines and by crowds. First, they demonstrate the principle that noise levels during the overnight period, usually defined as 11 p.m. to 7 a.m.,

Table 14.1
World Health Organization guideline values for community noise in specific environments

Environment	Critical health effect(s)	LAeq[1] (dB)	Time base (hours)	LAmax,[2] fast (dB)
Outdoor living area	Serious annoyance, daytime and evening	55	16	-
	Moderate annoyance, daytime and evening	50	16	-
Dwelling, indoors	Speech intelligibility and moderate annoyance, daytime and evening	35	16	
Inside bedrooms	Sleep disturbance, night-time	30	6	45
Outside bedrooms	Sleep disturbance, window open (outdoor values)	45	8	60
School classrooms and preschools, indoors	Speech intelligibility, disturbance of information extraction, message communication	35	during class	-
Preschool bedrooms, indoors	Sleep disturbance	30	sleeping time	45
School, playground outdoor	Annoyance (external source)	55	during play	-
Hospital, ward rooms, indoors	Sleep disturbance, night-time	30	8	40
	Sleep disturbance, daytime and evenings	30	16	-
Hospitals, treatment rooms, indoors	Interference with rest and recovery	(as low as possible)		
Industrial, commercial shopping, and traffic areas, indoors and outdoors	Hearing impairment	70	24	110

Environment/source	Critical health effect	LAeq[1]	Time base (hours)	LAmax, fast[2]
Ceremonies, festivals, and entertainment events	Hearing impairment (patrons <5 times/year)	100	4	110
Public addresses, indoors and outdoors	Hearing impairment	85	1	110
Music through headphones/earphones	Hearing impairment	85	1	110
Impulse sounds from toys, fireworks, and firearms	Hearing impairment (adults)	-	-	140 (peak sound pressure [not LAmax, fast], measured 100 mm from the ear)
	Hearing impairment (children)			
Outdoors in parkland and conservation areas	Disruption of tranquility	Existing quiet outdoor areas should be preserved and the ratio of intruding noise to natural background sound should be kept low.	-	120

[1] LAeq indicates a steady sound level that, over a specified period of time, emits the same total energy as a fluctuating sound. It stands for "equalized level measured in A-weighted decibels," and is used to measure sound pressure over time.

[2] LAmax, fast indicates the maximum value of sound pressure during a measurement period, measured in A-weighted decibels. The "fast" measurement system averages the sound over 0.125 of a second.

Source: Berglund, Lindvall, and Schwela (1999)

should be lower than daytime levels. The assumption arises from the tendency in the industrialized nations in much of the Northern hemisphere for most people to sleep in uninterrupted eight-hour cycles in the dark, with noise minimized to prevent interruption of sleep. That is certainly reasonable, but what about equatorial cultures that break the sleep cycle into two periods: early afternoon, when the heat drives everyone indoors, and a short cycle in darkness?

The recommendation for speech intelligibility indoors is designed to protect educational institutions from disruption, but it would be difficult to apply in a crowded bar or restaurant where the combination of conversation, background music, and hard-surfaced building materials make the content, if not the ritual, of communication irrelevant. The limit on loudness for outdoor events is above the threshold of potential hearing damage for individuals on or close to the stage if the event is electronically amplified: an assumption is being made that participants in such an event will have the judgment to use hearing protection, or will be advised or ordered to do so by their employers. As we saw in chapters 9 and 10, this is not necessarily the case. While the cost of effective protection is small – simple foam or soft plastic earplugs are usually adequate unless the exposure is frequent and of long duration – such protection is not available everywhere, and cultural barriers against its use may be significant.

In a broader perspective, the WHO guidelines are indicative of rising international and intercultural concerns with the issue of noise as an environmental pollutant. As more parts of the world industrialize, as cities grow in size and density and a greater proportion of the world's population lives in urban areas, the potential for noise annoyance and its attendant health concerns grows. Even beyond the health of individuals, the guidelines map a concern for the health of communities. In section 6.1, recommendations are made for the future of research into the effects of noise on health worldwide: these include development of standardized questionnaires that can be used in any culture, government-sponsored investigations of the economic consequences of exposure to excessive noise, and investigation of the "relationships between sound pressure levels and politically relevant variables," citing decreased productivity, workplace and school absenteeism, and "increased drug use and accidents" (Berglund, Lindvall, and Schwela 1999, 73). Preservation of quiet is also mentioned as essential to the emotional health of groups as well as individuals. Where options for quiet surroundings exist, the opportunity for recourse against stress is present. Where stress is reduced, so is conflict.

BYLAWS

Measurement Issues

Legal limits on ambient noise are dependent on accurate systems of measurement: how loud the noise is, what frequencies are present, what effects are produced by the noise, and what causes it are all matters for concern. When the parameters of the problem defy measurement or results are inconclusive, policy decisions may be inadequate. Since most measurement systems and resulting laws only concern loudness of sound, problems can arise from other physical components such as proximity, duration, and frequency spectrum. Variables, both objective and subjective, in the transmission and reception of sound complicate the process, but the most immediate barrier to accuracy is incomplete understanding of the measurement system.

Canadian physicist Jeremy Tatum, in an essay describing the limitations of municipal noise bylaws, points out that "the deciBel scale gives no indication at all as to how annoying an unwanted noise can be." He then poses the question of whether a sound measured at 55 dBA for five hours would be more or less annoying than one measured at 58 dBA – twice the level of sound pressure – for five seconds (Tatum 1996). Because laws are based on objective principles rather than subjective ones, and quantitative data are necessary for proof of harm, noise bylaws and their enforcement depend on precise measurement of whatever can be measured: primarily loudness (sound pressure level) and duration, with some allowance made for time of day and for community expectations.

Since there is no objective way to measure individual annoyance, its assessment must rely on qualitative measurements, consisting of subjective reports by individuals exposed to the noise. These may have some bearing on whether a legal nuisance is perceived, but they will rarely override quantitative data in a legal dispute. An ideal system for determining whether a community is harmed by noise would take both quantitative and qualitative information into account. It would start by providing sound-pressure readings at a variety of locations and times and under various wind conditions; it would also survey residents in the affected area about their social habits and their reactions to the noise, in order to determine their levels of sensitivity to noise annoyance.

As an example of the complexity of measurement, consider the ambient roar of urban traffic. The impact of traffic noise in residential areas and the interiors of residential buildings, like the exposure of employees in workplaces, is determined by "dose." Measurement of the dose is accomplished by taking readings with a sound-level meter along the street. This method may not be sufficient for accuracy, however, since interior and exterior soundscapes are different (Skånberg and Öhrström 2002). While it seems reasonable to assume that the interior of a building will be quieter, such complicating factors as position of the building in relation to others may change the outcome, as when high rises that face each other across a narrow street produce a vertical amplification tunnel. A fully accurate reading should be adjusted mathematically for the distance between street and building and for the floor level in the building (Rylander and Bjorkman 2002). Additional complications include the position and construction of windows, composition of building materials and furnishings, ambient noise levels within the building, time of day, and sensitivity of occupants. Because bylaw officers are rarely trained in the complexities of acoustics, what appears to residents to be an obvious problem may be difficult to resolve.

The measurement of low-frequency noise (LFN) is another example of the complexity of determining noise dose. Although dBA is the standard scale used to determine infractions of noise bylaws, it is inaccurate for determining the loudness of LFN, which is more accurately measured by the dBC scale. While the use of computer software for the identification of frequency ranges and the calculation and graphing of effects makes the process of measuring LFN far less difficult than it used to be, neither laws nor standard abatement industry practices outside northern Europe have yet caught up to the use of the dBC scale, in part because interpretation and communication of the measurements obtained often involve experienced judgment as well as measurement. As a result, noise complaints that involve LFN, which is produced by a variety of mechanical sources as well as heavily amplified bass from popular entertainment venues, are notoriously difficult to resolve. The legal validity of such complaints is often determined by which system of measurement is used.

Legal Issues

Because one person's preferred ambience may be another's unbearable noise, the law simply cannot address all possible situations encountered by communities and individuals. Neither can the budgets of communities for noise-control design, materials, and projects. Noise abatement law is always a balancing act between what is desired, what is physically possible, and what is affordable.

Since legislation depends on quantified descriptions, community noise bylaws depend on a combination of the questions asked by scientists, legislators, and community members and the technology available to provide measurements. Frustrated residents of a neighbourhood besieged by jackhammers, compressor stations, or late-night entertainment spots may feel strongly and rightfully that their quality of life is affected and that the government ought to do something about it, but if a sound-level meter indicates a level of loudness within the law, they have no direct recourse. Still, some guidelines for community standards are well established. One example, from a document specifying the regulations for an urban redevelopment project in San Francisco in 2004, will serve to illustrate some common assumptions about noise issues:

> Unnecessary, excessive, or offensive noise which disturbs the peace or quiet of any neighborhood or which causes discomfort or annoyance of any reasonable person of normal sensitivity residing or working in the area is prohibited. A noise level, which exceeds the ambient noise level by 5 dBA or more, as measured at an affected receptor's property line, is deemed a *prima facie* violation of the Ordinance.[1]

This statement, which is summarized in the document from the wording of specific bylaws, contains a number of assumptions that are worth examining. First, it suggests that noise becomes a problem in residential areas when it results from unnecessary activity or from sources that produce more noise than is necessary for their operation. This conclusion, designed to apply to limiting the impact of a construction project on residents, would forbid such actions as idling the motor of a construction crane when it is not in use. Theoretically, it could also apply – logically, though not legally – to the decision to run a gas-powered mower on a small residential lawn. "Offensive" noise may be the result of acoustic properties, or it may be the wrong style of music or a shouted profanity from the residence next door. Noise bylaws do not cover the content of speech or music, only their loudness and duration. A "reasonable" person is, presumably, one who will tolerate some degree of annoyance as part of everyday life, and "normal sensitivity" seems to presume undiminished hearing ability but also the absence of health conditions, such as migraine or chronic pain, that heighten sensory acuity. Residential noise bylaws are always less strict than those designed for hospital zones. Finally, the statement takes ambient noise levels into account because the loudness of an offending noise will be noticeable only when it exceeds the ambient level, and a noise that is not noticed does not offend: it can still cause some level of disturbance to the nervous system, but it will not motivate a complaint.

NEIGHBOURHOOD NOISE

Few community issues are as controversial as noise, or as likely to polarize residents. In the American Housing Surveys taken in 1999, 2005, and 2009 by the US Census Bureau,[2] neighbourhood noise consistently outranked crime and litter as a cause for complaints by residents.[3] Attitudes about what constitutes noise and how loud it should reasonably get can solidify cultural subgroups by occupation, ethnicity, age, gender, musical taste, and even parental status. While the stereotyped anti-noise complainant is elderly and female, the largest proportion of actual complainants comprises young parents of both genders, concerned that the neighbour's motorcycle will wake the baby. While the stereotyped anti-bylaw objector is a youth, actual objectors are usually business owners in the restaurant and entertainment industries. In any situation with music providing the basis for a complaint, the complainants may be people who don't appreciate that style of music. In order to be fair, bylaws must be written on the basis of quantifiable properties and time of day, rather than taking questions of taste into account.

What noise bylaws do take into account are the characteristics of the neighbourhood – population density, zoning (residential, industrial, or mixed) – as well as those of the noise. Table 14.2 provides a brief list of noise limits in municipal bylaws for residential districts in selected North American cities, showing that standards for maximum loudness fall within a fairly consistent range.

In Europe, bylaws follow the WHO directives adopted by the European Union and have ambient limits similar to the North American average. Individual countries and municipalities have leeway in determining limits for public events and even for when "night-time" designations begin and end: for example, regulations differ between Scandinavia and the Mediterranean as a result of culture as well as latitude. Brazil has regulations comparable to those in the EU. Elsewhere in the world, noise regulation is in its infancy. Published information from non-Western areas is scarce, and most of it is based on summaries done by journalists or graduate students in urban planning rather than by government-sponsored professional surveys. Major cities in India and China are now in the process of mapping and drafting bylaws based on the EU regulations, but expectations for loudness limits may be quite different. A noise map of

Table 14.2
Residential neighbourhood noise limits in selected North American cities

City (population in millions)	Daytime limit (dBA)	Night-time limit (dBA)
New York (18) – zone density dependent	60–65	50–55
Toronto (5)	50	45
Chicago (3)	55–62	unspecified
Vancouver (2)	55	45
Calgary (1.3)	65	50
San Francisco (.7)	55–60	50–55
Seattle (.6)	55–60	unspecified
Winnipeg (.6)	55	50

Beijing, compiled as part of an effort by the municipal government to initiate noise regulations, identifies zones adjacent to major highways where ambient outdoor noise is over 85 dBA and indoor noise averages 70 dBA, the former comparable to working in a factory and the latter to standing on a busy street. Recommendations for bylaws advise an outdoor limit of 75 dBA for residential areas, considerably higher than Western limits.[4] In other parts of the newly industrializing world, calls for research and regulation have not yet attracted the attention of governments: the international urban growth of the twenty-first century is racing ahead of the capacity of researchers and legislators to measure and control its effects.

Like all laws designed to serve complex communities with a variety of needs, noise bylaws represent a series of compromises. Consideration is taken of people who prefer quiet or require it for their home-based work, and of those who have a need for the stimulation of noise or produce it as a consequence of necessary actions. Standards are set by means of averages and of reasonable assumptions, but that does not mean that they take all important variables into account. The setting of nocturnal limits at 45–55 dBA is a case in point. The variation of 10 dBA, very substantial in terms of the increase in actual sound pressure, is allowed because ambient street noise in densely populated areas is higher in some cities, and some districts within cities, than in others, and ambient noise masks the audible bumps in the night that are more likely than level soundscapes to awaken residents. The presumed measurement is taken outdoors on a residential street, not in people's bedrooms. The loudness levels are exemplified as falling between "normal conversation" and ambient noise in an office. Such levels would be inoffensive coming from outdoors with the bedroom windows closed, but imagine them next to your bed, an effect easily approximated by the combination of open windows and traffic or loud conversation directly beneath them.

The noise limits presented here refer only to loudness, but many bylaws also cover other aspects of noise. One is specifying sources that are likely to produce major noise annoyance. Federal, regional, and municipal laws set specific limits on noise from construction projects, highways, airports, the hospitality and entertainment industries, and public festivals. Constraint of noise is often a matter of limiting duration and times of day that noise can occur from these sources, as well as limitations on loudness. Regulation by means of time is also important for minor sources in residential areas: running your gas-powered mower or leaf-blower at 5 a.m. is not legal, but the same actions at noon are. Leaving your dog outside to bark in the daytime might be legally, if not culturally, permissible;[5] doing so at night is not. Recently updated bylaws regarding protection of the environment in North America and Europe also provide guidelines for noise-reducing insulation materials to be used in the construction of apartments, factories, airports, and public buildings.

NOISE INDOORS: RESIDENCES

In *The Soundscape of Modernity*, her comprehensive study of the changes in urban soundscapes and architectural acoustics during the first decades of the twentieth century, historian Emily Thompson meticulously examined trends in registered complaints about noise in New York in 1930 (Thompson 2002, 157–68). The sources of noise annoyance – construction, vehicles, loudspeakers in the streets, music blaring from storefronts, and residences with open windows – are external to the complainants' domestic surroundings. They are urban outdoor noises encroaching on the private soundscapes of listeners indoors. The complaints were registered with the city because domestic privacy, while usually defined and demonstrated in visual terms, is largely an auditory concept.

Today, the indoor domestic soundscape is dense with auditory stimuli. Even in situations where people live alone, electronic media may provide a constant background of sonic activity at home as they do at work. While no apartment block would be designed or built with interior windows between units, allowing people to observe their neighbours visually, it is common for soundproofing to be so inadequate that conversations and the noises of plumbing, as well as media, are only slightly muffled. Noise abatement was rarely a major issue in buildings completed before the Second World War: their sturdy interior walls and lath-and-plaster construction prevented transmission of most speech and noise between units in apartment and office buildings. Building methods in the later half of the twentieth century were less effective. The postwar building boom encouraged speed and the use of cheap materials, resulting in thinner walls and hollow doors that transmit sound between residential units rather than attenuating it. Changes in architectural fashion led to an emphasis on open-plan and loft spaces, as well as the conversion of factory floors to urban residential units, eliminating auditory as well as visual barriers within units.

Residential noise abatement remains a challenge to architects, engineers, and the construction industry as well as legislators and communities. Tests used to determine the compliance of designs and materials to industry standards of noise abatement may be limited to lab trials that do not account for performance *in situ*, and clients are rarely familiar with the implications of test results (Weissenberger 2004). It is not always guaranteed that architectural training programs will include expertise in acoustical design: relatively few architects and engineers study acoustics even when the subject is available unless they plan to specialize in concert hall or theatre design or to work in the noise abatement industry. Unless clients are aware of the range of potential noise issues when a building is planned, the issues are easily ignored until they become problematic.

Precautions that would have been more effective and less expensive when planned at the design stage are often added when the building or complex is already a reality and the people using it start to complain. In order to provide a reasonable level of noise abatement in residential buildings, notice should be taken before construction starts of such variables as number of occupants, presence or absence of children, typical loudness and frequency levels of conversation for the population involved, number and type of appliances, entertainment devices and their typical use, layout of room adjacencies, structural and decorative materials used (wallboard grade, flooring, window glazing and covers), and ambient soundscapes, both interior (ventilation system, furnace) and exterior (traffic, zoning of neighbourhood, population density) to the building.

All of this requires planning ahead, with attention to a wide range of influences. Perfection of design and implementation is probably unattainable for all residents, since noise levels will also depend on where in the building a given apartment or condo unit is located, what the occupants' and neighbours' habits are, and each tenant's personal tolerance. While a quest for ideal domestic quiet may require the purchase of noise-cancelling headphones, a basic and reasonable desire to avoid unsolicited information about your neighbours' personal, sexual, and sanitary habits should be accommodated simply by the construction and insulation of your building.

Such accommodation, however, comes at a price. Current industry standards, which are not necessarily uniform, usually recognize six levels of effectiveness in wall partitions. The most effective grades are the most expensive, usually reserved for areas with high ambient noise and the potential for speech to encroach on partitions. More attention to prediction of noise problems in the planning stages of a project, as well as subsidies for insulation, would eliminate most problems before they occur. Planning ahead pays off: a well-soundproofed residence allows occupants to appreciate quiet, and also to be loud without disturbing others and causing complaints.

REGULATION OF OCCUPATIONAL NOISE

Regulation of workplace noise is covered by government departments of occupational health and safety, which concern themselves with limiting the loudness of noise to which workers are exposed, enforcement of hearing protection regulations, and compensation for workers whose hearing has been damaged by workplace noise. Such departments may be federal, as in the United States; provincial with federal guidelines, as in Canada; or intergovernmental, as in Europe. One of the responsibilities of government health-protection agencies is to provide legal limits for daily noise exposure without hearing protection. This would seem to be a matter of pure science, standardized and unambiguous, but it is not.

The US Occupational Safety and Health Administration (OSHA) standard notes that non-continuous exposure should be assessed as cumulative for the day; thus, three one-hour bouts of noise count as a three-hour exposure for that day. Whether the individual worker, the project supervisor, or the company is expected to keep track of exposure times is not specified, however, and therein lies one of the many complications in occupational health regulation. Another can be found in the determination of danger posed by sound pressure levels, as indicated by limits on length of exposure. American regulations for maximum safe exposure to

Table 14.3
US federal regulations for maximum safe exposure to continuous noise in work-places per day

Noise level	Maximum exposure time
90 dBA	8 hours
92	6 hours
95	4 hours
97	3 hours
100	2 hours
102	1.5 hours
105	1 hour
110	30 minutes
115	15 minutes

Source: US Department of Labor, Occupational Safety and Health Administration, n.d. It is worth noting that several online sources, including US government agencies and private consulting companies, differ on both sound pressure levels of common activities and length of safe exposure.

workplace noise, especially at higher levels, use different standards from those in Canada. Comparing tables 14.3 and 14.4 provides a sense of the differences.

Table 14.3 shows the noise exposure limits in the United States, where regulation is a federal concern, although methods of – and requirements for – measurement vary by state. Setting the maximum exposure at 90 dBA rather than the 85 dBA recommended by both the World Health Organization and the National Institute for Occupational Health and Safety[6] exposes workers to increasing risk of hearing damage if they retain the same job for years. Comparing the lists of permitted exposure times at each decibel level shows that US regulations reduce exposure time as the continuous sound gets louder according to the human ear's perception of rising sound pressure, at 5 dBA difference (table 14.3). The Canadian regulations, by contrast, reduce exposure time according to the measurable rate of sound pressure increase, at 3 dBA difference (table 14.4). In Canada, each successive incremental increase of 3 dBA from 85 dBA to 115 dBA *halves* safe exposure time. The recommendations in table 14.4 hardly come as a surprise: as explained in chapter 2, each successive 3-dBA increase results in a doubling of sound pressure, which necessitates halving of exposure time to maintain safety.

Since a revision of standards in 2016, all provinces except Quebec set the requirement for hearing protection at 85 dBA (table 14.5). Canadian federal

Table 14.4
Canadian provincial regulations for maximum safe exposure to continuous and peak noise in workplaces per day, 2016

Noise level	Maximum exposure time
85 dBA	8 hours
91	2 hours
94	1 hour
97	30 minutes
100	15 minutes
103	7.5 minutes
106	3.75 minutes
109	1.88 minutes
112	.94 minutes
115	28.12 seconds

Source: Canadian Centre for Occupational Health and Safety

guidelines are slightly different, with the national threshold set at 87 dBA, perhaps as a way of harmonizing with US and Quebec regulations for the benefit of cross-border industries.[7] Prevention, of course, saves individual workers from experiencing damage to their hearing, a significant consideration for a country with public health care. The preventive approach is effective, but only so long as both employers and workers comply with guidelines and regulations for the provision and use of hearing protection devices.

Along with government regulations on noise in the workplace, protection is provided in North America by workers' compensation boards (WCBs), which monitor workplaces, keep records of accidents and hazards, and process requests for compensation by employees. Assessments of risk, based on ongoing records of claims for injuries sustained at work, give an informative picture of the scope of the problem. According to WCB standards, occupations considered to produce high risk for noise-induced hearing loss (NIHL) include any that involve percussive action (hammering or processing of hard-surfaced materials) or working with machinery. Work Safe BC, the WCB for British Columbia, lists the following construction industry trades as being particularly risky for exposure to harmful levels of noise: bricklayer, carpenter, concrete worker (around a pump, vibrator, jackhammer, or powered finishing equipment), driller, drywaller (shooting track or boarding), electrician, form worker, foreman, framer, labourer, mobile equipment operator, pipefitter, plumber, roofer (shake, tar/gravel, membrane, shingle), sandblaster, steel erector, supervisor, truck driver, and welder/fabricator (Work Safe BC 2019). Its guidelines then go

Table 14.5
Noise exposure limits in Canadian jurisdictions, 2020

Jurisdiction	Continuous noise*	Impulse / impact noise*		
	Maximum permitted exposure level for 8 hours: dB(A)	Exchange rate dB(A) +	Maximum peak pressure level dB(peak)	Maximum number of Impacts
Canada (federal)	87	3	-	-
British Columbia	85	3	140	-
Alberta	85	3	-	-
Saskatchewan	85	3	-	-
Manitoba	85	3	-	-
Ontario	85	3	-	-
Quebec	90	5	140	100
New Brunswick	85	3	140	-
Nova Scotia	85	3	-	-
Prince Edward Island	85	3	-	-
Newfoundland and Labrador	85	3	-	-
Northwest Territories	85**	***	140	100
Nunavut	85**	***	140	-
Yukon Territories	85	3	140	90

* Asterisks in chart refer to "exchange rates" of 3 or 5 dBA: see p. 233. The permitted rates differ among jurisdictions in Northwest Territories and Nunavut rather than being standardized.
Source: Canadian Centre for Occupational Health and Safety (CCOHS), *Noise: Occupational Exposure Limits in Canada* (2020), https://www.ccohs.ca/oshanswers/phys_agents/exposure_can.html.

on to mention particular risks to farm workers who operate machinery, to workers in the hospitality industry, and to musicians and their technical support workers. All workers in the designated categories are at high risk for hearing loss, but also for the somatic effects of sustained noise – discussed in chapter 6 – that are not mitigated by protecting the ears. Hearing protection is essential, but so are better education about risks, better insulation of work areas, and design of quieter machinery.

Since loud noise is inevitable in industrial workplaces, construction sites, and a host of occupations, unions often demand hearing protection. In most cases employers comply, if only because the cost of protective equipment is less than the potential cost of employees' compensation claims. As early as 1989, public health researchers in Alberta warned that compensation claims for hearing damage, then estimated at over Can$14,000 per case, would increase drastically in number and cost in the near future (Alleyne et al. 1989). A similar study conducted in the state of

Washington found that compensation claims had doubled between 1984 and 1991, largely due to increased reporting but indicative of a growing and expensive trend (Daniell, Fulton-Kehoe, et al. 2002).

While the law can specify that hearing protection be used in the workplace, compliance with the law is another matter. Further studies by the Washington research team and a similar group in Michigan focused on the question of whether industries were complying with regulations for hearing protection in high-risk workplaces and concluded that compliance rates were low: almost half of companies surveyed did not achieve full compliance with OSHA regulations in the 1990s (Reilly, Rosenman, and Kalinowski 1998). Even when companies met the standard, only 62 per cent of workers in the Washington survey routinely used the mandated hearing protection (Daniell, Fulton-Kehoe, et al. 2002; Daniell et al. 2006). This means that a substantial number of workers in high-risk industries, mostly in construction and manufacturing, have not been adequately protected.

Compliance with regulations for noise reduction and protection of hearing rests with workers as well as workplaces. Why, then, has compliance been relatively low? A number of factors contribute: lack of knowledge about the hazards of noise, belief that hearing loss "can't happen to me, and there must be a cure for it anyway,"[8] association of noise with productivity and action, and cultural beliefs – particularly for males – about the importance of ignoring pain and disability. Even where industrial programs for hearing conservation are successful, workers may remain at risk. Hazardous levels of sound in the workplace are determined by means of calculating exposure that is equalized over a period of time. In the case of noise that varies in intensity, equalized measurements may not be adequate. Research suggests that accounting for pressure at peaks or spikes of intensity will provide better protection for workers (Seixas et al. 2005; Merluzzi, Pira, and Riboldi 2006). However, even setting a specific decibel level as a cut-off point for mandatory hearing protection might not provide adequate protection. A 2007 study of ten-year hearing-loss rates for more than 6,000 employees at a US aluminum manufacturing company found that those who worked with noise levels above 85 dBA had lower rates of hearing loss than those at 80 to 85 dBA. The research team concluded that mandatory protection above 85 dBA was having its intended effect, but that workers who were not required to wear protection were now at higher risk than those who were. Their suggestion? Extend legally mandated protection to sound pressure levels at and just below 85 dBA (Rabinowitz et al. 2007).

Industrial hearing protection, like workplace safety regulations and medical insurance, results from the recognition of workers' rights in societies with a long-standing history of industrialization and of struggle for such recognition. In places where industrial culture is new, or where the notion of workers' rights has not been recognized because of repressive political cultures, NIHL – along with other forms of occupational injury – is likely to be normalized rather than treated as a problem to be solved. Consider, also, the ethical repercussions of exposing millions of workers in developing countries to the risk of hearing impairment. In places where factory jobs are in demand despite low wages, protest against unsafe working conditions is unlikely. Where literacy levels are also low, the loss of ability to hear speech frequencies through NIHL can doom workers to drastically reduced quality of life, cutting them off from full participation in their social surroundings. In municipal, national, industrial, and multinational policies, attention to the problem as a socio-legal and economic issue, rather than a series of isolated individual cases, is essential.

Hearing and Healing:
Soundscapes in Health Care

Over recent decades, the proliferation of medical devices and their use even outside of surgery and intensive care has grown dramatically ... When these different devices come together at the point of care, the environment can become a cacophony of bells, buzzes, chirps, and flashing lights from central stations and patient rooms throughout the unit – to communicate threshold-based deviations in patients' status.

Solet and Barach 2012, 86

The meanings of being a patient in a sound-intensive environment were interpreted as never knowing what to expect next regarding noise, but also of being situated in the middle of an uncontrollable barrage of noise, unable to take cover or disappear.

L. Johansson, Bergbom, and Lindahl 2012, 108

Places of healing are supposed to be quiet, with clean and pleasant surroundings and efficient nurses gliding through the corridors on soft-soled shoes, murmuring gently to patients and each other. Doctors appear at bedsides with good news, patients wake refreshed from sleep, food is plentiful and nutritious. Friends and family members arrive with flowers, everything is bathed in gentle light, the soundtrack is full of soft violins ... *Wait*. Does this resemble any hospital you have encountered outside of a Hollywood movie?

SOUNDSCAPES IN HOSPITAL CARE

The soundscape of the average modern hospital is a cacophony of overlapping beeps, blats, and pings from monitors; roaring ventilation systems; raucous bursts of conversation from visitors, patients, and the nursing station; carts rumbling down hallways; televisions and phones. Sleep, essential to the healing process, is available primarily through medication. Even medicated sleep can be elusive, however, when mechanical alarms sound in the patient's room at an average of six times per hour, signalling nurses and technicians to perform tests and administer medications as well as alerting the nursing station to emergencies (Görges, Markewitz, and Westenskow 2009). While the procedures are necessary, the use of auditory signals can be counterproductive: sleep disturbance for patients is echoed by stress for staff, who are expected to respond immediately to alarms and call buttons even when there are not enough of them available to do so.

Despite growing awareness of the noise problem by hospital staff and administrators, it has not improved quickly. An 2005 on-site study of noise in a US hospital stated that levels had been rising since 1960 with an average of 4 dB every 10 years, and that, for hospital staff, "noise may impact mental efficiency, [and] short term memory and contribute to stress" (Busch-Vishniac et al. 2005). Crowding is also an important variable for patients and staff: a 2012 study of medical literature about noise in intensive care units stated what should be obvious: "a potentially important clue for future work involves the finding that as the number of patients and staff of the ICU increases, noise levels appear to also increase" (Konkani and Oakley 2012).

Hard-surfaced floors and walls, chosen for ease of cleaning, reflect and amplify noise. In her innovative book on the effects of environment on health and healing, physician Esther Sternberg draws attention to the importance of architectural design and materials to patients' processes of recovery from illness, injury, or surgery. She discusses the paradox posed by sterile surfaces: while necessary for reducing the risk of bacterial infection, they render the atmosphere of hospitals aesthetically sterile as well: devoid of the textures, colours, and comforting familiar objects that can enhance the process of healing by providing emotional sustenance (Sternberg 2009, 218–19). Such surfaces enforce a mechanistic aesthetic in visual terms and serve as amplifiers for the noises produced by machinery, making it difficult for patients to relax, sleep, or feel a sense of comfort. In effect, they delay the process of healing.

The World Health Organization's recommendation for noise exposure in hospitals is set at 30 dBA for ambient levels, equivalent to hushed indoor conversation, both day and night; the US Environmental Protection Agency recommends 35 dBA. Most North American hospitals set ambient daytime levels at 40 to 45 dBA, the intensity of conversation at typical levels, with peak levels of 65 dBA not to exceed 10 per cent of a twenty-four-hour period. Yet studies of hospital noise show that the recommended decibel levels are often exceeded (Busch-Vishniac et al. 2005; Darcy, Hancock, and Ware 2008; Connor and Ortiz 2009; Elliott, McKinley, and Eager 2010; Pope 2010; Eggertson 2012; Cordova et al. 2013; Eivazzadeh et al. 2017).

One study of noise levels in four nursing units of a US hospital found that all units ranged from 61 to 67 dBA: these were not peak levels but averaged from samples taken throughout the day. The author concluded that all units were "as noisy as a busy office," identifying twelve sources of noise in each five-minute sampling period and recommending more extensive use of sound-level meters in further research (Pope 2010). Worldwide, daytime noise levels in hospitals are reported to have grown from 57 dBA in 1960 to 72 in 2012, while night-time levels have risen from 42 dBA to 60 (Eggertson 2012). Psychiatric units can produce ambient levels of 76 dBA, equivalent to heavy traffic, with peak levels at 85 to 90 dBA (Holmberg and Coon 1999). This is loud enough to compromise the hearing of patients and staff with repeated exposure and to cause trauma to noise-sensitive patients.

The noisiest places in most hospitals are those containing patients most in need of relaxation and sleep: intensive care units (ICUs) for newborns and for patients recovering from major surgery, including open-heart surgery (Topf 1992; Gabor, Cooper, and Hanly2001; Epstein 2004a; Drouot et al. 2008). One study that combined sound-level measurements taken in the cardiac intensive care unit in a Turkish hospital with patient questionnaires designed to gauge levels of noise annoyance and sleep disturbance revealed noise levels that ranged from 49 to 89 dBA, with an average of 65 dBA. Patients located closest to the nursing stations reported the most disturbance. Monitor noises and conversations, as well as sounds made by patients entering the ICU from surgery, were reported as the most disturbing sources (Akansel and Kaymakçi 2008). More recently, monitoring and recording of nocturnal noise in an Australian ICU showed averaged levels of 52.85 dBA and a maximum at an astonishing 98.3 dBA, with peak levels above 70 dBA happening routinely ten times per hour (Delaney et al. 2017).

Even newborns may not be sufficiently protected. Research conducted at a US neonatal intensive care unit (NICU) concluded that upper limits for noise levels, set at 45 dBA, were exceeded 70 per cent of the time (A. Williams, van Drongelen, and Lasky 2007). The effects on premature infants can be serious. In noisy conditions, they are at elevated risk for irregular heart rate and blood pressure, sleep apnea, and respiratory problems, while prolonged exposure can affect brain development (Brown 2009; Wachman and Lahav 2011).

The most likely sources of elevated NICU noise? Monitors, staff conversation, and even the incubators that keep "preemies" alive until their bodies adjust to life outside the womb. But consistent reduction of NICU noise may require an attitude of vigilance that goes beyond quieting the obvious sources. Protocols used by researchers in India included attaching rubber pads to the feet of furniture, handling metal trays gently to minimize clashing sounds, reducing the volume of alarms, and appointing monitors among the staff to remind others to speak softly. The average level was reduced to 60 dBA, still far above the goal of 50 dBA (Ramesh et al. 2009).

Noise in operating rooms is also problematic, especially because the need for sterile conditions may preclude the use of sound-absorbing materials for infrastructure. Sound pressure levels reach 100 to 120 dBA (West et al. 2008) with the use of bone saws in orthopedic surgery, while equipment used in several contexts, including surgeries on the head and neck, exposes patients and staff to 131 dBA (Fritsch, Chacko, and Patterson 2010), well above the limit for safety.[1]

Even the crucial need for clear communication between surgeons and their teams can be compromised by a combination of machine noises, conversation, and music. A survey of published studies on operating room noise conducted in 2014 verified expected conclusions – that machine noise and irrelevant speech during surgeries distract surgeons from fully concentrating on their tasks. The same survey also revealed some surprising results: patients under anesthetic are at risk of hearing damage during prolonged surgeries, music can be distracting to novice surgeons, and the hearing acuity of anesthesiologists is often compromised by occupational noise (Katz 2014). Another study of anesthesiologists found that noise distraction was held responsible for a 17 per cent decline in their ability to detect subtle changes in oxygen saturation, a situation that could potentially endanger patients (Stevenson, Schlesinger, and Wallace 2013). Suggestions for improvement include elimination of irrelevant conversation – analogous to the protocol for pilots – better insulation, and the use of plastic bowls and trays to eliminate the clanging of metal instruments against metal containers.

Music in the Operating Room: A Mixed Blessing?

Many hospitals permit the use of music during surgeries for the benefit of the surgical team; it has the potential to benefit patients as well. Music has been found effective for reducing anxiety, which commonly includes fears about disfigurement and death, in patients immediately before elective surgery, even if it does not necessarily affect physical results (S. Wang et al. 2002). One controlled study reporting on the effects of live classical piano music on patients during ophthalmic surgery found that, while the control group (which was not provided with music) experienced rising blood pressure, heart rate, and respiratory rate, these rates decrease in the group exposed to the music, which was played on a digital piano in the pre-operative waiting area immediately before the surgery (Camara, Ruszkowski, and Worak 2008).

While most hospitals have neither the space nor the budget to provide pianos of any sort in operating rooms, recordings are always an option. Whether music in or near the operating room is used for the benefit of staff or patients, or both, it probably needs to be a shared experience: members of a surgical team are likely to perform more effectively when they respond to the same rhythm, and patient preferences should be considered when the surgery does not involve full anesthetic.

Although music has potential to contribute to a calm and alert state in surgeons and their teams, its use in operating rooms is now controversial because of its potential to function as noise when played too loudly or superimposed over a background of mechanical noise. Investigators in the United Kingdom who videotaped twenty surgeries over a six-month period in 2012–13 discovered that instructions from surgeons to members of their teams were five times more likely to need repetition when music was present. Each repetition wasted precious seconds of operating time – over a full minute in some cases. Their conclusion? "Music played in the operating theatre can interfere with team communication, yet is seldom recognized as a potential safety hazard" (Weldon et al. 2015, 2763). Because decisions about style and loudness of music are usually made by the operating team's lead surgeon, the rest of the team may feel annoyed by the style of music or simply unable to hear vital communication the first time because the music adds to a level of ambient noise that already exceeds WHO guidelines because of the machinery and ventilation systems in the room (Kleebauer 2015; Shambo, Umadhay, and Pedoto 2015; Weldon et al. 2015). The researchers, concerned about an absence of standards for best-practice policies about music during surgery, recommended that "frank discussions between clinicians, managers, patients and governing bodies should be encouraged for recommendations and guidance to be developed" (Weldon et al. 2015, 2763).

Operating rooms in old buildings need special attention to noise for another reason: antiquated metal cabinets for storing equipment can be hazardous if their doors squeal or rattle during a delicate surgical procedure, causing the surgeon to startle even slightly. In the words of one chief of surgery in a large US hospital, the "machine-gun" rattling of equipment cabinet doors in a cardiac surgical suite was a disaster waiting to happen: "if that cabinet gets opened at just the wrong time, the involuntary movement of the surgeon could kill a patient" (Dubbs 2004, 18). Scrupulous attention to monitoring noise, identifying hazards, fixing malfunctions, and alerting staff to potential problems – no matter how trivial they seem – is crucial to maintaining patient safety.

Machines That Go "Ping!"

The alarms from monitors tracking patients' vital signs are essential components of their care in emergency rooms, surgical suites, and intensive care units, but they are also interrupters of sleep and healing. Some of the problems with hospital noise are a direct result of the sounds made by equipment that monitors patients, and how that equipment is designed. For example, pulse-oximeter probes that monitor blood oxygen rely on a sensor attached to the patient's finger by a loose-fitting plastic clip, which is easily dislodged by movements of the hand. Use of the monitor is routine in emergency and intensive care, and a significant amount of alarm noise is caused by the sensor clips slipping off and activating loud alarms from the monitor when the patient is not actually in danger from low oxygen intake. While the decibel levels of the alarms may not be problematic, they activate a startle response that can wake dozing patients and contribute to stress. Improving the design of the sensors so that they do not fall off the fingers so easily would reduce the alarm component frequently found in patient care, and thereby reduce some of the ambient noise in emergency and intensive care units. This might contribute to decreasing recovery times and increasing quality of patient care.

Nurses might also be grateful for a reduction in "alarm fatigue." Constantly discerning whether a given alarm is signalling a dislodged monitor or an actual emergency without having time to check the patient until they finish another task is potentially aggravating, frustrating, and

244   Sound and Noise

tiring. It is also potentially hazardous to the patient, as multiple false alarms run the risk of desensitizing nurses to genuine hypoxia alarms (Eggertson 2012). One US study estimates that the rate of false alarms in critical care is typically over 80 per cent, and may be as high as 94 per cent (Görges et al. 2009). Consequent desensitization of nursing staff – as well as high levels of ambient noise that make alarms difficult to distinguish – can be responsible for hospital deaths (Solet and Barach 2012). Another US study tracked alarm signals for a month in one post-surgical unit of Johns Hopkins Hospital, and reported that 16,934 alarms sounded in eighteen days (Graham and Cvach 2010). That's an average of 940 per day. Even if the responsibility for answering them is spread over a large number of staff, and not all of them are audible in every part of the unit, that number may well represent a major source of stress to both patients and staff.

Auditory vigilance, which enables quick response to auditory cues, succumbs to fatigue when hospital staff are bombarded with too many signals of too many types. If the ambient noise of a nursing station is filled with constant alarms and most do not stand out against the background roar of air systems and computer fans, the evolutionary tendency to startle at sudden sounds is overridden by complacency: reaction is slow because urgency is not perceived. Recommendations for increasing the effectiveness of alarms include more careful design of their acoustic signals, better cooperation between manufacturers of monitoring systems and the hospital workers who use them, more accurate risk assessment for patients, and better training for staff to distinguish levels of urgency conveyed by alarms (Solet and Barach 2012).

Additional solutions seem obvious – alter the design of monitors to modify auditory signals and intensify visual ones, use sound-absorbing materials for floors and ceilings, educate staff about noise issues, provide ambient soft music where appropriate, and post signs requesting quiet. In fact, such innovative measures are not so easy to implement. Supplying new monitors and remodelling rooms are major expenditures for health-care systems already under financial strain. Education of staff is not always effective, particularly when turnover is high (Taylor-Ford et al. 2008). Relaxing music? According to whose preferences – those of individual patients, nurses, unit or hospital administrators, the executives of commercial Internet music channels? When the same carefully selected soundtrack is played for the seventeenth time in one day, at a volume that can override the ambient noises of the unit, is it still relaxing? Providing patients with headphone devices, or allowing them to bring their own, seems feasible, but setting the volume at a level audible over the ambient noise might not enhance health or relaxation.

Research into ways to reduce the impacts of noise is ongoing, and results are sometimes surprising. Experimental use of bedside privacy curtains equipped with pockets to hold sound-absorbing materials was found in a US hospital to improve rest and sleep periods slightly, but the improvement was not implemented because of increased costs (Locke and Pope 2017). It is often the case that hospital staff and administrators are aware of what can theoretically be done to improve conditions but do not have access to sufficient funds. Meanwhile, low-cost solutions, like educational programs for staff, do not usually make much difference (Konkani and Oakley 2012) unless compliance is high and mechanical sounds are also well controlled – an unlikely combination. However, designation of "quiet times" – reducing levels of noise and light while forbidding visitors for a set period of time during the day – can be effective (Gardner et al. 2009; Dennis et al. 2010). So can the use of inexpensive earplugs (Wallace et al. 1999). White noise can enhance sleep by reducing the contrast between relatively quiet ambient levels at night and intermittent loud monitor noises. For patients who are averse to white noise, the Internet provides hundreds of soothing soundscapes ranging from waterfalls to rainforest birds to gentle wind chimes. Even fans can provide masking to offset the startling effects of external noise.

While sleep disturbance is known to be a significant problem in intensive care units, research is divided about the impact on sleep of peak noise levels, usually caused by intermittent sounds that are considerably louder than the ambient background. Interruption of sleep includes full awakening as a result of a startle response as well as variations in the quality – the "depth" – of sleep. Two studies that measured sleep quality in relation to ICU noise concluded that noise is responsible for only 30 per cent of sleep disturbance, and that noise affects the quality of sleep, more than its continuity (Freedman et al. 2001; Gabor et al. 2003). A third, attempting to resolve controversy about the impact of noise, measured arousals with a polysomnograph – a monitor designed to record quality of sleep – and compared responses to two conditions: peak noise levels against a background of relative quiet, and against a baseline of soft noise provided by a white-noise generator. The authors found that white noise reduced the number of arousals, suggesting that sleep disturbance may be caused more by the contrast between noise and quiet than by the absolute loudness of noise (Stanchina et al. 2005). Thus, a background of white noise may be beneficial in situations where quiet cannot be maintained.[2] Such backgrounds can be provided by ventilation systems, but only if there is little variation in their sound and they are not loud enough to become part of the problem.

SOUNDSCAPES IN HOSPICE AND PALLIATIVE CARE

What about situations in which the goal of care is not healing, but end-of-life transition? Is noise a problem in hospices and palliative care units, which care for the dying? What sounds and soundscapes might give comfort?

Palliative care, whether in hospital or hospice, is concerned with treatment of pain and the effects of terminal illness, but also with enhancing the quality of life for the dying. Calm and quiet surroundings are central to care, but so are the options for communication and enrichment that are provided in many facilities. The concept is a very old one with modern revisions. For centuries, such care was provided by families, close-knit communities, or religious organizations. The primary goal was to ease the transition into a final state of rest or an expected afterlife by sheltering the patient from any excitement. Innovations in medical care for terminal cancer patients in the mid-twentieth century began a reform that saw palliative care as an enrichment of whatever life remained: moving beyond the provision of physical comfort, care facilities branched into attempts to provide for terminal patients' emotional needs. Conversation, music, art therapy, spiritual practices, and even occasional parties entered the quiet spaces of palliative wards (D. Clark 2007).

While there is not yet much research on the effects of soundscapes or the sonic preferences of hospice patients, it is assumed in Western societies that a "good" death is a peaceful one, and that the state of peace includes quiet. This paradigm is occasionally questioned, however, by some commentaries on the sociology of end-of-life care. One analysis of reactions reports the comments of two patients newly arrived at a palliative unit. The first is surprised at the presence of any significant noise:

> I was expecting when I came to the palliative care unit, that it would
> be more tranquil. But they [staff] want to allow each person as full a
> life as possible ... So you have to take the needs of the other patients
> into account ... Sometimes you hear a radio and the volume is too
> loud ... I had to ask one person here two or three times to turn down

his radio. And when the nurses go and see the patients at night or at 6 in the morning ... they speak all along the corridor, loudly. They leave the doors open and speak loudly to patients. You find yourself awake at 2 a.m ... I was not expecting that. (S.R. Cohen et al. 2001, 367)

The second patient is both surprised at and appreciative of a soundscape that includes multiple activities: "There is a good mood here ... I thought all would be in hushed tones, but to the contrary ... there are activities here, volunteers, a music therapist, a library, you can listen to cassettes. It surprised me: it is not exactly what I imagined" (ibid., 368).

Another critique points out that non-Western cultures, with differing assumptions about what constitutes appropriate action for the end of life, call into question the idea that quiet must be assumed and enforced. Citing an example of a large Bengali family surrounding the bed of their dying relative in a very quiet English hospice with loudly recited prayers, the author draws attention to the necessity for hospice staff to understand diverse cultural needs and to plan for designated "auditory spaces" that can be made available to meet those needs (Gunaratnam 2009). She goes on to mention all the unacknowledged sources of ambient noise that underlie palliative care: mechanical equipment, meal deliveries, staff rounds, visitors, coughs, groans, and laboured breathing.

The sounds of grief are also part of the ambience. They can vary tremendously with culture and family dynamics, forming a spectrum from soft weeping to shouting and screams (Gunaratam 2009). Given this, the planning of care for the dying and their families should always include attention to the auditory ambience. Providing a tranquil soundscape that includes private spaces for loud expressions of grief and for ceremonial practices is a major challenge that requires the active cooperation of architects, building contractors, and financial authorities with facility administrators and health care staff. Because dying and death bring both patients and visitors into tremendously vulnerable emotional states, providing a comforting atmosphere – visual, spatial, and auditory – whenever possible should be a major priority.

HOSPITAL SOUNDSCAPES:
THE PATIENT'S EXPERIENCE

As measurement of hospital noise becomes more accurate and its effects are better understood, a promising new line of inquiry is including patients in the research. Their reports of reactions to noise, once dismissed

as insignificant complaints, are helping investigators to improve conditions. One such study, which logged patient reactions to sound for each
day of their hospital stays, found that they showed responses – turning
their heads toward the sound or opening their eyes – an average of 35
times per day (Meriläinen, Kyngäs, and Ala-Kokko 2010). For those whose
recovery requires extended periods of sleep, sounds that catch attention
are counterproductive.

Monitoring of patients' physical responses provides one method for
determining both positive and negative reactions. Another is supplied by
questionnaires that gather information about emotional reactions. They
are designed to measure the patient's degree of disturbance or relaxation
as well as likes and dislikes. This line of research is showing that understanding the sources of noise helps patients feel less negative about it, and
that being asked how they feel raises their sense of security (Mackrill,
Cain, and Jennings 2013 and 2014). In fact, some small degree of noise is
reported as comforting, at least in the daytime. The sounds of staff going
about their work with soft-soled shoes and modulated voices help to reassure patients that they are being cared for. This reassurance is particularly
significant when patients' views are limited by bed curtains or room
dividers: when visual signals are restricted, sound becomes a crucial
source of information about the proximity of help as well as information
about hospital routines and customs (Mackrill, Cain, and Jennings 2013).

SOUNDSCAPES THAT AID HEALING

Recommendations for improving hospital soundscapes begin with taking
a noise inventory: identify all sources of unpleasant or startling sounds
and measure how loud they are, how disturbing, and how their sounds
combine. Include non-medical equipment like vending and ice machines,
carts, phones, and vacuum cleaners. Identify equipment that needs repair
or replacement and places where staff tend to carry on loud conversations.
Develop and implement plans to reduce noise where it is identified as
bothersome or distracting. Measure before-and-after results, and keep
monitoring periodically (Mazer 2002).

Reduction of noise can reduce stress and annoyance for patients, but
improvement does not need to stop there. Recordings of soothing natural soundscapes (forest, ocean, rain, rivers, birdsong) can serve as safe and
low-cost interventions to enhance relaxation without raising controversies about musical taste. Brief live performances of calming music can
humanize the hospital atmosphere as well as providing performance

opportunities for local musicians and advanced music students. In any case, keeping noise levels controlled will enhance the effectiveness of any health-promoting soundscapes. Reduction of annoyance equals promotion of healing.

Preliminary research into uses of music for post-surgical recovery also shows efficacy in promoting relaxation and in reduction of post-operative pain, both in isolated interventions (Shertzer and Keck 2001; Tse, Chan, and Benzie 2005; Cepeda et al. 2006; Cutshall et al. 2011) and in combination with therapeutic touch (Kshettry 2006). Although it is not yet known whether all personality types can benefit from music – because responses are complicated by variables like anxiety (Bradshaw et al. 2011) and the nature of the patient's medical condition – music has repeatedly been found effective in reducing anxiety during minor surgical procedures and subsequent recovery (Cadigan et al. 2001; Evans 2002; Mok and Wong 2003). Responses of cancer patients also suggest significant reduction of anxiety with music (Bradt et al. 2016) and improvement of body awareness and self-confidence with dance therapy (Bradt, Shim, and Goodill 2015).

Far more research is needed. So is standardization of reporting. While medical and nursing studies are meticulous about describing patient numbers and characteristics, tests administered, statistical data, and design parameters, they are often vague about the music. For example:

- What style of music and specific repertoire was used?
- Was it live or recorded?
- How fast was the piece played, and was it in a major or minor key or a non-Western mode?
- What instruments and/or voices were included, and what technology was involved (CD or digital file, device used for playing it, quality of speakers)?
- How did the listeners receive it (open-air listening/earphones)?

Keeping track of such details, and publishing them, may open new levels of knowledge about whether the content of music used in health-care interventions, and the technologies used to convey it, are significant to reported outcomes. Including someone knowledgeable about the structures and variables of music – a music therapist or musicologist – on the research team can help ensure that this aspect of the investigation is reported accurately, since health care personnel are unlikely to have all the requisite skills.

In the meantime, science is gradually catching up with what musicians have known for centuries: "music induces an arousal effect, predominantly related to the tempo. Slow or meditative music can induce a relaxing effect; relaxation is particularly evident during a pause. Music, especially in trained subjects, may first concentrate attention during faster rhythms, then induce relaxation during pauses or slower rhythms" (Bernardi, Porta, and Sleight 2006). Focusing research on aspects of rhythm – speed (tempo), patterning, complexity, repetition or variation – may well prove fruitful in answering persistent questions such as the following:

- Does music affect the parasympathetic nervous system purely through entrainment of respiration and heartbeat to tempo, or are other factors involved?
- To what extent is conditioning – familiarity with or preference for a particular type of music – significant in producing a beneficial effect?
- Are all responses culturally determined by familiarity, or are some universal?

Given the influences of ambient sound on patient comfort, hospital staff would benefit from greater knowledge of how to use sound to enhance healing responses, as well as how to decrease noise. If you are a health care practitioner conducting research in patient reactions to sound, consider styles of music drawn from the vast repertoire of "world music" now on the Internet. They can be used in formal experimental situations to test the validity of theories about familiarity and about the effects of rhythm or scales distinct from other factors, but also informally to find just the right soundtrack for an individual patient or client. Start by experimenting on yourself: sample West and South African percussion ensembles, Turkish, Iranian, or Indian classical instrumental music, and traditional Chinese compositions designed to enhance healing processes. Be aware, when working with unfamiliar styles of music, of whether the effect is calming or stimulating, intriguing or irritating, and of which category will be appropriate for the condition being treated. Does the patient need to be cheered up and motivated, calmed while awake, or lulled to sleep?[3]

QUIET TIME

Let us return to the notion of hospitals as quiet places, with calm surroundings that allow patients to recover from injury, surgery, or illness. Can a daily period of imposed quiet time – no visitors, media, or interruptions – benefit patients' mood, sleep, and rate of recovery? Nurses in critical care units began to ask that question early in the twenty-first century (Lower, Bonsack, and Guion 2003); in 2008, a team of Australian researchers spent five months attempting to answer it definitively. They observed almost 300 patients in the acute care orthopedic wards of two major hospitals: one implemented a daily quiet time of 1.5 hours in the afternoon, limiting both visitors and staff activities, while the other provided a control group without quiet time. The team found that the quiet time reduced ambient noise levels by more than 10 dBA (three to four times less sound intensity) and significantly increased both afternoon sleep and patient satisfaction (Gardner et al. 2009). The same beneficial results were observed in similar trials in US hospitals (Boehm and Morast 2009; Dennis et al. 2010), leading to implementation of afternoon quiet periods in many of them.[4]

Afternoon quiet times show promise for improving patient outcomes, but reduction of hospital noise at night is also essential. Recommendations from a protocol developed in a Taiwan hospital's surgical ICU give important information about effective techniques. These include closing room doors at 11 p.m., dimming room lights, decreasing the ring volume on telephones and bedside monitors, and training staff to limit conversation at night. The research team also recommended that repetitive night-time alarms be reduced by making sure that intravenous supply bags would not run out of fluid while patients slept, and by responding to alarms within one minute (S. Li et al. 2011). Similar conclusions were drawn by a US research team, which, in addition, recommended suspending public paging in favour of using individual pagers carried by the nurses, and offering to provide patients with earplugs (Murphy, Bernardo, and Dalton 2013). All of these suggestions require that hospital staff be trained in awareness of noise issues and exercise vigilance on behalf of patients (Konkani and Oakley 2012).

Creating a Culture of Listening

Noises are the sounds we have learned to ignore. Noise pollution today is being resisted by noise abatement. This is a negative approach. We must seek a way to make environmental acoustics a positive study program. Which sounds do we want to preserve, encourage, multiply? When we know this, the boring or destructive sounds will be conspicuous enough and we will know why we must eliminate them. Only a total appreciation of the acoustic environment can give us the resources for improving the orchestration of the world soundscape.

Schafer 1994, 4

Silence is an enabling condition that opens up the possibility of unpro-grammed, unplanned and unprogrammable happenings ... Silence has to remain available in the soundscape, in the landscape, and in the mindscape. Allowing openness to the unplannable, to the unprogrammed, is the core of the strength of silence. It is also the core of our sanity, not only individually, but collectively.

Franklin 2000, 2–4

There is growing recognition – indicated by the proliferation of organi-zations in North America and Europe dedicated to reducing community noise, as well as a spate of recent conferences and publications – that along with an excess of noise comes an insufficiency of quiet. If excessive or unrelenting noise acts as a physical and social pollutant, and with-drawing attention from it is a reasonable attempt at self-protection, what will it take for us to come out from behind the earplugs and earbuds? When, and under what conditions, might it be safe to listen to the world as openly and carefully as our ancestors did?

As author Garrett Keizer states in his reflections on the culture of noise, "We should always be wary of drawing pat moral analogies between noise and evil, quiet and good," pointing out that serial killers and mass murderers are usually described as quiet people by their neighbours (Keizer 2010, 16). The purpose here is not to advocate an inflexible standard of quiet either as a means of combating noise or as a moral principle; it is simply to explore a conceptual nexus that might prove useful, or even inspiring, in appropriate circumstances. If noise is overstimulating our ears and our culture, periods of quiet would allow for recovery.

The need for some degree of moderation in listening is not simply a matter of taste, but of science. The goal is awareness, and using it to give the ears, the nervous system, and the adrenal glands – the source of the stress hormone cortisol – an occasional break. A steady diet of anything that raises blood pressure or cortisol levels – be it music, sodium, or the constant urgencies of urban life – requires regulation, for reasons of physical health but also for the health of communities. Just as constant stress to the body can eventually produce illness in lab rats and in people, constant stress to attention and cognition can result in people who are angry, anxious, restless, and/or afraid. Large numbers of people in such states affect the health of communities and bodies politic.

Recognizing the problems associated with noise, then, is only a first step. What can be done to solve them, and to prevent noise pollution from affecting more parts of the world, more communities and environments? Moving beyond the practical remediation of noise annoyance and solutions arising from noise abatement technologies, what are the roles of sound and silence in the future of our society and its interactions with others? What are their practical and symbolic implications for facilitating learning, healing, communication, co-existence, and even peace?

Several paths radiate from these questions. The first revisits noise pollution and its abatement as a component of a larger process of awareness. The second explores the importance of music and silence in healing processes, both physical and social. The third – in the next chapter – carries the implications of music, silence, and conscious listening into the realms of ethics and social change. Each path branches and spirals: you can follow it beyond these pages as far as you choose. Like the mythical quests of ancient folktales, these journeys will lead you to unexpected destinations and surprising resonances.

ABATING NOISE, ABETTING AWARENESS

When noise is corrected with abatement technologies, we tend to presume that the problem is solved, that a reduction in decibel levels is all that can be expected. Noise that does not wake us up at night or interrupt common activities is tolerable, and complaints that persist beyond that point tend to be written off as obsessive. But what if there were a goal to reach beyond reduction of annoyance? Expanding the question beyond health care is an important exercise. What might comprise nourishing, or even healing, soundscapes in our everyday lives?

General answers are not difficult to find, although the specifics vary tremendously. For some, the ideal soundscape is filled with music and available at the touch of a button. For others a walk in a forest, on a beach, or in a large urban park will bring satisfaction and release from stress. For a third group, the quest for quiet is paramount. Each preference bears further examination, because while the hazards of noise have been amply discussed and documented here, analyzing the problem is not enough. Solutions are needed, and they can be found by examining cultural decisions as well as available technologies.

If being conscious of what we eat can contribute to physical health, can conscious listening contribute to emotional well-being? If we are constantly exposed in urban life and manufacturing work to mechanical noise, in clerical work to the white noise of ventilation fans and the hum of fluorescent lights, and in leisure time to the bombardment of advertising and to soundtracks filled with shouting and machine-gun fire, where are our sanctuaries?

In her influential essay "Silence and the Notion of the Commons," first presented as a speech at the founding conference of the World Forum for Acoustic Ecology in 1993, Canadian physicist and philosopher Ursula Franklin referred to silence as an enabling condition, a matrix from which thought and action can proceed. Tracing the decline of silence and acoustic neutrality in public spaces back to the rise of the recording industry and its alliance with advertising, she emphasized the contrast between intentional silence as a source of creative discovery and the act of silencing brought about by the constant presence of noise. Pointing out that the soundscape is a common domain increasingly colonized by sonic manipulation, she raised the question: "Who on earth has given anybody the right to manipulate the sound environment?" (Franklin 2006, 162).

Franklin's concern is well founded. The designation of a *commons*, originally attached to land held in common ownership by a small village in order to provide grazing rights for families that owned just a few cows and sheep, rather than entire herds, later came to indicate places where residents could gather freely to discuss and debate their concerns. By definition, a commons is shared space, not to be dominated by any individual or party. Its use is governed by tradition and mutual agreement, but because any imbalances in the established pattern can lead to misuse, it requires defining and defending by the community. When the members of the responsible community are unaware of the value of the commons and of potential threats to its neutrality, it becomes subject to degradation and appropriation.

In equating the public soundscape with a commons, Franklin draws attention to the role of noise, and particularly advertising, in colonizing public awareness and expectations. So pervasive is the presence of manipulative sound – speech and music designed to attract, distract, and persuade, embedded in media, entertainment, and Internet sites as well as overt advertising – that many individuals in North America are thoroughly conditioned to need it and to avoid silence as if it were a potent threat to their sanity. Are you old enough to remember a time when television news broadcasts did not have startling and stimulating musical soundtracks? When public buildings and their elevators did not broadcast Muzak or its offshoots? If you are not, find people who are, and ask them what they have noticed. If you are, consider these questions yourself:

- To what extent have we been conditioned to expect, and even to need, constant auditory stimulation?
- To what extent does the need for constant stimulation support a market for music and information as trivial commodities?
- What effects does such conditioning have on our abilities and desires to learn and communicate, to interact consciously and intentionally as a society?

These are not questions that can be definitively answered for entire populations, in part because the research involved would be astoundingly complex, but also because the answers would probably vary according to each individual's interpretation of the questions. They are raised for the purpose of bringing awareness to the subtleties of turning the twenty-first century's technological soundscape into a focus for sociological and even philosophical scrutiny, for connecting it to its own history and to whatever lost Edens of non-mechanized quiet are still directly accessible or at least within living memory. Such connections fall within the domain of the acoustic ecology movement, but they should not be kept there exclusively: as stated in previous chapters and in many other sources, cacophony in the commons is a practical concern for science, law, and public attention. Just as the unchecked growth of cities and their suburbs is now being reconsidered and classified as potentially harmful urban sprawl, are we now ready for a reassessment of urban, suburban, and technological soundscapes?

More questions easily follow:

- Does the growing replacement of mass broadcasting by personally chosen MP3 soundtracks to every waking moment represent an advance in our cultural relationship with sound or a decline?
- Is the soundtracking of everyday life an exaltation or an addiction?
- Does it inspire creativity, or mask it with distractions?

Again, the answers, given the variety of ways in which individual minds can learn and operate, will probably produce a broad spectrum of responses rather than a single trend. The questions are still worth asking. If they lead to recognition of distraction and addiction, we can take steps to modify our dependence. If the practices under scrutiny prove to be beneficial, whether for all or for a few, encouragement – with warnings about sound pressure levels – is warranted. At this time, however, the culture of perpetual soundtracking is a huge unregulated social experiment with potential effects on the cognitive and social development of millions of children and teens, as well as the future of their hearing acuity. Could its effects on emotional and social health be analogous to the effects of diet on physical health?

DIGESTING SOUND

A steady diet of junk food is now known to be harmful to physical, and possibly even mental, health. Could there be a comparable correspondence between auditory input and emotional health? Are there such things as healthy and unhealthy listening?

Since the components of a soundscape – indoor, outdoor, or electronic – are registered by the auditory system whether or not they are given attention, the ability to choose personal soundtracks may carry a certain responsibility. What we consciously choose to listen to can either exhaust the nervous system by producing further alarms that raise levels of cortisol, or replenish it by stimulating the production of endorphins, the hormones associated with pleasure and relaxation. The soundscapes under your control – on recordings or in places you choose to visit – may be important parts of your everyday "diet."

The characteristics of sound that cause annoyance and physical damage have been presented in previous chapters; they might begin a list of junk food parallels in the auditory realm.

Prescription for Auditory Health (Draft 1)

Avoid the following: prolonged exposure to excessively loud, abrupt, repetitive, harsh, and/or unpredictable sounds.

The list of toxins can be expanded at will: The vocal hysteria of commercials? Cars with disabled mufflers? Power mowers? Leaf blowers? Snowblowers? Any style of music you don't like? Beyond your personal list, are there categories of sound analogous to the excessive salt, sugar, and chemical flavours – attractive and addictive – that characterize nutritional junk that we can apply to the auditory system? It is likely that there are, although actual evidence that can be generalized is still scarce because there has not yet been enough study. Personal taste may well be significant, especially where music is concerned. It is almost a given that any style of music favoured by the young, or imposed on them by the commercial interests of the recording industry, will be regarded as toxic by their parents. That particular battle can be avoided by remembering that it is the volume at which music is played and its duration at excessive loudness – not its content – that will cause physical harm, and that susceptibility to emotional harm varies with the individual.

The difference between music designed to stimulate and music designed to calm is also important, since discords – clashes of sound that irritate within a musical phrase – are stimulants to the auditory system that command attention as surely as a bite of jalapeño. The alternating pulse of *startle-release*, *startle-release* has been one of the driving forces for music in the Western world for centuries, and each historical development in musical style and technology – ancient military trumpets, medieval polyphony, seventeenth-century opera, nineteenth-century symphonic orchestras, twentieth-century jazz and rock bands, current rap and hip-hop – has produced protests about auditory and moral perils.[1]

If certain styles of music are deemed hazardous to health, does the association go beyond decibel levels? The question of subtler forms of harm is still open because understanding of the psychological effects of music is still in its infancy. Assumptions about the effects of particular styles have not been verified by research, at least not yet. This is in large part because two different models are used to assess both the healing and potentially harmful properties of music. In medical research, the gold standard for benefit is stress reduction, with its attendant properties of lowered blood pressure and decreased production of stress hormones. The literature of music therapy, however, does not limit the definition of healing to decreasing stress. Its practitioners regard stimulation of activity, increases in communication skills and sociability, and acquisition of musical skills – as well as relaxation – as cherished goals. Music therapists believe that any musical style preferred by a client can be beneficial to them, and that assigning "therapeutic" music can be counterproductive if the client has no positive emotional connection with the music being presented, either because it is unfamiliar or because it has unpleasant associations for that individual. In their model, even heavy metal can potentially have therapeutic uses as a goad to emotional expression in those who identify with its culture, although in some listeners it can lead to feelings of isolation and anxiety (McFerran 2014).

Even in the absence of definitive research, some preliminary evidence can be gathered from a 2005 Canadian study that explored the chemistry of physiological stress, including increased heart rate, breathing rate, and production of cortisol, in players of video games. Medical researchers at the University of Montreal assigned volunteers to play either while listening to the game's soundtrack or in silence, and then measured cortisol by testing saliva samples. The group who was able to hear the music and sound effects of the game's soundtrack had consistently higher cortisol levels, indicating greater stress. The authors concluded that "the auditory input contributes significantly to the stress response found during video game playing" (Hébert et al. 2005, 2371).

Stress-inducing soundtracks undoubtedly make a positive contribution to the excitement of playing, and whether or not the game eventually produces a harmful level of stress will depend on the player and the frequency and duration of play. Adrenalin rushes can be pleasurable and even addictive, as any athlete or sports fan can attest. The intention here is not to demonize video games or any other form of entertainment that includes bombastic auditory components: *chacun à son goût*.[2] Stress is a stimulant, and whether you crave it or avoid it will depend on your culture, gender, age, and personality as well as your individual preferences. To the soundtracks of video games can easily be added those of horror movies, crime thrillers, cop shows, truck commercials, and *"HURRY!!! SALE ENDS TOMORROW!!!"* ads, each with its carefully measured doses of cortisol-raising auditory attractants and repellents. As with exposure to noise in the workplace, duration as well as loudness is significant. So is context.

COMMERCIAL SOUNDSCAPES:
THE QUESTION OF CONTEXT

Lunchtime on a university campus, at a restaurant-chain outlet with real cooked food: tasty, low on chemical additives, advertised as a healthful option. A place to gather with colleagues. On this particular day we bring our food to a table just as the speaker overhead starts to blast a grunge-rock classic, complete with singer screaming in apparent agony, his voice hoarse with strain, against a sheen of engineered discord over thudding bass. None of us finish our meals. The sound of a fellow human in deep distress, however ritualized and even familiar, is sufficiently disturbing to shut down appetite. Is this really a healthy place to eat? The staff are constrained by a legal contract that binds the entire chain of restaurants to an Internet music service, and believe that they have no choice about the content of the soundscape that results. The variety of preferences in music and atmosphere brought to the place by diners at any given time, many of whom wear personal music player (PMP) earphones to mask the corporate soundtrack, is accommodated onsite only by one option: whatever was popular with American teens a decade or two ago.

What is the intended audience for the subscription soundtrack services chosen by corporate executives? Are the soundscapes of club dance floors – including those relegated to history – appropriate in places where nobody is dancing? Is it really a good idea to convert songs that convey shock and stress into background tracks for malls and restaurants? If such tracks are actually intended to stimulate diners and shoppers to speed up and move out quickly, perhaps they should include machine-gun fire and explosions, to which we are so accustomed by movies and TV that their real-time presence on news reports may fail to shock when it should.

Advising on matters of personal taste in leisure time may be unrealistic as well as intrusive, but there are two conditions in which attention to soundscapes is especially important: eating and sleeping. Auditory "junk food" may be particularly harmful in combination with the ingestion of actual food. Pausing to ingest a meal while on the run from a predator was not something our hominid ancestors would ever have done, because they would never have survived long enough to become anyone's ancestors. The human digestive system, like those of other species, does not function well under conditions of stress. If our soundscapes, ambient or chosen, are keeping us in a chronic state of fight-or-flight, are appetite and absorption of nutrients compromised?[3]

Quality of sleep may also be a casualty of constant stimulation. If a thriller movie soundtrack turned up loud or a boom-shriek song on your PMP precedes your bedtime, try an experiment: does listening to something more mellow make sleep easier and more refreshing? Do you wake less anxious, less irritable? Take at least a month to track results, and share them. Challenge your friends on social media to give it a try. Personal experiments, while not scientifically conclusive, are worthwhile for their benefits to the individual, but also because they may in time stimulate controlled research that is greatly needed.

Prescription for Auditory Health (Draft 2)

- Be aware of what you hear: practise conscious listening.
- Avoid prolonged and frequent exposure to excessively loud, abrupt, repetitive, discordant, and/or unpredictable sounds, especially when eating and before sleep.
- Alternate periods of intense listening activity with periods of quiet.

MUSIC AS A HEALING ACTIVITY

At this point, enough attention has been given to hazards. *Where is the nourishment?* If noise is a pollutant and a potential toxin, what are its antidotes? Where are the auditory equivalents of vitamins, protein, fibre, and flavonoids? Again, much depends on individual conditioning and taste, but general suggestions for acoustic nutrition can be drawn from research in the fields of music therapy and neuroscience as well as acoustic ecology. Music, natural sounds, and silence all contribute to finding answers.

Health care in the twenty-first century is increasingly based in scientific evidence: standard interventions are studied for their effectiveness and compared in order to determine "best practices." Because some of the most effective practices are derived from ancient traditions, the field is now moving toward a holistic model that looks to a variety of traditional cultures for effective techniques to enhance relaxation – massage, acupuncture, yoga – and incorporates nutrition, exercise, and emotional well-being into the medical framework of biochemistry and medication. Decades of research now validate the role of arts-based therapies, with practices based on a combination of venerable traditions and modern neuroscience, in healing trauma, dealing with pain, and enriching the lives of people with debilitating chronic illness or disability. Therapeutic systems that incorporate drawing and painting, storytelling, dance, and drama, as well as music, are now recognized professions with practitioners working in hospitals, hospices, care homes, schools, and communities all over the world.

The association of music with healing and the maintenance of health is a long-standing one, probably dating back to prehistoric shamanic healing practices and now gaining scientific approval as evidence of efficacy is gathered, tested, and published. The chanting Indigenous shaman, the biblical story of David playing the harp to comfort Saul in his state of agitation, the ancient Greek and Hindu physicians whose understanding of music and sound as therapy was transmitted and expanded by their scientific heirs in the medieval Islamic world – are all now being validated by computerized scans of the brain in high-tech research facilities. Music and environmental sounds enhance healing by activating the body's natural responses to stress: emotional release, physical activity, and relaxation.

Enhancement of healing processes through music can be accomplished in several ways. Because activation of the auditory system also stimulates emotion and memory, music encourages release of suppressed emotions and consequent release of physical tension. Because respiration, heart rate, gait, and gesture entrain to heard rhythmic patterns by means of auditory-tactile connections in the brain, strongly rhythmic music can motivate people with injuries or disabilities to participate in physiotherapy and therapeutic dance programs. For the same reason, soothing music at a slow pace can activate the parasympathetic nervous system that governs relaxation, sleep, and spontaneous healing. Mothers in every culture who sing or croon to their infants demonstrate this connection, and researchers are investigating its potential for speeding recovery from surgeries (Tse, Chan, and Benzie 2005; Bernardi, Porta, and Sleight 2006). So powerful are the responses that deep memories and even lost abilities can be triggered by music. It has been known for decades by dementia caregivers, and more recently verified by research, that Alzheimer's patients who have lost the ability to speak are often able to sing along with songs familiar from their youth (Dassa and Amir 2014; Palisson et al. 2015; Jacobsen et al. 2015).[4] Research on Parkinson's disease concludes that the uncertain gait of patients can be strengthened by practice with walking to rhythmic music, by dance, or even by unvocalized mental singing while walking (Rochester et al. 2005; Hackney et al. 2007; Hackney and Earhart

2010; Satoh and Kuzuhara 2008; Arias and Cudeiro 2010). As neuroscientists investigate the effects of music on the brain, therapeutic protocols employing music or derived from elements of it – for example, rhythmic or melodic speech – are coming into use as recognized rehabilitative procedures for brain injuries, aphasia, and stroke (S. Kim 2010; Schlaug et al. 2010; Amengual et al. 2013).

An overview of medical literature validating the uses of music for healing processes shows both the promise and the limitations of current research. Most recently published studies end with statements about the need for additional research on larger numbers of patients, and for controlled studies in which the outcomes for patients given the protocol are compared with those of a control group that does not receive it. Far more study is needed in order to produce thoroughly convincing evidence for the medical community, which does not accept anecdotal or qualitative research. The medical standard of quantitative research – which requires large numbers of research subjects, statistical significance, and results that are consistent over many repetitions of the experiment – may be achieved over time if sufficient attention and resources are directed to it. In the meantime, several decades of qualitative studies from the literature of music therapy give preliminary evidence of efficacy for people with conditions such as autism and Alzheimer's disease, as well as a host of physical disabilities.

The methods used by professional music therapists include listening to live or recorded music and discussing the feelings and memories that result, moving to music in order to improve coordination and confidence, playing simple instruments that enable non-verbal communication, and writing personal songs to express emotions. One of their techniques applies the "iso-mood" principle, which recognizes the need to match the client's state of mind before attempting to alter it. Thus, clients who are grieving may be encouraged to listen to slow-paced music in a minor key, deemed to be in sympathy with what they are experiencing, until it becomes possible to respond to more upbeat listening. Those who are agitated and restless may be given a listening prescription that is fast-paced, loud, and stimulating at first, then gradually introduced to slower and more calming musical fare. Music listening can be used prescriptively, based on the therapist's knowledge of examples associated with specific moods, kept within the client's range of known preferences, or chosen during a session in dialogue with the client.

Even more than listening, participation in music can promote emotional and social health. Training in basic musical skills – recognition of auditory patterns in rhythm and harmony, pitch discrimination and patterning, coordination of breath and of manual and facial muscles, development of auditory imagination and "inner ear" – actually changes rates of activity and even the shape and density of neural networks in the brain, enabling improvements in the cognitive and executive functions that govern learning, concentration, and decision making (Trainor, Shahin, and Roberts 2003 and 2009; Levitin 2006; Pallesen et al. 2010). The research literature of music education is full of evidence that learning music enhances comprehension of mathematical concepts, and drumming is used as a curriculum option in some US schools – as well as in traditional African cultures – for teaching mathematical ratios and fractions (Huylebrouck 1996; Stevens, Sharp, and Nelson 2001). New research in neuroplasticity, the ability of the brain to change its own neural pathways in response to changing conditions, suggests that benefits gained in these areas are fully available to seniors and other adults as well as children.

When the social benefits of musical training are also taken into account, the ability to play something other than a PMP device seems to be a good idea. While professional-level skills and talent are rare and require extreme dedication, the proliferation of televised "Idol" shows and YouTube performers attests to the lure of amateur musicianship, restoring it to a prominence largely lost when broadcasting was invented. Might the healing of the modern soundscape include greater encouragement for garage and kitchen bands, backyard drum circles, and community-centre basement choirs, appreciated less for their musical perfection than for the personal and communal thrill they bring to participants?

Try this experiment at your next gathering of family and/or friends at home. Distribute wooden and metal spoons, bowls, and cups of various sizes, bottles containing different amounts of water, sheets of foil and paper. Ask participants to put their keys and combs on the table as well. Anything within reach, as well as hands, feet, and voices, is now considered a musical instrument. *Tap, clap, rattle, crumple, shake.* Alternate loud and soft, dense and sparse, group and soloists: invent dinner table jazz. Let the family dog join in. Observe the (very few) rules of the game: don't pound hard enough to break anything or annoy anyone, and stop what you're doing occasionally to listen to others. Then, watch the faces around the table. Is there anyone without a smile?

Music stimulates and soothes, as it has done in every culture throughout history. Before there was history, even before there was human cognition, there were the sounds of the natural world that gradually shaped our species' auditory perceptions. The next steps in the search for acoustic health lead outdoors.

THE ROLE OF NATURAL SOUNDSCAPES

The ambient soundscapes of industry and technology dominate the experience of billions of people worldwide. The industrial soundscape is loud and endangers hearing; information technology is comparatively quiet but fragments concentration. Both produce noise annoyance by their ambient sounds, but also by removal of attention from natural soundscapes, the accustomed spectrum of sounds in which our species evolved and developed its cultures and from which it is still capable of drawing comfort and reassurance. Such acoustic surroundings gave our ancestors both information and a matrix for language and music, foundations of culture. Losing connection with them neglects our roots and leaves us confused about our common heritage as a species.

The usual notion of an ideal soundscape – always under control, filled with pleasing and informative sounds, free of harsh noises and auditory clutter – is a social construct of the individual human listener and the human community rather than a reality in the natural world (Redström 2007). Just as a forest is more complex and chaotic than a park, a natural soundscape is diverse and unmanageable. It contains the voices of crows as well as songbirds; it encompasses gentle breezes and thunder. To our prehistoric ancestors, the diversity of voices carried critical information about weather conditions, potential dangers, and the presence and ripeness of food sources as announced by birds and demonstrated by the calls and movements of animals. It was a rich and complex text that required close attention. To the urban-dwelling weekend hiker or camper today, largely free from the need to listen for signals indicating food and peril, the same wild soundscape can be relaxing music, an antidote to the accustomed ambience of road traffic, aircraft, and lawnmowers.

Access to wild soundscapes today is beneficial, but not necessarily easy. It requires expenditures of time and money beyond the reach of many people, and the numbers of visitors now descending on some national parks and wilderness preserves threaten to destroy their designation as wild areas as well as their local flora. Traffic competes with birdsong as the dominant element in many park soundscapes, while chatter and litter abound on hiking trails. What measures are needed to preserve wild sounds and silences?

As is so often the case, awareness is the first step. Among the practices of the acoustic ecology movement is the preservation and documentation of the sounds of places outside of human habitation. For the past four decades, Dr Bernie Krause, bioacoustician and director of the Wild Sanctuary Project (http://www.wildsanctuary.com), has engineered and archived thousands of field recordings of ambient natural sounds. He estimates that half of the sites he recorded early in his career have since been destroyed or degraded by human interventions. As wild places are altered by tourism, resource extraction, agriculture, and urban expansion, their sounds as well as their native species – indigenous populations of humans and their languages included – will become matters of archival record rather than components of living environments. Recording and film provide documentation, but preservation requires public will and political support for legislation that will establish and protect extensive wild park systems along with private land trusts.

Still, the benefits of natural sounds need not be limited to wilderness encounters or exotic field recordings. Garden design is easily adapted to incorporate water and attract birds. Historical and current examples abound. R. Murray Schafer writes with delight about the "polynoise of water," the "deviousness of wind," and their incorporation into gardens designed to turn natural noise into outdoor music (Schafer 1992, 237–52). He recommends that architectural designers and city planners develop more imagination, envisioning and auralizing staggered urban roofs with drainpipes engineered to turn rain or melting snow runoff into musical expression, as well as clusters of wind chimes placed in urban parks to diffuse the sounds of traffic. The growing field of sound art – sculpture that includes auditory components, futuristic climbing frames that double as musical instruments, outdoor installations of electronic sensors that trigger auditory surprises or sample and amplify the sounds of observers – also provides options for transforming parks and playgrounds into places of sonic enchantment. Fountains, incorporated into the design of public parks and private estates since antiquity, have always been valued for their calming auditory effects as well as their visual beauty.

Community planning and architectural design can benefit from the imagination as well as the advice of the acoustic ecology movement. Schafer offers dozens of sound awareness exercises in his writings, and many are apt for translation into action. Along with clever and often whimsical suggestions for expanding auditory awareness and imagination, he urges individuals to recognize sounds that will enhance the atmosphere of the home (wind chimes, door knockers) and those that should be eliminated (squeaky doors, rattling fans) (Schafer 1992, 125–7).

Public buildings also come under scrutiny: for Schafer, the modern enclosed building is markedly inferior in auditory atmosphere to its historical predecessor. He has found fault with buildings that privilege the artificiality of sealed windows and white noise over access to natural soundscapes and the human activity of footsteps and speech (Schafer 1992, 222). But technology can be transformed from enemy to ally, bringing fountains indoors to the lobbies of corporate office buildings and the corridors of hospitals. Where nourishing natural soundscapes are inaccessible, recordings can at least provide an echo of their efficacy: flowing water, ocean waves, rainfall, or songbirds broadcast indoors can serve to ease the tensions of a workday when escaping to actual beach or forest is not an option. Even the neighbourhood you live in right now, or the one surrounding your workplace or school, can be explored and recorded for information and inspiration, whether its dominant sonic ambience is wind in trees or buses in traffic. All you need to do is leave the building and take a soundwalk.

Soundwalking

The practice of soundwalking, first developed by Schafer and his composition students in the 1970s, is easily followed alone or in small groups. Hildegard Westerkamp, who has refined the practice extensively, is clear that it is a crucial activity for environmental awareness: "To be in the present as a listener is a revolutionary act. We absolutely need it, to be grounded in that way ... Listening will help us reconnect to the environment. If we can understand what listening can do to reconnect us to our environment, we can understand what's happening to our environment ... we would be enriched, hugely."[5]

Westerkamp's colleague Andra McCartney, a sound artist, posted the following example of such awareness by a participant on her "Soundwalking Interactions" website:

I went to visit the Pier 21 museum in Halifax, which focuses on immigration to Canada, and has a special exhibit this year for Canada 150. One activity it asks of museum-goers is to write about their first day in Canada, which I did. It was August 1968. I remember clearly walking by a vacant lot by the highway between our high-rise apartment and the restaurant where we had dinner that day. The lot was filled with chirping crickets, and that sound along with the pervasive sound of a multi-lane highway, seemed very Canadian to me as an immigrant from the UK. Both highway and insect sounds were unusual to my ears, although it feels strange to write that now, having become so accustomed to them in the intervening half century. In both cases, it was a buzzing, layered, sound caused by many individuals. And it turns out that my perception of the cricket sounds as unusual, and more Canadian than English, has an ecological reason. At that time in the UK, crickets were almost extinct because of industrial farming practices. So I would not have heard a field filled with crickets like that before.[6]

Soundwalks can have an educational focus and are easily incorporated into school activities or sound awareness workshops, but they can also serve as recreational breaks in a workday or just an excuse to get outside. The most basic soundwalk is simply this: a walk in silence, the object of which is to hear as many ambient sounds as you can. Lists can be written or not, as you choose. No talking by participants is allowed, so leaders of group walks use hand signals to indicate changes in route or stops and starts. More advanced varieties, designed for participants with backgrounds in music, search for rhythmic and melodic elements in ambient sounds, often with the goal of creating soundscape compositions that integrate recorded natural and/or mechanical sounds with music composed to contextualize them.[7]

When I lead soundwalks for students in my acoustic communications class at the University of Calgary, we tour the campus with stops at places that contain sounds that might otherwise go unnoticed: the rattling buzz of vending machines in a hallway, a back-of-building air vent's roar, practice room corridors in the Music Department (*piano + flute + singer + trombone, each in a different key*), quiet concentration in a chemistry lab (*clink, shuffle*), intense activity on an indoor basketball court (*thud, thud, grunt, THUD*). Afterward we discuss findings and reflect on what was noticed. Students enjoy the process; so do children and the occasional harried colleagues who can be coaxed into participating.

Soundwalks can be agenda-free voyages of discovery or focused on a theme like wind, water, birds, or machinery. The process can be about finding sounds that are attractive or about locating sources of sonic pollution. Practice will expand your auditory powers of observation as well as your knowledge of local conditions. Try it in surroundings that are not familiar. Try it on your next vacation trip, hike, or visit to an environment quieter or noisier than the one to which you are accustomed. The process will expand what you know about listening and give you a more nuanced appreciation of your surroundings. It will also bring to mind the value of keeping silent as a way to enhance your powers of observation.

THE USES OF SILENCE

True silence – a state not found in nature, despite long-standing traditions associating uninhabited wilderness with it – is an absence of sound, and thus an absence of motion. A study of absences requires first the recognition of what is missing, then the determination of *why* it is missing. Is the absence a matter of intention or omission? Is the condition of silence desirable, uncomfortable, repressive, regenerative, reverent? What does it tell the observer, the listener, about the soundscape and the conditions that produce it?

Noise orients us in space and time. Conditions approaching silence, free of the auditory clutter produced by media and mechanical infrastructure, can produce reactions of anxiety as well as calm. The absence of communicated information – *who is nearby, what is going on around me, what is my proximity to danger or protection or companionship?* – can be disorienting or liberating. Silence can free the mind to find contentment or drive it to panic, to relish solitude as a pathway to peace or to flee it as a peril to sanity. If the noise of a crowd signals communal activity and high emotion, silence denotes the absence of community and consequent irrelevance of emotion and identity. Perfect silence is perfect stillness, the absence of vibration and consequently of sound. The ultimate silence, and the focus of existential fear, is death, which permits neither motion nor communication. Is this fear a component of the aversion to quiet that many people in Western culture feel? Is it grounded in the fear of losing significance in a culture that glorifies activity and speed, or is there a deeper root to the fear of silence?

Writers Sara Maitland and George Michelsen Foy, on personal quests for silence, have recounted experiences of anxiety and fear in the presence of it. Maitland (2008, 80–115) describes being snowed in during a Northern English winter. Despite consciously welcoming the quiet and isolation, she found herself succumbing to depression, hallucinations, and attacks of panic that resembled those described in the journals of polar explorers, isolated survivors of shipwrecks, and volunteers in sensory-deprivation experiments. Foy (2010, 58–60) recalls a solitary trek through the Algerian desert and the "alien" grandeur of its silence, which induced fear. Throughout history, solitary confinement, banishment, and shunning, all conditions that limit or eliminate human communication, have been used as extreme forms of punishment. They evoke a basic threat to existence in a species whose survival in harsh climates and conditions depended, for hundreds of thousands of years, on the skills, assistance, companionship, reproductive availability, and sheer physical warmth and strength of their companions.

Paradoxically, the fear of silence can also be activated by excesses of activity and noise. Urban soundscapes surround us with the fleeting auditory traces of strangers' lives – traffic, feet on concrete or apartment building corridors, fragments of conversation overheard – without direct social contact. The feeling of being alone in a crowd can provoke anxiety. The early ancestors lurking in the human subconscious hear the crowd and search it for kin, for tribal solidarity. If none is found, they anticipate the state of the outcast abandoned in silence: unprotected and vulnerable to harsh weather, enemies, and predators. Sound artist Wreford Miller, examining attitudes toward silence within the parameters of the acoustic ecology movement, calls attention to the common uses of broadcast media as a defence against it, pointing out that "the disembodied voices of electroacoustic media accompany most of us throughout our lives, allowing us to gloss over the paradox of increasing isolation and crowding" by "reliev[ing] us of our own silences" (Miller 1993, 83). Even in the midst of noise, silence becomes a metaphor for exile, for impending danger, and for absences of agency, communication, and meaning.

There is substantial literature on silence as a metaphor in three social contexts: political, educational, and spiritual. In the first of these, it is overwhelmingly regarded as a negative quality, an indication of deliberate repression, isolation, or inattentive neglect. Cultures and individuals are silenced when their wisdom is ignored, their voices unheeded; the act of silencing is a form of oppression that has been practised historically on women in many cultures, including our own (Luckyj 2002), and more recently – or currently – on children and the aged, and on racial, religious, linguistic, sexual, and other cultural minorities. The equation of silencing with oppression is carried over into the majority of commentaries on silence in education, where it comes to represent the exclusion of cultural traditions and individual experiences from the canon of knowledge that is considered worthy of preservation. Thus, the activity of silencing becomes a social injustice – an act of aggression – when enforced against an individual or group that is other than oneself or one's culture. The plethora of titles that include the word "silence" in databases of literature on education is testimony both to the prevalence of oppression and to the variety of ways in which it is perceived. But silencing is not always an imposed condition. Self-silencing in political and educational terms implies a condition in which silence is maintained as a protest or as protection against revealing information. It is a cocoon from which identity can emerge or a veil that shadows it, shielding the individual or cultural group from hostile reactions or preserving secrecy until conditions are favourable for revealing a truth.

Sara Maitland (2008, 223–57), tracing the history of appreciation for silence, cites its uses as a foundation for spirituality in various cultures and as a component of creative solitude in the Romantic movement of nineteenth-century Europe. Wreford Miller (1993, 93–5) associates three activities with it: resistance, regeneration, and religious inquiry. In the literature of spirituality, silence can be a vehicle for inspiration and enlightenment, enabling the cessation of mental chatter, anxiety, and desire. Silence and its approximations facilitate interiority: contemplation, focused thinking, brainstorming. They reduce or eliminate distraction and allow the mind to focus on a single task.

The state of silence can be dependent in part upon individual expectations. For some, the craving for it is absolute, and mere quiet will not suffice. The Public Transit Authority in the state of New Jersey, following precedents set in parts of Europe, has introduced "quiet cars" on its commuter trains for the benefit of riders who seek to escape from other people's cellphones, conversations, and music, but success has been questionable. An allowance for quiet conversation on the designated train cars irks some travellers enough to motivate complaints to conductors or even requests to shut down the automated announcements that identify stations (Mallozzi 2011). It is likely that such punctilious passengers are involved in a favourite pastime of solitary commuters: reading, which involves the translation of narrative or dialogue into an audio-visual process and is therefore especially apt for disruption by actual speech. It is for this reason that the traditional practice of enforcing quiet in libraries developed, and that the interior of a train can be regarded by passengers as "silent" as long as nobody is talking.

Silence carries signals and can be an active component of communication: the reflective pause for constructing a response in speech or the editing of emotion or information, the signal given that the topic is distasteful or the statement confusing. In music, it is a necessary platform for the development of rhetorical structures. Musicians are taught that music consists of rests – the pauses between notes or phrases – as well as the notes and phrases themselves. Rests define phrases, and must be integrated into the player's understanding of the piece. They are often the places where one player is silent so that another player's phrase can be heard, enabling dialogues in the abstract languages of melody and rhythm, providing moments for absorption and reflection, clarifying the importance of each voice in the community of the ensemble. For the player, rests are a place to breathe or relax muscles and to gather energy for the next phrase. For the listener, both pauses and seamless phrases are clues to the structure and meaning of the music.

Silence also provides a platform for the refinement of communication. Like music, spoken dialogue requires it. So does the process of arriving at consensus.

Silence as a Matrix for Communication

On a Sunday morning in October, a group gathers in a sunlit modern room, its floor-to-ceiling windows overlooking bright sky and trees loud with colour, red-gold-bronze against startling blue. The people vary in age and origin and probably in habits of speech, although that is unknown at first because they enter the room so quietly. They sit in a state of introverted reflection for some time, then – one by one, man or woman, youth or elder – a few rise unsummoned and speak about what they have noticed, experienced, felt, or been moved to question since the previous Sunday. Their speeches – brief, gentle, and full of warmth – rise out of companionable silence.

As a university student in Toronto in 1980, I had the opportunity to attend services with the Religious Society of Friends – commonly known as Quakers – at their Meeting House, and, in the process, received an experiential education in the value of silence. Since my focus of study at the time was musicology, the history and cultural functions of music, there was a challenge involved: how to open my well-trained and focused auditory awareness to an absence of purposeful sound? Could silence, usually perceived as an absence, become an active presence?

For the Religious Society of Friends, silence is the foundation of meaningful communication. The movement, which dates back to seventeenth-century England, represents a protest against both the rigidity of Puritan beliefs and the ceremony of the state-sponsored Anglican Church. Founded on the idea that internal focus will reveal the "Light of God," the movement practised communal silence in worship as a way of stilling the mind in order to get past the distraction of random thought and arrive at revealed truth. Worship services have diversified since the seventeenth century, but most involve extensive use of silence even today. Some congregations choose to keep silent for the entire service, usually an hour long, while others use silence as a generative matrix for meaningful commentary from members of the congregation on events of the week, experiences and memories, or internal spiritual and emotional states. There is often no minister or facilitator to direct the comments. They arise spontaneously from the supporting platform of silence as the profound melodic certainties of jazz soloists arise spontaneously from the undemanding sonic support of their fellow players. Each comment elicits thought from others, so that a theme may develop for the service without any planning, or it may drift comfortably among discrete reflections.

My connection with the group led to an invitation to observe a traditional "Quaker business meeting," so designated by the congregation. The issue under discussion was whether space in the Meeting House should be rented out to appropriate community groups – those with a focus on social justice or community improvement, in keeping with the Friends' mandate – when it was not being used for their own activities. While I do not recall the outcome of the decision, the process of arriving at it was thoroughly memorable and has served since as an example in university courses of alternative structures for communication and decision making. The community does not vote in such cases, nor does it have elected or appointed authorities who make the decision for the group. The model for the Quaker business meeting is neither democracy nor autocracy, but the active use of silence as a means for arriving at genuine consensus.

The worship service is a model for the business of the congregation. When an issue requiring a decision arises, a business meeting is announced to the congregation after a worship service. All members who have an interest in the issue are invited to attend. When the meeting is convened, initial comments are given supporting each side of the issue. This first round of comments is followed by a period of silent meditation on what has been said. Another round of comments follows, then another silence. This alternation continues – even if the time required exceeds the time allotted for the initial meeting and requires a second meeting or even a third – until all members involved in the process come to an agreement. And they do. There is no disgruntled minority, outvoted and resentful of it. There is no lone argumentative individualist who keeps objecting against the rest of the group just for the sake of doing so. Because harmony is valued above efficiency, there is time to arrive at genuine consensus, which reinforces the intentions of the community with purity and clarity. Consensus comes about *because* of the use of silence.

Accustomed to taking space between thinking and speaking, to using silence as a way to integrate matters of the head, the heart, and the spirit, members of the Society of Friends have been able to develop a culture in which speech is neither frivolous, competitive, nor combative. Speech is deliberate, and comes from a stance of deep conviction. This culture enabled the sect to survive persecution in its early history and to gain respect, both social and legal, for its deeply rooted pacifist beliefs.

PEACE. AND QUIET?

Peace is incomprehensible without some degree of quiet. Speech and music can lead us there, but the relation of peace to quiet is central. The state of peace – whether political, emotional, or mystical – is free of urgency. It is first enabled by the slowing or cessation of activity, whether physical or mental. Wars end and peace is declared when the fighting stops and the guns are silent. Mediation reduces the urgency of emotional, legal, and political conflicts by carefully defining shared goals and reminding antagonists that finding a point of agreement will bring the conflict to a shared stasis, a point from which beneficial actions and attitudes can proceed and cultural healing can begin. Beyond the need to voice anxieties and objections, beyond the shouting and accusation, is a state of reflective silence.

Professional mediators are well aware of the potential for using silent pauses in confrontational negotiations. In an article outlining the positive and negative uses of silence in mediation, Canadian occupational safety consultant Renée Gendron points out that "silence can be an opportunity for mediators to allow the participants time to reflect. It can provide opportunities during a specific session or in the overall process, for parties to consider options presented and discussed" (Gendron 2011, 4). Like the Quaker practices employed in both spiritual and task-oriented gatherings, periods of silence defuse confrontational dialogue. Pauses in the discussion allow participants whose cultures have differing levels of comfort with silence to reflect on their views and to question other participants in order to arrive at an understanding of the communication patterns inherent in the situation: this in turn will ease tensions and enable compromise.

Silence also has value for easing tensions in everyday life. Internal quiet is the direct goal of much meditation, which facilitates mental and emotional peace by slowing the reactive chatter of the active mind and directing its focus toward a consistent goal or object, acting on the nervous system to enable states of relaxation and inspiration that proceed from stillness. Noise is antithetical to these states; quiet enables us to access them.

The path to peace, whether personal or cultural, is paved with listening. If a global state of political peace – professed goal of international organizations, elected leaders, schoolchildren, and beauty pageant contestants alike – is ever to be achieved in an increasingly crowded and conflicted world, it will begin quietly. The shaping of such a future is in our hands – or, more realistically, in our ears, minds, and voices – through knowing when to speak and how to listen.

Tuning a Future

The perception of sound can most easily be studied from the artifacts and contexts of music ... The history of music, with its cultural variants, provides a repertoire from which deductions can be made about what different eras were expected to hear, and equally what they missed, for in the study of any sound-scape what is missed is just as important as what is listened to, perhaps more so.

Schafer 1993, 118

Improvisation for us is more than an artistic conceit, more than the sponta-neous creation of notes by musicians or words and gestures by actors. In its most fully realized forms, improvisation is the creation and development of new, unexpected, and productive cocreative relations among people. It teaches us to make "a way" out of "no way" by cultivating the capacity to discern hidden elements of possibility, hope, and promise in even the most discour-aging circumstances.

Fischlin, Heble, and Lipsitz 2013, xii

What changes, what concepts and processes, will help us reach a culture of listening? Such questions are no longer merely rhetorical. Recognition of the importance of soundscapes to the health of populations and com-munities is already producing tested recommendations for reducing spe-cific harms and enhancing symbolic harmonies.

Imagine, for a moment, a future soundscape free of common annoy-ances, with quiet traffic on streets, quiet appliances in homes, and better soundproofing everywhere in the built environment. Consumer demands for these innovations have brought prices down to reasonable levels, so that extreme wealth is no longer a requisite for protection from noise-induced stress. No, the everyday ambience is not silent. There are faint machine hums and clacks, there are whooshes from passing vehicles, there

are sounds of weather and nature and conversations. There are still barking dogs and excited children and thunder. There are multiple options for
entertainment at whatever level of loudness you prefer because your neighbours won't hear your choices unless they choose to do so – both their
soundproofing and your noise courtesy have taken turns for the better.
Public transport options are quieter and more efficient, traffic management has improved, and new discoveries in sound-cancelling materials and
devices have even assured residents living near factories, highways, airports,
nightclubs, and sports arenas that their everyday activities and sleep will
rarely be interrupted. A hypothetical future, of course – but steps are
already being taken to arrive there.

RE-SOUNDING THE CITY

Current predictions about the future of cities almost invariably fall into
two categories. There is the mass-market movie theme of dystopian crime,
collapse, and chaos, with one heroic white American family or handsome
honest cop saving a corner of the world from itself. There is also the exul-
tant tech magazine article, lauding the efforts of the industry's heroes to
bring about the perfection of smart vehicles, energy systems, lighting systems, surveillance systems, and communicative appliances – the Internet
of Things that will save us from ever having to remember to buy milk or
refill the ice cube tray, "if we can get the software right."[1] Less public attention has thus far been given to a third option. It does not require heroes
because it relies on all of us to re-imagine the city as a commons: an
"open-source city" guided into improvements by communication
between experts and residents. In this paradigm, local needs shape the
progress of urban renewal.

Among the ideas now circulating in publications and conferences for
architects, city planners, and experts on public health is the concept of
human-centred urban design as crucial to the success or failure of urban
environments. An ideal city has walkable neighbourhoods, efficient and
affordable public transportation, thriving local businesses, inclusive cultural centres and events, accessible and affordable health care, well-man-
aged services, public safety, and access to natural areas that provide space
and quiet. Every city has, in reality, areas that are neither safe nor spacious
nor quiet. Is a less loud future feasible? Obstacles abound: existing infrastructure is difficult and expensive to change, new technologies need time
for sufficient testing of efficacy and safety, there are many competing priorities – and where is all that money supposed to come from, anyway?

It is fairly certain that future urban soundscapes will contain far fewer combustion engines, whether because of further development of electric and hydrogen fuel-cell technologies, or because of greater attention to the dangers of climate change associated with fossil fuels. Electric cars are not entirely noise-free: traffic noise includes the interaction of tire surfaces with road surfaces, wind noise, and horns. Their motors are quiet enough, though, that new regulations in Europe are calling for some form of engine noise to be added because pedestrians and cyclists are 40 per cent more likely to be hit by electric and hybrid cars than by conventional ones.[2] This statistic exemplifies the complexity of finding solutions, but complexity is not the same as impossibility. Many remedies are already available to those who can afford them; more will come into general use with time and attention.

A brief survey of current actions will provide examples. The European Union is measuring transportation noise in cities and formulating plans to reduce it: the first step is a program of strategic noise mapping carried out in districts with heavy traffic. It goes beyond tracking decibel levels to cover reported impacts, breaches of regulations, and numbers of people affected. Mapping processes can now be implemented and disseminated on mobile phones. As of 2017, studies analyzing road surfaces, tire construction, and traffic patterns are contributing to the analysis of practical plans for reduction of traffic noise: their implementation is beginning with new neighbourhoods and with road repairs.

Even low-income districts are now receiving attention from planners determined to reduce the impacts of noise that endanger the health of their residents.[3] Among their methods are easing traffic congestion, improving the quality of housing, and re-envisioning neighbourhood infrastructure. The most immediately feasible and necessary changes involve food markets stocked with fresh produce to eliminate "food deserts," and increasing the number of neighbourhood parks. Ideally, parks will be substantial areas with water features, walking paths, community gardens, and plenty of trees, but even small green spaces on the corners of city blocks can make a difference.[4] As well as giving residents a break from unpleasant conditions, parks and green spaces reduce the impacts of traffic noise. In fact, perception of noise annoyance can be reduced even by the sight of trees (Gidlöf-Gunnarsson and Öhrström 2007) and by the visual relationships of "buildings, vegetation, and sky" (Liu, Behm, and Luo 2014). While actual reduction of traffic noise is the eventual goal, reduction of perceived annoyance can provide a significant stopgap measure. Encouraging residents to care for trees and plant yard or balcony gardens will make city-planting projects even more meaningful. A study published in 2006 found that a majority of residents in Huangzhou, China would be willing to pay extra taxes, scaled to household income, in order to preserve urban green spaces because of their perceived aesthetic and stress-reducing benefits (Chen, Bao, and Zhy 2006);[5] a smaller-scale US study in 2016 found the same reaction from a substantial minority of 47 per cent in a Chicago neighbourhood (Kocisky 2016).

Current research based in the ideas of acoustic ecology is also providing the groundwork for new ways to approach reduction of urban noise. Applying the concept of soundscapes to city districts is leading planners to the realization that while the physical abatement of noise is a necessary component of improving urban life, it does not necessarily produce meaningful and lasting change. Recognizing the diversity of soundscape elements – noise, natural sounds, landscape, land use, built environments, population density, spoken languages, preferences in music and conversational tone – in different locations and cultures leads to a new paradigm: paying attention to sonic preferences as well as annoyances. Since these concepts are "culturally loaded," surveying residents about their attitudes and sensitivities can lead to more precise and effective understanding of community needs (Kang et al. 2016). Through the use of interviews, questionnaires, soundwalks, soundmapping, and direct observation, researchers are showing the way toward better understanding of sonic environments and ecologies. Understanding then leads to successful planning of urban soundscape interventions that take cultural preferences into account. The city of Sheffield, UK, provides an example. The addition of sound-barrier walls and fountains that mask mechanical noise to the surroundings of a major train station reduced noise complaints and enhanced feelings of social cohesion through designs incorporating the city's major historical industry, steel (Kang et al. 2016).

Architect and planner Antonella Radicchi applies similar principles to finding locally recognized "quiet areas" in European cities, integrating conventional noise measurement procedures with soundscape design. Her definition of an everyday quiet area is "a small, public, quiet spot embedded in the city fabric, at a walking distance from the places we work and live, where social interaction and spoken communication are not only undisturbed, but even favored" (Radicchi, Henckel, and Memmel 2017, 4). Suitable spaces are identified through the work of resident volunteers trained to gather data – beginning with decibel levels and soundscape recordings – using a mobile app called Hush City, designed for the project, to map quiet spaces rather than noisy ones. Information, which includes volunteer, resident, and researcher observations as well as collected data, is then analyzed by a research team and featured in education campaigns and community discussions. The hope is that nearby residents will enjoy these spaces, which include cafés with outdoor seating as well as parks, public gardens, and even quiet streets. Preliminary data sets from nine European cities and two North American ones suggest that many residents feel calm and secure in these spaces while walking, conversing, reading, resting, taking coffee or smoke breaks, or just listening (Radicchi 2017).

Radicchi's work joins that of World Health Organization researcher Brigitte Schulte-Fortkamp and urban planner Jian Kang as an example of a new way to approach urban soundscape design. First implemented in a district of Berlin in 2017, this design is part of an innovative paradigm for the "smart city." The term as conventionally used refers to the development and use of increasingly sophisticated technology – from driverless cars and automated thermostats to artificial intelligence and massive media platforms – to shape the future of everyday life as well as systems for commerce and information. Another way to define "smart" in the urban context, however, is to concentrate on "a socially constructed set of activities," including design on a human scale, that pay attention to human needs before corporate ambitions (Radicchi, Henckel, and Memmel 2017, 1). This approach regards the city as a commons, and goes beyond spatial definitions to encompass local culture, shared management, and consensus-based problem-solving. It foresees a process for urban noise remediation that begins with resident participation in gathering both measurable and anecdotal information, moves to sharing that information with city councils and planning departments, and results in better appreciation of quiet areas, and eventual planning of new ones, as a public health measure.

The inclusion of residents' responses in planning processes can extend to emotional as well as auditory aspects of sonic awareness. A trio of investigators at the University of Warwick, UK, recommends expanding the vocabulary of soundscape analysis as a way to identify positive ambiences. They move beyond physical and auditory descriptors (for example, loud, harsh, rumbling, continuous) by adding terms related to how soundscapes make people feel, examining such factors as familiarity and appeal as well as annoyance and intrusiveness (Cain, Jennings, and Poxon 2013). Their findings include recommendations for planners: use of 3-D sound labs to approximate real conditions before decisions are made, use of variables like the frequencies produced by air conditioners and traffic patterns to modify existing soundscapes, and encouraging technological changes such as electric vehicles and sound-absorbing pavements that will make urban noise less intrusive.

It seems, then, that the long-standing desire of acoustic ecologists to "improve" urban soundscapes is now being advanced not so much by reverting to nature as by making use of innovative technologies. Attention to accuracy of measurement and description leads to better understanding of variables, which leads to recognition of the role of listeners, which leads to the development of effective problem-solving options. Rather than counting entirely on new technologies to save urban residents from noise annoyance and stress, however, let's take a moment to re-examine a vital step in the process of change: awareness of the cultural symbolism of noise.

RE-SOUNDING A CULTURE OF COMMUNICATION

Along with its physical manifestations, noise is often a voice of conflict, a symbol of disorder or dissent. Consider, for a moment, the structure of an argument: two or more interests colliding as opinions harden, as the determination to win becomes more and more urgent, as speech escalates to shouting. Because everything is loud, nothing is heard; fight-or-flight mode eclipses the ability to interpret messages accurately because attention is on defensive responses. Now consider disarming the verbal combatants with the system described in chapter 16: using silences to remove the heat of urgency, leaving time for breathing, listening, relaxing of muscles and attitudes, softening of voices into reasonable dialogue. When reaction time lengthens and stress declines, confrontation gets a chance to become communication. When communication prevails, compromises and reasoned decisions become possible.

An increasingly polarized worldwide political climate in the past few years has made widespread comprehension of this point a matter of some urgency, but the need to lower the temperature of international communication is nothing new. As Ursula Franklin noted in an essay entitled "Peace as an Ongoing Issue," conflict-based models consistently dominate international relations as well as the rhetoric of political systems, leading to use of the world's economic resources for conflict and defence rather than the social and environmental justice that could make conflicts less common. Describing the legacies of the twentieth century's Cold War, she cites "the lack of non-military perspectives in public discourse and public policy" as an impediment to global social progress, pointing out that insistence on winning frames any situation as a conflict, and is therefore a "losing proposition" (Franklin 2006, 110). Even the vocabulary of health care in the popular imagination incorporates conflict-based images: we battle with cancer and wage wars on drugs. As in an ongoing argument, the tension grows without resolution. Noise escalates and nobody can hear much valid information through it.

If noise mutes reason, can quiet restore it? Can we imagine sound-scapes of order and agreement? This idea is introduced with questions for a reason: examples are not immediately easy to find because consensus-based decisions cannot, by definition, be enforced by an external power. Forced imposition of order – silencing of dissent – is oppressive, a situation that generates protest. Its opposite is rational consensus. Order derived from agreement is fluid and flexible, changing when necessary to meet the needs of a household, a neighbourhood, or an entire society. It is a form of harmony, linked to the deep-rooted paradigm of sound as a creative force.

In ancient myths about the origin of life, sound was considered the primal mechanism for bringing aspects of the world and its inhabitants into existence, moving matter from idea to vibration to physical presence. The repeated phrase "And God said" in the biblical account of Genesis emphasizes speech as a motivating force. In one of the many Hindu descriptions of world origin, it was the vibration of the syllable *om* that filled the initial cosmic void and awakened the sleeping god Vishnu, who called for the work of creation to begin. Song is featured in a number of Indigenous American creation stories: among the Hopi, Spider Woman sang to animate the first beings and charged them with sending resonance through the newly formed earth; for the Sioux, it was the Creator's singing that brought rain and rivers. If such explanations for the origin of matter have been displaced by the physics of atomic structure, they still resonate as potent metaphors for the vibratory realities of the subatomic realm as well as the mechanics of sound. The basis for these creation stories is a series of interlocking scientific truths: sound is vibration, vibration is motion, motion produces action.

Just as the natural motion of wind produces a measurable force, a spoken idea can initiate cultural change. In the science of acoustics, sound waves are enhanced when two or more sources produce wave forms that are identical – peak to peak and trough to trough – or mathematically concordant, and muted when wave forms are opposing: peak to trough. If we translate the physical reality into a symbolic concept of social order, we arrive at consensus as a form of harmonic enhancement: when there is general agreement about a course of action, it happens. It may not happen quickly or smoothly, or produce the expected results, but there is change. When the agreement is concordant – when overlapping areas of need are discovered and acted upon, and the participants are satisfied that the result is fair to all concerned, if not always perfect – it is more likely to be stable and even beneficial.

Let us then consider reviving a persistent ancient notion: that harmony is more than notes in a chord, and that the sounds produced by a community or a society are indicative of its health. Many cultures throughout history and even pre-history – aboriginal Australia and Africa, ancient Greece and India, medieval France, Renaissance Italy, and Elizabethan England among them – devised elaborate theories about the role of music in originating and maintaining cosmic, social, emotional, and somatic harmony. Divine beings sang the world into form, tuned and plucked cosmic strings to set the planets in motion, and inspired the verities of mathematics, medicine, architecture, and audible music as interrelated aspects of the same reality. Traditional healers used songs and drumming to stimulate relaxation and healing responses long before anything was known in scientific terms about psychoneuroimmunology.[6] The ideal state of health and the ideal political state echoed each other: the balanced flow characteristic of a beautiful melodic line also governed the human circulatory system and the traffic in a city; purposeful rhythms inspired physical strength, wise decisions, and rational governance.

A brief excursion into the structures underlying music will bring its mathematical properties to light and give a glimpse into the correspondences that inspired ancient theories. Like architecture, music depends upon physics as well as aesthetics: structures in both are supported by the science of ratio and proportion. For example, the note that musicians in the Western world call "A 440" – the basis for tuning instruments in a modern orchestra with each other – vibrates at 440 Hz. Halving the vibratory rate to 220 Hz will produce the same note at the octave below, while doubling it to 880 Hz will give the same note at the octave above. All three notes will be perceived by the brain as versions of the note designated as "A" in Western scales, deemed "high" or "low" according to their vibratory speed. This 2:1 ratio holds true for any note at any vibratory rate: doubling or halving the rate will always produce an octave from the note you start on, whatever it is. Some ratios, like 2:3 and 3:4, will produce sounds concordant with the original note; others will produce discords. All musical intervals, which govern the structures of melody and harmony, can be understood this way, as elements of a mathematical infrastructure that supports the actual sounds we hear.

Ratios also govern the rhythmic patterns that define units of time and illustrate the mathematics of periodicity, represented by clocks and calendars as well as by drumming and counting beats. As a result of mathematical infrastructure, the acoustic relationship of notes in a chord or beats in a rhythmic pattern, unlike a system of beliefs, does not change with ideology, invasion, revolution, or revelation. The ratios involved remain constant regardless of differences in instrumentation, interpretation, or speed. No such certainty is available to non-musical forms of community. Political and religious orthodoxies are subject to differences in interpretation and to revision, even with the strongest intention to maintain a pure tradition. The passage of time results in cultural changes that render the meanings of words and phrases obsolete; translations between dissimilar languages produce ambiguities. Invasions, migrations, and religious conversions have altered the traditions of every culture and faith; sects and factions arise when interpretations fall into dispute. Wars are fought as a result of such differences, especially when their origins are poorly understood.

What if there were a better way, a less conflictive model? Along with the use of silence as a time-out from disagreement, what if we were to adopt the principles of harmony and rhythm as guidelines for humane and effective social change, for economic justice and intercultural communication? Musical rhetoric in Western cultures demands that the end of a song or piece be in some way satisfying, whether through harmonic resolution to a "home" chord, relaxation of a driving rhythm, or repetition of an effective concluding phrase. Harmony includes both concord and discord in its development, but the struggles between them eventually resolve in a statement that encompasses both, whether in pure resolution or in expansion of the initial vocabulary.[7] Rhythm teaches the value of time and timing, of knowing *when* to act as well as what to do. As both a metaphor and a literal activity, music can expand the imagery available for the teaching of leadership and ethics.

SOUNDING SOCIAL JUSTICE

Along with militaristic imagery, the dominant metaphors and models for political and economic leadership in North American society today are derived from sports: the level playing field, being at the top of your game, supporting the home team, stepping up to the plate, scoring points, defending turf. They enhance the themes of conflict and competition, emphasizing the notion in popular culture that statecraft and the production and trading of goods are about being bigger, faster, stronger, tougher, and *louder* than ever. Unlimited growth is seen as unlimited power; success is measured by increasing impact on social and physical environments; human interactions are characterized as moves in a vast and inevitable game. Given the exigencies of climate change and unstable globalized economies, as well as the natural physical limits on unchecked economic and population growth, isn't it time for some alternative models?

- Rather than outpacing the competition or fighting for turf, rather than acting as if there were no limits, can we create harmonies of scale and scope?
- Can we learn to communicate clearly in meaningful counterpoint about what is necessary and what is not?
- Can we recognize the rhythms of rising and falling power, and learn to find contentment in the small and local when the struggle for worldwide domination comes at too high a cost?

Moving beyond metaphor to a more literal interpretation, music can inspire cooperation even through the practice of skills required by its formal structures. While competitions of musical skill are almost as firmly entrenched as competitive sports in Western popular culture, performing ensembles from a variety of musical traditions can serve as models for teaching cooperative activities. Classical string quartets and chamber ensembles, along with marching bands, choirs, and close-harmony ("barbershop") vocals – like partnered dancing and figure skating – demonstrate precise teamwork at its most exhilarating. Jazz ensembles model creative improvisation and the epiphanies that come from letting each member of the group have a turn at being the soloist. Drum circles based on Indigenous traditions in Africa or North America emphasize inclusion, open communication, and the sheer joy of making noise that is purposeful, rhythmic, and shared. Choirs teach the values of consensus and cooperation by encouraging individual singers to blend their sonorities, listening to each other and matching tone qualities in order to produce a collective voice that thrills participants as well as listeners.

Musical metaphors for social action are already available, with an intriguing example rising from the alliance of African-American activism with the musical lineage of jazz. In their collaborative 2013 book *The Fierce Urgency of Now*, musicians Daniel Fischlin, Ajay Heble, and George Lipsitz (2013) examine the structural and emotional traditions of jazz as a model for social justice and social change. Unwilling to leave democratic reform movements at the stage of endless complaint – "feel[ing] good about feeling bad" (xxx) – the authors present cooperative creativity as an alternative to systems based on fixed ideologies. Rather than fiery slogans that generate strong feelings, they promote quietly effective local actions that rise out of community dialogue and the recognition of genuine needs (xxx–xxxi). Describing improvisation – the foundation of jazz – as a way to break through the hesitations, critiques, and ideological disputes that prevent social justice movements from advancing, the authors pose a series of questions drawn from their experience as performers and teachers:

Who are you here and now?
What do you have to say?
How are you going to join this dialogue?
How will you speak your mind and also listen to how others speak their
*   minds?*
How will you contribute to the outcomes of the improvisation?
How far are you prepared to go to experiment with alternative forms of
*   expression that make sense in the particular moment? …*
Can you integrate your ideas with the larger group in ways that make new
*   meaning?*
What resources can you bring to the table to make a difference in the
*   improvisatory moment?* (71)

Through these questions, they centre social change in the structural principles of a rehearsal process in which the outcome of plans is discovered rather than enforced. While the intention of perfecting the outcome is shared, the ultimate form it takes is not predictable. It grows out of the time participants spend together and the extent to which they inspire each other. It is inventive, communicative, and free of hierarchies.

The idea of music as a structural model for social interaction and social justice can return us to the ancient belief that it is far more than transient and commodified entertainment. As a component of purposeful human soundscapes, music can bond communities and create channels for communication where conventional opportunities are inadequate, ignored, or conflicted. The basics of its structure are not hard to learn:

- Play when it's your turn; listen when it's someone else's.
- Know the difference.
- Cooperate with others and appreciate their contributions.
- Help others to improve their skills.
- Be sincere.
- Practise constantly to gain expertise, and take every opportunity to learn from players who are more expert than you are.

What, then, might characterize a model of interaction based on noise, rather than music? First, the over-layering of too many signals, too many voices clamouring for attention, the loss of meaning in an excess of stimuli. To which we can add, attraction to loudness because it's the only thing that captures fragmented attention. Constant competition, constant pressure to outshout rivals, to gain approval, adulation, wealth, status. Constant repetition substituted for significance.

Does this sound familiar?

Do you recognize this pattern in broadcast and print media, social media, entertainment? And what about political campaigns? When you examine candidates, do the slogans mask the issues? Is media attention given to the most valid reasoning, or to the loudest lies? When shouting overcomes meaning, noise wins. When reasoned dialogue is submerged by fear and hysteria, nobody wins.

A culture that privileges noise loses the perception of subtlety. If all that we hear about are the most privileged major players in a worldwide game of politics and economics, we miss information about the lives and needs of the vast majority who are barred from playing. When seemingly insoluble problems loom, the best answers can come from quiet sources: local businesses and charities, small innovative companies, family farms and local markets.

Let us visit another example, then, by returning to the analogy mentioned in chapter 16 between listening habits and diet.[8] The North American addiction to the convenience of processed food, beginning with post-war acceptance of cake mixes and packaged breakfast cereals, grew for decades into the fast-food industry that now dominates urban habits of consumption and sustains industrialized agriculture. In those decades, little attention was given to the nutritional implications of processed food, so that the current epidemic of malnourished obesity in "developed" cultures came as a surprise to most. Now that the problem is getting attention, now that there is extensive literature, both popular and scientific, focusing on the dangers to health and detriments to productivity that result from the habits associated with industrialized fast food as well as its ingredients, questions are being raised and "slow food" is gaining credibility.

The slow food movement was founded in 1986 in Italy as a protest against the encroachment of the McDonald's fast-food chain on traditional Italian food culture. It concentrates on nutrient-rich "from scratch" home cooking and on eating fresh, locally grown ingredients, as well as home and community gardening. Slow food is a dialectical reaction against processed industrial pseudo-food, but also against the social habits of speed and inattention to detail that characterize the driven life. It is convivial, individualistic, and nostalgic, but the intention behind it is not to drive women back to their kitchens as full-time managers of the food supply. Slow food encourages everyone – women, men, children, communities – to learn about food production, selection, and preparation, and to regard food as a beloved art form instead of a necessary annoyance. Rather than a call to recreate the past, it is a plan to learn from the past about how to improve the future.

Parallels between slow food and acoustic ecology are easy to find. Among the practices supported by slow food enthusiasts worldwide is the preservation of traditional farming methods and of heritage seeds and livestock breeds. These efforts are comparable to the acoustic ecology movement's desire to preserve or restore quiet places and respect for quiet in order to sustain population health by combatting noise-induced stress. Through advocacy of conscious eating and conscious listening, both movements are promoting awareness of emotional nutrients as well as physical ones. Thus, the slow food movement can provide one more potential model for the re-evaluation of our soundscapes. If the phenomenon of malnourished obesity has a parallel in the ubiquity of ambient noise and earbud culture, is its antidote to be found in learning to balance constant sonic stimuli with periods of quiet? In soundwalking habits taught early and maintained throughout life? In the development of social courtesies around the uses of cellphones in public? In insisting, *loudly* if necessary, that machines and ventilation systems be designed and manufactured to be as quiet as possible? In recognizing soundscapes as shared and precious resources?

Can we listen to noise, find meaningful signals within it, and reduce the volume on what is neither meaningful nor supportive for our lives, our cultures, and our futures?

HEARING A FUTURE

The auralization of historical soundscapes in chapter 4 has its counterpart in imagined futures. As concerns about noise continue to shape urban planning and bylaw policy, as the healing properties of sound are recognized and given implementation, as the soundscapes of speech become recognized as aspects of communication, a culture of listening can be built. Step into this future of unspecified immediacy and notice what is missing as well as what holds your attention: find the silences along with the sounds.

Soundscape 1:
An Urban Residential Street

Step outside of your home in the morning and notice the wind rustling the leaves of natural trees while activating the leaf-like collection structures (*rustle, whoosh*) of the neighbourhood's wind turbines. Children chatter and squeal as an electric school bus stops to pick them up at the corner; the traffic of small electric cars, buses, and bicycles down the street is barely noticed over the wind and the birds. The wooden wind-chime on your balcony provides a soft clacking counterpoint to the mellow bronze bell tones of the one across the street. Your walk to the public bus stop is accompanied by music you have chosen for its energizing properties, fuel for a brisk pace, but you turn it off and slow your steps while passing the grounds of the hospital in order to listen to the sound-sculptures as their sensors turn your footsteps into soft chiming percussion riffs in a variety of tones and trills.

On the bus, most people stare intently at their personal communication devices; some gaze out the windows, a few participate in small bursts of conversation. For the benefit of those using earphones, stops are signalled by flashing coloured lights as well as the *ping* of an auditory signal. Feet shuffle, packages and bags are shifted on laps, traffic whooshes softly past, doors open and close: all are audible because the electric engine is very quiet.

Soundscape 2:
An Office Building and Its Surroundings

As you enter your workplace, you pause for a moment to appreciate the fountain in the lobby – audiovisual reward and synesthetic pleasure doubling as an air ionization system – before reaching your office, where a choice of relaxing and motivating music channels, environmental record-ings, or near silence awaits your decision to push a button. If you prefer, a small window can be opened to admit air and sound from the street below. The air circulation system in the building is so well insulated that the distracting HVAC roar you recall from your first job years ago has become a "back in the day" story to tell co-workers.

You choose to take your morning break with a few of those co-workers today, gathering in the office next to yours to chat and share Internet pho-tos, making a plan for lunch together. The corner restaurant chosen uses sound-reducing structures and technologies, absorbent floor and ceiling materials, and stain-repelling carpeted panels on the walls to reduce the clatters, roars, and hisses of food and beverage preparation. Thanks to pre-cisely engineered insulation and a sound system playing upbeat instru-mentals softly, conversation at your table is easily heard while the surrounding tables maintain their privacy. You return from lunch calm and invigorated.

Soundscape 3:
A Hospital

On the way home from work, you visit a friend who is recovering from surgery, bringing a card with an embedded chip that plays a song she loves. Sounds of running water, birds, and soft bells on the ambient soundtrack in the corridors are enhanced intermittently by the pattering rhythmic dialogues of *tabla* drums from India or by the gentle majestic plucked-string sonorities of Chinese zithers. At certain times of the day, live performers may visit as well: players of Celtic harp or Brazilian jazz guitar, a classical flute and guitar duo, an octet of singers from the local university whose director composes songs designed to promote healing, the fruit of years of research into music's effects on the nervous system and the biochemistry of recovery. A nurse walks briskly past, engineered soft-soled shoes nearly silent on the sound-absorbing floor.

At the nursing station, displays of light on a panel alert staff members to patients' needs; vibrating wristbands worn by nurses also transmit sig-nals from the monitors so that alarms are rarely needed. The station is partially surrounded by noise barriers that reduce transmission of sound along corridors and into patients' rooms.

Walk along the corridors in different units. You will hear soft music or nature sounds but also volunteers and family members reading to patients in a variety of languages. In some rooms, the occupants listen to personal devices or television soundtracks through earphones; in others, the soundscape is shared. Intensive care, surgical recovery, cancer, psychiatric, palliative, pediatric, and maternity units all have their own variations on the soothing ambient soundscape, designed according to the findings of research and the taste of current patients.

For those who prefer silence, designated rooms and noise-cancelling headphones are available on request. In any case, periods of silence are provided each afternoon, enabling sleep for patients and completion of record keeping for staff.

Soundscape 4:
A Town or City Council Meeting

In the evening you attend a meeting at your community centre. There is a contentious issue on the agenda and you have a strong opinion about it, as does everyone in the crowd that fills the hall and confronts the council members, who sit in a row at the front of the room. It's about property taxes or parking regulations, business licences, dog licences, building permits, or access to recreational facilities; the issue has come up before and remained unresolved because opposing factions have not been willing to compromise. The issue has been declared unsuitable for definitive voting because opinion is evenly divided. You enter the room with some apprehension: last time this subject was up for discussion, everyone left feeling frustrated.

Before the meeting begins, two professional mediators walk in and sit in front of the council, gazing warmly at the crowd without speaking as a trio of musicians begins to play from the back of the hall: instruments chase each other through quick spiralling phrases, then slow vibrant ones, then a single note held and pulsed, fading into silence. The crowd is quiet. A mediator with a rich and deeply resonant voice instructs the crowd to consider the issue and its potential outcomes; silence is held for three minutes while they do so. Discussion begins, bracketed by measured periods of music and silence. Faces in the crowd relax, heads nod, smiles appear.

The second mediator, busy with a tablet computer through the discussion, projects and summarizes notes along with sketched diagrams of potential outcomes. Another break for silence is taken while the crowd and council consider results. A preliminary vote is taken by the council; silence follows, then responses from the crowd. The mediators summarize and explain the implications of the result. The council votes again; responses show that a large majority of the crowd is in agreement. A reasonable consensus has been achieved. Dissenters are respectfully told how and where they can register complaints, but few are expected. You smile at passing neighbours and leave the hall with a sense of relief.

Soundscape 5:
An Elementary School

Today you have taken a flex day from work in order to visit your children's school, R. Murray Schafer Elementary. In one classroom, a math teacher brings out several hand-drums for a unit on fractions. She holds a slow steady beat on one drum, then asks a small group of children to divide the beat in half. When they settle into their pattern, she asks the next group to divide the halves into halves, producing a fast patter. Then the slow beat starts afresh, and another group divides it into thirds. Patterns are repeated, exchanged, and superimposed until each child has had a chance to practise all of the patterns, then screens and writing materials come out for a lesson on translating the rhythms into numerical fractions. The next day's lesson will see the children mentally calculating the fractions, then translating the results into drumming patterns and pie charts.

In another room, older children in a social studies class track Google maps of mountain villages in South America, identifying terrain and the outlines of towns and agricultural lands, commenting on land use and density of habitation while checking statistics on exports and population. Another group looks at architecture through Internet street views and observes a live camera feed of an open-air market. Their earphones provide samples of ambient sounds from the streets. Later, the children share snacks made from a traditional Andean recipe while listening to music from Peru. The teacher tells stories about the summer he spent in Argentina improving his Spanish, and guides the children in repeating some phrases, which they use to make up a song.

Down the hall, a speech class is in progress. A colourful chart of speech-related anatomy – the respiratory and vocal systems – is projected on a wall. As the children take turns reading aloud, a teacher coaches them in the skills of voice development: *take a deeper breath, relax your jaw, that's good! That sounds way better!* The eager and the shy are provided with suggestions, encouragement, opportunities to shine. A game follows, taking turns around the circle with call-and-response chanting and clapping in rhythm: *WHO-took-the / COOK-ies-from-the / (pause) COOK-ie / jar?* As each child playfully accuses the next and is met with denial, smiles break out.

In the gym, a game of rhythm-ball is in progress at the volleyball net. You see what seems to be volleyball, but nobody minds when the ball hits the floor and nobody bothers to keep score. The object of the game is to send the ball back and forth over the net in rhythm, listening for interactions of the ball with hands and floor. (*Thud-tha-THUD-thud. Thud. Thud. THUD-thud. Thud-THUD.*) Nobody actually wins or loses, but greatest appreciation is given to players who can control and vary the pattern, syncopating the impacts or producing rhythms that invoke speech. These talented players are further encouraged toward the end of the gym period by being asked to form a team with each other while the other children watch and listen to their extended phrases. An after-school practice session is arranged for the skilled group to act as mentors for those who are less expert but interested in improving.

In the next period, the gym will resound with high-energy rock or salsa-based dance tracks programmed for a group of children who need vigorous exercise breaks in order to concentrate on their studies. They will be followed by a different group that jumps and stretches to reggae and jazz, testing their balance with lateral moves and off-beat footwork.

Lunch in the school cafeteria today is accompanied by recorded voices of birds, whales, and wolves and other environmental sounds woven into relaxing music. On alternate days, the soundtrack includes voices of children from different cultures, one continent per week, speaking and singing songs in their local languages. Once a month after school, an interest group meets to research the places represented by the songs and send Internet messages to similar groups in schools all over the world, exchanging greetings, songs, and recipes. Friendships and fascinations thus generated can serve as the basis for extended social media contacts or even family vacation trips.

Soundscape 6:
An International Mediation Centre

That evening you check the computer for news feeds. One headline story is the resolution of a land dispute between two continually warring nations. The dispute is complex, involving a long history of conflict and deeply held convictions about social justice. It has been going on for so long that generations of inhabitants have lived out their lives in the soundscape of gunfire and home-made explosives.

The Internet video feed brings you into the scene of negotiations: a large, light-filled room with windows overlooking a wooded area, its location outside the disputants' countries. A place of calm and beauty, it is equipped with potted plants and flowers, a small cascading fountain set into one wall, and a series of tables arranged in a semicircle. At one end of the room is a platform holding several drums of hourglass shape with goatskin or synthetic heads bound to their wooden frames by lengths of cord: the "talking drums" of West African tradition, used to imitate and enhance speech. There is also a large cylindrical drum with four leather-padded beaters; its traditional use in Indigenous North American ceremonies involves the synchronization of four players representing the directional perspectives of North, South, East, and West.

The camera pans over the room to show people taking their seats at the tables: heads of state with their translators and secretaries, mediators, assistants, designated reporters, and technical crew. With the professional mediation team are two ethnomusicologists equipped with expertise in both popular and traditional musical styles from a wide variety of cultures, as well as the latest research information from neuroscience and psychology on the effects of such music on mood and brain function. When the participants have all been seated and initial speeches given, they summon the heads of state to the drum platform and guide them through the first exercise: using the large Indigenous North American drum to synchronize a slow steady pattern at the pace of a resting heartbeat. When the pattern settles into steadiness, the room vibrating with its confident weight, the participants make eye contact and nod in rhythm. They rise from the drums at a signal and join hands in a brief silent gesture before returning to their tables.

At designated breaks in the formal dialogue throughout the next few days, the musicologists will play soundtracks of instrumental music from the disputants' cultures, carefully alternated to avoid the impression of bias, along with live mellow jazz improvisations on piano based on tunes from both traditions. The recordings are infused with musical patterns designed to induce calm and rationality. When anger or frustration breaks out, when discussions reach an impasse or accusations are made, the African talking drums are distributed and used to give wordless voice to the disputants' concerns. Held at the players' left sides, their tones rise and fall with the pressing and releasing of the cords that fasten their goatskin heads, imitating the inflections of speech and the intensity of argument. Gradually the players begin to echo each other's patterns: tentative at first, then more confident as a language of agreement is built. As the proceedings stretch on for days, as offer and counter-offer are formulated and discussed, breaks are taken to absorb and create rhythm. Breaks for silent contemplation follow the exchange of written documents. After ten days, the negotiations conclude with a binding agreement announced on global news services and heralded as the beginning of a new era in international communication.

ECHO: LISTENING TO NOISE, HEARING SIGNALS

Along with its physical realities and its long history of cultural manifestations, noise is a symbol. It represents urgent and emphatic communication, but also the clutter that obscures our ability to communicate, to concentrate, to notice what is important. In the sciences, the term is used to describe clusters of irrelevant data that obscure the clarity of evidence. In broadcasting, it represents unfocused signals. Noise is also, paradoxically, a voice of culture: what we permit and forbid, what we endure, stifle, protest, condemn, rejoice in, amplify, and silence. It can remind us of where we are located, who is around us, what is valued and what is not. It can make us forget that attention is a precious and limited resource, that hearing and listening are not the same. Noise can motivate us to seek refuge from the auditory assaults of the culture we inhabit, and to build defences against what cannot be avoided.

Technological remedies for mechanical noise have improved considerably since the World Health Organization issued its first set of guidelines in 1999. Noise-cancelling headphones, once available only to the military and the aviation industry at prohibitive cost, are now easily available to commercial airline travellers and the patrons of crowded cafés. Decades of evidence about occupational noise hazards are finally producing a well-established culture of concern among governments for the auditory health of industrial workers. Building construction codes are constantly updated as materials designed for sound absorption are invented or re-formulated to produce improved results. Exporting such new technologies and the regulations that mandate their use to industries in the global economy is essential. So is continuing medical research on the physical and psychological effects of new technologies for communication and entertainment. Technological improvements and medical data, however, are not enough: as stated here in so many contexts, public awareness and public discussion are needed.

A broad discussion is now happening around issues related to noise, carried by sources ranging from academic journals publishing histories of noise complaints and anti-noise legislation to media broadcasts, podcasts,

and Internet sites giving advice about auditory safety for teens and factory workers. It echoes debates reacting to urban growth, economic competition, and globalization. Throughout the twentieth century, we in the
"developed world" – the economically dominant Western and Northern
cultures that define modernity – mechanized and technologized our
sources of physical and aesthetic nutrition as well as our industries producing food and entertainment, until they were no longer under our control. Corporate conglomerates have for decades dictated to us, through
their advertising, what we should eat, wear, watch, and listen to, if we
wanted to be given the respect due to consumers who "keep up." The
course of progress appeared to be a projectile travelling one way into an
ideal future of effortless pleasure, until the effort and the cost finally
became apparent: destruction of resources, confinement of lives, and production of immeasurable waste.

*What if that projectile changed its trajectory and become a pendulum? What
if it recognized ebbs and flows like the inexorable powers of the natural world:
tides, seasons, breath?*

Progress does not have to be measured in competitive advantage, constant growth, and increasing loudness. When the inclination of a culture
impels its participants toward more noise, more speed, more wealth, more
pleasure, more distraction, when it is pushed and harried to greater and
greater excesses, we can begin to question whether the goal is worth the
stress it induces. We can decide to live with sound and noise by recognizing their benefits and controlling the harms they cause. We can schedule,
construct, and search for pockets of quiet to recover from sonic assaults.
We can rediscover the art of listening.

*Consider this book, along with the many sources that inform it, as a call to
pay attention: a siren in both senses of the word. Let it shock you awake and
lure you to listen.*

Notes

PREFACE

1 Examples of recordings are available at
https://aporee.org/maps/work/projects.php?project=corona.
2 Email from Ioanna Etmektsoglou in Greece, 7 May 2020.
3 Email from Hildegard Westerkamp in British Columbia, Canada, 31 March
2020.

CHAPTER ONE

1 The term *earlid* appears to have been coined by R. Murray Schafer in his pioneering work on acoustic ecology, *The Tuning of the World* (1977c), and is much
used in the literature of acoustic ecology as a symbol for the physical inability to
block out sound.
2 A *decibel* (dB) is a unit for measuring the loudness of a sound: see the discussion
in chapter 2.
3 The abbreviation dBA indicates the "decibel A" mathematical scale, a system that
approximates the range of human hearing (see chapter 2). The statistics are from
environmental psychologist and noise expert Arline Bronzaft, interviewed by
Andrea Bartz in *Harper's Bazaar*, 25 July 2017.
4 There are, of course, exceptions to this observation, and urban planning has a lot
to do with them: see Wang and Kang (2011).
5 See, for example, Bakalar (2006) and Bartz (2017).
6 According to a report published in 2018 by Statistics Canada, more than one
million adults across the country have reported a noise-induced hearing-related
disability. It is stated that hearing loss costs the Canadian economy more than
$10.6 billion each year. Heather Ferguson, president of the Hearing Foundation

of Canada, estimates that actual numbers may be as high as three million, since hearing loss is believed to be under-reported (Hearing Foundation of Canada, accessed May 2020, http://www.hearingfoundation.ca/statistics/).

7 See, for example, Fuente and Hickson (2011).

CHAPTER TWO

1 Research questions are even raised occasionally about responses of plants to sound waves, at least as a reaction to vibration. See, for example, Woodlief, Royster, and Huang (1969). The authors reported a 40 per cent reduction in the growth rate of tobacco plants exposed to random noise. See also Coghaln (2007).

2 Terminology for loudness (*sound energy, wave amplitude, sound pressure, volume*) can vary according to what profession is doing the measuring and what is being measured. For example, physicists, engineers, environmental assessors, and musicians may use different modes of analysis and different terms to describe the same phenomena.

3 For an explanation of frequency measurements, see the section on frequency/pitch below. The alternative C-weighted decibel scale (dBC), mathematically adjusted to the entire frequency spectrum and used in the detection of low-frequency noise (infrasound), is discussed in chapters 6, 11, and 14.

4 See chapter 11.

5 The organ of Corti is filled with endolymph rather than perilymph, and the chemical difference between the two fluids is believed to be the mechanism for the conversion of vibration to electrical signalling in the brain.

6 The authors speculate that noise-induced synaptic loss results in "hidden hearing loss," which is not necessarily measurable with standard audiometric testing.

7 This condition is referred to as "temporary threshold shift" (TTS). As further research is done, evidence is showing that these temporary shifts in hearing accumulate potential long-term damage.

8 National Organization for Hearing Research Foundation website, http://nohrfoundation.org/regeneration, accessed 15 December 2015. As of 2019, the website is no longer active.

9 Impaired hearing comes in many varieties. Some result from conditions that are genetic, prenatal, or present from birth, and unrelated to noise. Others are caused by tumours, infections, or head injuries. Some noise-induced conditions are temporary and minimally invasive, while others cause major disability and emotional repercussions. Most noise-related varieties of hearing loss can be avoided with attention and due care.

10 Traynor's comments speculating on how demanding Beethoven would have

been as a patient with today's practitioners of medicine provide interesting commentary on what it's like to treat those with special concerns when it comes to alleviating hearing loss.

11 "Message from William Shatner to You," American Tinnitus Association, https://www.ata.org/managing-your-tinnitus/patient-stories/message-william-shatner-you.

12 For further information on treatments, see the website of the American Tinnitus Association, https://www.ata.org/.

CHAPTER THREE

1 The major third (5:4) and minor third (6:5) were not considered fully consonant by music theorists until the seventeenth century, when the development of new methods of tuning musical instruments changed interval ratios from the earlier Pythagorean system. To hear the intervals as they are defined today, see the Wikipedia article "Interval (music)," at https://en.wikipedia.org/wiki/Interval_(music)#Frequency_ratios, and scroll to the chart in the section titled "Main intervals," which has an audio column with recorded examples.

2 The early works of Hildegard Westerkamp and Barry Truax, both of whom began as students of Schafer, are representative of this style. Westerkamp has continued to develop the content and technological resources of acoustic ecology extensively: her recent work focuses on the creation of rich evocative soundscapes with the layering of multiple recorded tracks on speakers surrounding the audience. Close your eyes and you're in a rainstorm with birds from a dreamscape jungle, or on a train crossing a Canadian prairie in another dimension. Truax, a computer scientist as well as a composer of electroacoustic music, is the author of two foundational texts: *Handbook for Acoustic Ecology* (1978) and *Acoustic Communication* (2nd ed., 2001).

3 For more information about soundwalking, see chapter 16.

4 See the bibliography at https://www.leonardo.info/isast/spec.projects/acousticecologybib.html.

5 The academic field known as critical studies stems from the observation that many cultures today have endured a long history of colonial domination that has led to long-standing political and economic injustices. Revealing and criticizing these injustices is the central work of the discipline's scholars.

6 For an explanation of the term *interdiscipline*, see my article "Growing an Interdisciplinary Hybrid: The Case of Acoustic Ecology" (Epstein 2003).

7 These programs include options at the School for Contemporary Arts and the Sonic Research Studio at Simon Fraser University (Canada), the Natural Resources Institute at the University of Manitoba (Canada), the Discovery Park

Center for the Environment at Purdue University (US), the University of Huddersfield (UK), and the Berlin University of the Arts (Universität der Künste Berlin, Germany). Relevant courses can also be found in some university departments of communications studies, environmental studies, or music.

8 See the *Sounding Out!* blog at https://soundstudiesblog.com.

CHAPTER FOUR

1 There are still questions in the medical literature about whether presbycusis is universal. Thurston's article mentions a field study in the 1960s of an isolated tribe in rural Sudan whose elders showed no sign of it, but questions were raised later about the accuracy of reported ages.

2 The source of this information is a report on a meeting of the American Society for the Hard of Hearing in June 1936 from the newsletter of the Society for Science and the Public, and the blog post "Ancient Noise Ordinances" (31 July 2010) on the blog *InertiaWins! The Iron Law of Politics* (http://inertiawins.com /2010/07/31/ancient-noise-ordinances/), which references historian Donald Kagan's *Pericles of Athens and the Birth of Democracy* (New York: Simon and Schuster, 1998), 125.

3 The poet Juvenal complained in Satire 3.235 about the nocturnal clatter in Rome. As a satirist, he may not have been an entirely credible source of information, but his complaint gives life to legal regulations about cart and wagon traffic.

4 For a set of speculative interpretations of auditory culture in prehistory and history, focused particularly on sound in architectural spaces, see Blesser and Salter (2006), 67–126.

5 The eighteenth-century Scottish writer Tobias Smollett provides an example from his time in London: "I go to bed after midnight, jaded and restless from the dissipations of the day – I start every hour from my sleep, at the horrid noise of the watchmen bawling the hour through every street, and thundering at every door; a set of useless fellows, who serve no other purpose but that of disturbing the repose of the inhabitants; and by five o'clock I start out of bed, in consequence of the still more dreadful alarm made by the country carts, and noisy rustics bellowing green pease under my window" Smollett ([1771] 1966), 136–7. Retrieved from the archive of literary references to sound of the World Soundscape Project, http://www.sfu.ca/sonic-studio/srs/time.html. Like Juvenal, Smollett was a satirist.

CHAPTER FIVE

1 See Redström (2007).

2 In electronic and audio engineering, "coloured" noise spectra are defined by their distinct intersections between frequency and loudness, and named by analogy with their counterparts in the visual spectrum. While white noise is defined as having equal pressure (perceived by the human ear as loudness) in all octaves, the pressure of pink noise decreases in octave bands as the frequency increases. It is used as a reference signal in audio engineering. Recorded examples can be found at https://en.wikipedia.org/wiki/Colors_of_noise.

3 See the website of the Hyperacusis Network, at http://www.hyperacusis.net/hyperacusis/home/default.asp.

4 There is considerable literature about this phenomenon in the bibliographies on the websites of the American and Canadian Associations of Music Therapists (AAMT and CAMT). The principle raises the spectre of children lulled to sleep with the Rolling Stones or Metallica demanding the same repertoire in senior care homes of the future.

5 Levitin (2006, 61) suggests that the cerebellum and basal ganglia are involved in the perception of periodicity, which may form the basis for musical rhythm. Patel (2008) explores rhythmic patterning in music and speech, speculating that there is little relationship between them in terms of brain function.

6 Another question is easily raised: Could the instinct for recognition of rhythm also be present in some other mammals? Horses, with their distinctively rhythmic gaits, would seem to be an obvious species to investigate for this if some way could be found to do it. Although the precision of dressage riding and of the Royal Canadian Mounted Police Musical Ride and similar exhibition events that involve coordinated riding to music are based on riders' signals to their horses rather than equine musical sense, some individual horses reportedly have more aptitude than others for this type of work.

7 Listen to Gioacchino Rossini's "Duetto buffo di due gatti" ("Funny Duet of Two Cats," or "Cat Duet"), a nineteenth-century operatic parody of cat vocalizations for two sopranos. Several versions are on YouTube.

8 Rhythm of a sort may be present in the form of patterned repetition: one dog in my neighbourhood barks in patterns of four calls with a *decrescendo*, another in sets of three with the first beat stressed, although neither pattern is perfectly consistent.

9 A preliminary project that uses brain imaging to study human responses to the vocalizations of cats and monkeys suggests that affective responses are not limited to the hearer's species: see Belin et al. 2007.

10 An *ostinato* (plural: *ostinati*) is a repeating pattern in the background of a piece of music, persisting for the duration of the piece. It is usually an accompaniment to a foreground melody.

11 The motorcycle has an association with adventure and the romance of travel as well, although its sheer loudness is polarizing, inspiring either attraction or annoyance. Its appeal may be connected to the association of noise with power and with images of youth and masculinity.

12 A study of the effects of recorded machine-gun noise, now familiar to more than one generation through innumerable film and television soundtracks, on cortisol levels is long overdue.

13 LFN is also a component of thunder and ocean currents, and is involved in the vocalizations of elephants, whales, and some reptiles and burrowing mammals. It is basically the result of long slow vibrational waves carrying through earth or water.

14 The study is flawed by ambiguities in descriptive terms but is useful for its raising of questions about the effects of frequency.

15 The authors state that, even though the connection of symptoms to electromagnetic fields is not yet proven, the symptoms are real; they encourage further research on the condition.

16 This experiment is reported in Hviid (2010) as having taken place in Texas in 1991, but no further information about it is available.

17 See Granlund-Lind and Lind (2004), with reference to the work of Swedish medical researcher Åke Bergman.

18 Removing the mixture of metals using a very specific protocol and replacing some or all with ceramic fillings has been found in some cases to produce relief or reduction of symptoms. See "Hospital Accommodations of Electrically Hypersensitive Patients in Sweden" (2007), at http://www.eiwellspring.org/ehs/HospitalAccommodationsOfEHSPatientsInSweden.htm.

19 *Neuroplasticity* refers to the potential for creating new neural pathways in the brain and nervous system as a response to changed conditions. Learning and practicing a new physical skill like playing tennis or the piano is one example; new theories are now suggesting – not always with sufficient proof – that the principle has medical applications as well. For neural reprogramming, see, for example, the website for the Dynamic Neural Retraining System, at http://www.dnrsystem.com/dnrs_dvd.html. While personal testimonies report success, no actual studies have yet been done on the system.

CHAPTER SIX

1 For more information relevant to industrial workers, see chapter 10.
2 The abbreviation mmHg refers to the measurement of blood pressure, expressed as a ratio of "millimetres over mercury": systolic pressure from the heartbeat and diastolic pressure on the relaxed heart muscle between beats.
3 For technical and medical analysis of the phenomenon and theories about its causes, see K. Kraus and Conlon (2012) and U. Kraus et al. (2013). Both studies are considered by their authors to be at early stages in an ongoing process of explaining the physiology of noise-induced heart rate variability.
4 This micro-level of variability is not the same as heart arrhythmia, which can often be felt and heard. Arrhythmia requires medical monitoring or treatment.
5 For discussion of the relation between sleep disturbance and noise annoyance, see chapter 5.
6 Further discussion of social and legislative issues appears in chapters 13 and 14.

CHAPTER SEVEN

1 This is because smell and taste represent different but adjacent reception sites for the same sensation, triggered by the chemistry of food sources and environmental stimuli. In evolutionary terms, smell is a warning system that tells us what is and is not safe and pleasant to eat.
2 "Normal" synesthesia is essentially virtual: one sense summons another in the memory and/or imagination. True genetic synesthesia involves measurable physical triggering of senses linked in specific pairings, often sound with colour.
3 For example, think – right now – of red.
4 To experience and stimulate the gustatory imagination, read recipes and look at pictures of food. Like the auditory imaginations of musicians, who can hear music internally without physically listening to it, the gustatory imaginations of chefs and expert cooks are more thoroughly developed than those of the general population.
5 Analogous hints at the plasticity of human sensory perception are suggested by an excerpt from Helen Keller's observations on the importance of smell for her negotiation of everyday experience. See Drobnick (2006, 181–3). Sensory deficits require the brain to compensate through alternate sensory processing.
6 It is now suspected that handwriting should be taught to children even if they use computers at an early age, because it embeds letter recognition in the memory better than typing: see Longcamp, Zerbato-Poudou, and Velay (2005).
7 This may be due in part to the constant decision making involved in taking

notes by hand, in addition to the physical processes of handwriting and the tendency of laptops to offer distractions.

8 In rare cases, tinnitus can be modified by eye position (Lockwood et al. 2001).

9 Medieval monastic libraries, for example, were probably full of whispers and mutters as each monk sounded out the sonorities of the text before him, like modern undergraduates taking a linguistics exam. For further background, see Ong (1982).

10 The insufficiency of defective auditory processing as a *sole* cause for reading problems is easily shown by the reading proficiency of hearing-impaired individuals. Children with profound deafness learn by means of "increased reliance on the articulatory component of speech when the auditory component is absent" (MacSweeney et al. 2008, 1369).

11 The studies, carried out in Japan and Europe, specify a direct link between the diagnosis of ADHD and low levels of the neurotransmitter dopamine in the brain: since the diagnosis of ADHD is controversial in North America, the results summarized here may not be generally applicable to all children labelled with the condition.

12 White noise and its uses are discussed in chapter 10.

13 For example, noise is also recognized as detrimental to sensory processing and social interaction in autism: see Teder-Sälejärvi et al. (2005).

14 See also Al Mannai and Everatt (2005); Ho (2006); Shu, Peng, and McBride-Chang (2008); and J. Williams and Lynch (2010). Phonological processing patterns for learning to read are apparently consistent regardless of the child's first language and the characteristics of its alphabet.

15 Learning-style theories may be particularly applicable to children and teens. Most adults develop a mix of styles, making the categories less relevant (Lujan and DiCarlo 2006).

16 The *Classroom Noise Booklet* is still an excellent layperson's guide to acoustics, how sound functions in architectural spaces, and why it matters. It is available as a pdf file from the Acoustical Society of America website: http://asa.aip.org.

17 For the association in adults between uncontrollable noise and learned helplessness, see Hatfield et al. (2002).

18 Noise audits can be performed by noise abatement companies. They usually require several days of measurements taken in classrooms while school activities are in session, so advance planning is required. So is funding.

19 For an example of thorough reporting of a wide range of variables, see Haines et al. (2001).

CHAPTER EIGHT

1 For example, one commercial site that claims its abdominal speakers will affect brain development uses the term "pre-natal education," which normally refers to education of parents-to-be on birth and infant care, as an advertising slogan (https://www.amazon.com/BabyPlus-Prenatal-Education-Heartbeat-Promote/dp/B00140KS9I). While babies respond to music played before their birth for a few months afterward, nothing is known about whether their intellect benefits later: the varied influences (genetics, education, opportunities, experiences, familial and cultural attitudes, and so on) are interdependent and too complex to identify separately.

2 It can easily be argued that exposing animals and birds to such stresses is not ethical either, but medical research has had to rely on doing so for centuries. Computer models are now replacing animal experimentation in many fields of inquiry; it may not be possible for all of them.

3 Because the Hohmann team's interpretation of results from prior studies included evidence drawn from a wide variety of methodologies, its conclusions are sometimes questionable, which they acknowledge. As in many other cases, more research is needed.

4 For web posts on television commercials for vintage toys, see Mike Cane's "TV Commercials: 1960s Toys," at http://mikecane2008.wordpress.com/2008/05/18/tv-commercials-1960s-toys (accesses 3 June 2012) and John Behrens's web page for commercials from the 1950s to the 1970s, at http://www.jonbehrensfilms.com/toys.html (accessed 2015).

5 I consulted the Sight and Hearing Association's website for the "Annual Noisy Toys List" at http://www.sightandhearing.org/Services/NoisyToysList©.aspx for the years 2004, 2008, 2011, 2014, 2018, and 2019 (SHA, "Annual Noisy Toys List").

6 SHA, "Annual Noisy Toys List," 2019.

7 SHA, "Annual Noisy Toys List," 2014.

8 SHA, "Annual Noisy Toys List," 2013.

9 SHA, "Annual Noisy Toys List," 2011.

10 Directive 2009/48/EC of the European Parliament of 18 June 2009 on the Safety of Toys, Official Journal of the European Union L170, 2009, s. 17.

11 Ibid., Annex II, s. 10.

12 For additional discussion of the "sound education" concept, see chapter 17 below.

CHAPTER NINE

1 Statistics published by the US Centers for Disease Control in 2014 on the health of children under age eighteen did not include data on deafness or hearing loss. The 2012 survey of adult health gives a figure of 15 per cent for hearing impairment from mild to severe; this does not distinguish between hereditary, congenital, and noise-induced hearing loss. A follow-up survey of children In 2009 tracks only infants born with hearing impairments; results for hearing loss will not be available until at least the end of 2020, and probably later.

2 The percentage may be misleading, since attendees who did not experience tinnitus would be less likely to fill out and return the questionnaire than those who did: this is a methodological flaw that the authors should have accounted for. Still, even if the actual rate were only half what is reported, it would – and should – be alarming.

3 My experience with teaching acoustic ecology to undergraduates at the University of Calgary supports this conclusion. In any group of sixty students, six to ten will indicate that they are unpleasantly surprised to discover that they already have experienced some degree of tinnitus, that their hearing is probably compromised, and that their history of concert and/or rave attendance is the most likely cause. They also report that encouragement of hearing protection – and distribution of earplugs – is more likely to happen at raves than at live rock or hip-hop concerts. It is difficult to determine whether any of those who do not speak up in class are also experiencing tinnitus, so the actual number may be somewhat higher.

4 See, for example, Lena and Peterson 2008; ter Bogt et al. 2011; and Arsel and Bean 2013.

5 Such delays are common for studies of mass-market phenomena. It takes at least that much time for markets to grow, for observations to be made and questions raised about effects, and for studies to be designed, funded and published.

6 Exposure time was not specified, although most aerobics classes are approximately one hour in duration.

7 People with highly developed skills in visual arts, music, mathematics, dance, or sports may verbalize thought less than others, substituting images, sound patterns, motion, or abstract reasoning for verbalized self-communication. I once attended a lecture sitting next to someone who made intricate multi-coloured abstract drawings while listening. When I admired the drawings and asked how she managed to recall the content of the lecture without notes, she replied: "These *are* my notes." Studying new languages and travelling shed further light on the process: internal voices can become bilingual.

8 See, for example, D. Williams 2008.

9 Does this reflect your experience? Does your music have to be extremely loud to distract you from the terror of what you've been thinking? Are the lyrics of your favourite songs fixated on despair and destruction? *If so, tell someone: friend, family member, teacher, counsellor, doctor, elder.*

CHAPTER TEN

1 It's worth noting that many of the experiments conducted in the 1970s and cited by Cohen and Spacapan (1984) involved the use of continuous noise at 90–95 dBA, now known to damage hearing acuity with prolonged exposure.

2 For summaries of research, see Shek and Schubert (2009) and A. Smith, Waters, and Jones (2010).

3 The researchers were students in my Introduction to Acoustic Ecology course at the University of Calgary who also held clerical jobs in corporate head offices.

4 Variants include "pink," "red," "grey," "brown," and "purple" noise, defined by their frequency components. Each evokes a particular response from the basilar membrane and has the potential to provoke specific reactions from listeners.

5 Cited in Banbury and Berry (2005) as part of a critique on earlier findings.

6 A 2012 article in the *Wall Street Journal* on clerical workplace music began with the complaint of an open-plan office worker that she was unable to get the attention of co-workers because of their ubiquitous use of earphones (Shellenbarger 2012).

7 Listening to songs in a language you don't understand, at low to moderate volume, can also be effective.

8 Some symphony musicians now wear specially designed hearing protection devices that screen out particular frequencies while allowing others to be heard normally. This does not entirely eliminate the problem, but does mitigate it.

9 A 2015 article in the *Atlantic* magazine shows that acoustic shock can be a hazard for classical musicians as well as call-centre workers. In "A Musician Afraid of Sound," a symphony cellist describes the career-ending onset of acoustic shock brought on by the heavy amplification of a "pops" concert: Horvath (2015).

10 In 2000, the *San Francisco Chronicle* started a trend in restaurant reviews by including noise ratings, giving negative reviews when ambient noise levels reached 80 dBA. At the time, this was a fairly unusual level; it has since become common for trendy restaurants in large cities (Gill 2008); see also Harden (2008) for an account of the same phenomenon in Australia.

11 On the other hand, soft music – whatever its content – may be more effective than loud music as a motivator for restaurant diners to stay longer and order more: see Lammers 2003.

CHAPTER ELEVEN

1 Dutch historian Karin Bijsterveld's book *Mechanical Sound* (2008), an instructive history of noise control, abatement, and complaints derived from published and archival records of the twentieth century, gives a fascinating account of the issues and controversies involved.

2 In Porto Velho the limits of regulation led to a strong initiative for community education about noise pollution, which has been successful in reducing levels (Paul et al. 2013). Education should be considered a passive form of noise abatement.

3 If you try this at home, use something that really clangs: neither a concert hand-bell, which has built-in frequency dampers, nor a tiny ornament will produce the full effect.

4 A selection of abstracts from the website *Engineering Village* (http://www .engineeringvillage2.org/home.url?acw=&utt=) is summarized in this chapter.

5 I am grateful for information provided by Terry Thompson, director of environmental services at the Calgary International Airport.

6 This information was conveyed in a presentation by Canadian acoustic engineer David DeGagne at the 2007 Spring Noise Conference of the Alberta Acoustics and Noise Association (now the Alberta Association for Noise, Vibration and Sound), and is used with his permission.

CHAPTER TWELVE

1 Comments posted on the FIFA website, accessed 14 June 2010: http://www.fifa.com/confederationscup/news/newsid=1073689.html.

2 Anthropologist Steven Connor asserts that cinema is driven by six grades of "violent sound tactations": "the kiss, the punch, the cut, the shot, the crash, and the explosion" (2004, 167).

3 For definitions and more detail, see chapter 3.

4 See, for example, Prochnik (2010); Keizer (2010), and Foy (2010).

5 See Labelle (2010) and M. Smith (2004).

6 Numerous examples from studies conducted in the 1970s are cited and analyzed in Cohen and Weinstein (1981, 47–9).

7 See *Walden, or Life in the Woods* (1854), available in multiple modern editions.

8 *Bruits: essai sur l'economie politique de la musique* (1977); English translation by B. Massumi, 1985.

9 Its origin is the text of a 1973 Chilean song of solidarity and protest, "¡El pueblo unido jamás será vencido!"

10 Studies of the effectiveness of repetition in advertising and other persuasive

communication now recognize that its persuasive effect diminishes with length of use. While it initially enhances perceived credibility, fatigue and scepticism eventually set in: see Cacioppo and Petty (1989) and Lehnert, Till, and Carlson (2013). However, since most test subjects for such studies are university students, there is still a question to be raised about whether the eventual shift to scepticism also occurs in less educated populations, which in turn raises questions about repetition in political messaging.

11 See "Panama: Operation Just Cause – General Manuel Noriega and How the Rock Music Came to Be" at http://www.psywarrior.com/rockmusic.html, accessed February 2008, January 2016). The site author, Ed Rouse, a retired officer and "Psyops" (psychological operations) expert in the US Army, claims that the music was not primarily intended to grind down Noriega's resistance: it was used to mask the sound of security-sensitive conversation by CIA operatives surrounding the building where Noriega had taken refuge. Media broadcasts at the time of the incident stated that the music was used intentionally to cause psychological irritation.

12 See Alex Gibney's 2007 documentary film *Taxi to the Dark Side*.

13 Rumours circulating sporadically since the Second World War about the use of low-frequency noise as a weapon are apparently unfounded: besides being impossible to control, it would affect its users as well as their opponents. For a summary of the science of sonic weapons as it was known in 2001, see Vaisman (2001–02). For an extended analysis of the cultural implications of sonic warfare, see Goodman (2009).

14 "Troops in Iraq Get High-tech Noisemaker," an Associated Press article posted 3 March 2004 to the Military.com website, http://www.military.com, and "Sonic Weapons in Iraq," condensed article posted 3 March 2004 on the DefenseTech website, http://www.defensetech.org/archives/000807.html.

15 Jardin 2005. The truck-mounted system was also capable of projecting public address messages clearly over long distances; this may have been its intended use. It was apparently not put into action in either capacity.

16 Postings on DefenseTech.org: http://www.defensetech.org/archives/001741.html.

17 Posts on Twitter during the protest on 9 July 2016 in Baton Rouge described the "hellish scream" of the LRAD and warned local participants to keep away from it. The device's model was not specified.

18 The sound was recorded on a video taken during a night of protest and posted by *Slate Magazine*: see Newman (2014).

19 Some discussions of the potential for deafness mention the use of earplugs to mitigate risk (Altmann 2001); this is not feasible for accidental exposure. In any case, technical analyses of LRAD sound do not agree about whether earplugs have an effect in reducing the sound pressure: see Altmann (2001) and Vinokur (2004).

20 The idea of the soundscape as a commons was first advanced at the 1993 "Tuning of the World" conference on acoustic ecology in Banff, Alberta, by physicist and philosopher Ursula Franklin in a talk entitled "Silence and the Notion of the Commons." Further expansion of Dr Franklin's concept appears in chapter 17.

CHAPTER THIRTEEN

1 Because temperatures above 25 degrees Celsius – 77 degrees Fahrenheit – have usually been rare in Calgary as a result of its climate and altitude, relatively few homes had air conditioning until the summers of 2016–18, when climate change became apparent in longer and more extreme periods of heat and drought.

2 The gathering and verification of information about the asphalt plant and its operation were undertaken by a group of senior undergraduate students in my "Introduction to Acoustic Ecology" course at the University of Calgary in September 2007. I am grateful for their assistance, and for the care they took in documenting the unpublished class project entitled "NW Calgary Asphalt Plant: A Study in Noise Pollution."

3 Nausea, nosebleeds, and throat irritation were also reported, but these were the effects of airborne particulates rather than noise.

4 What authorities considered to be a "substantial number" may be debatable as well. For example, inability to sleep because of the noise from the asphalt plant itself and the nightly traffic of trucks carrying aggregate from the site was reported more often by women and children than by men. Does this indicate a correlation of gender and/or age with noise sensitivity, or were they just more likely to register complaints?

5 According to WHO guidelines, sleep disturbance can occur at 30 dBA, especially if the noise contains low-frequency components.

6 For an explanation of the difference, and of the parameters of measurement, see chapter 2.

7 Summer festivals in Calgary often apply for permits to exceed the night-time limits: since most of them take place in a park on an island in a river running through the city, the surrounding expanse of water amplifies the sound, leading to complaints. Also exempt is the annual Calgary Stampede, a ten-day summer festival of traditional rodeo culture and sports that includes amplified grandstand shows and an open-air midway active late into the night.

8 A reading of 65 dBA represents more than twice the sound intensity of 61 dBA: see chapter 2.

9 Spy Hill Lands Development Project Open House #3, 21 October 2003: Summary of Questions and Answers.

10 Government of Alberta, Ministry of Transport, "Noise Attenuation Guidelines for Provincial Highways under Provincial Jurisdiction within Cities and Urban Areas" (26 July 2006).

11 The owners' plan to use noise abatement measures was voluntary, since the site was exempted from municipal regulations, and served as an example of good community citizenship.

12 Ironically, the jump in oil prices in 2008 temporarily changed the focus of provincial planning from roads to public transportation.

13 YouTube clips of the group can be found at "NYC Drum Circle Marcus Garvey Park Harlem," 11 October 2007, http://www.youtube.com/watch?v=ZORFFIsEcBk and "Drum Circle Marcus Garvey Park NYC," 31 August 2008, http://www.youtube.com/watch?v=2jIpDS-ozPk&feature=related.

14 Information on the resolution of the dispute was provided by Steve Simon of the New York City Parks Department in response to my query. The story was not covered in news sources beyond 2008.

15 Electromagnetic field levels of 3 to 50 milliGauss were measured around the affected homes (Cowan 2008).

16 See the discussion of otoacoustic emissions in the context of hypersensitivity to noise in chapter 5. Thermoacoustic effects are explained by the physics of oscillating waves, which can produce either sound or heat depending on the circumstance. Sound and heat can even occur together; an example is found in glassblowing.

17 CBC interview with lead researcher Colin Novak, University of Windsor: "Windsor's Mysterious Hum Is Real, Says Researcher," CBC News, 1 August 2013, http://www.cbc.ca/news/canada/windsor/windsor-s-mysterious-hum-is-real-says-researcher-1.1414596.

18 For information about the Ranchlands (Calgary) Hum, see the blog http://unidentifiednoiseincalgary.blogspot.com/.

19 The team included myself, two acoustic engineers, and several volunteers from the community.

20 A map of existing and planned wind farms in the United States is available at http://www.caller.com/data/economy/windfarms/. It shows greatest concentration in the central plains states from the Dakotas to Texas and in the Northeast near the Canadian border.

21 A CTV (Canada) news clip posted on YouTube includes a brief interview with Dr McMurtry: "Wind Turbines and Health Problems," 6 October 2008, http://www.youtube.com/watch?v=lmoOe8J6qT8&feature=related. Conflicting

evidence about wind farm noise is given at "Wind Turbine Syndrome Dismissed Visceral Vibratory Vestibular Disturbance Illness Brain Damage," 29 September 2008, http://www.youtube.com/watch?v=40xl-Jggya4&feature=related, and at "Wind Turbine Noise: Suncor Wind Farm Ripley," 13 May 2008, http://www.youtube.com/watch?v=mablINxg3zE.

22 An initial report about an ongoing German study of this phenomenon appeared in the *Daily Mail* online in July 2015 (Zolfagharifard 2015). The study was published a month later: see Weichenberger et al. 2015.

23 Definitions of LFN and infrasound, equated in physics and engineering, are distinct in audiology: the boundary is drawn between frequencies barely audible by humans (LFN) and those below the audible limit, presumed to be 20 Hz (infrasound).

24 For recordings of wind turbine noise, see (hear!) the following: "How Noisy Is a Wind Farm," 6 July 2011, https://www.youtube.com/watch?v=zKgN2G9dodc; "Wind Turbine Noise," 27 November 2012, https://www.youtube.com/watch?v=79aARWPalt4; and "The Sound of Wind Turbines in 2014," 25 February 2014, https://www.youtube.com/watch?v=nq9QNlbnG6c. A recording "Wind Turbine Syndrome: A Matter of Bad Prevention," 23 December 2013, at https://www.youtube.com/watch?v=Rm1b11YCwWg, accessed in 2014, mentions a study to be published by the Danish government in 2015, but it does not seem to be available.

25 *Westlaw Canada* XVIII, item 589, https://www.westlawnextcanada.com/academic/.

CHAPTER FOURTEEN

1 City of San Francisco (2004), Bayview Hunters Point Redevelopment Project and Rezoning EIR, III-I, p. 7.

2 Information about noise complaints was not published beyond 2009.

3 . American Housing Survey, 2020, https://www.census.gov/programs-surveys/ahs/about.html.

4 Accessed from http://www.hudong.com/wiki/%E5%8C%97%E4%BA%AC%E5%8C%BA%E5%9F%9F%E5%99%AA%E5%A3%B0%E5%9C%B0%E5%9B%BE, in 2011. I am grateful to Naiwen Liu for translating this item.

5 Within reason, of course: if the dog is neglected, animal protection bylaws come into play. For that matter, running an electric-powered or manual mower, raking your leaves rather than blowing them, and walking your dog regularly can cause your neighbours to adore you.

6 See https://www.osha.gov/laws-regs/regulations/standardnumber/1910/1910.95. In 2020, the NIOSH website displays a statement that it is no longer being maintained.

7 Website of the Canadian Centre for Occupational Health and Safety, accessed 28 May 2020: http://CCOHS.ca.

8 There isn't. Hearing aids are a treatment, not a cure. Higher-tech treatments may well be developed in the future, but they are likely to remain expensive and intrusive. Prevention really is the best strategy.

CHAPTER FIFTEEN

1 Do any hospitals supply earplugs to patients whose surgeries generate these sound pressure levels? There is no literature on the subject yet, either because it isn't being done or because nobody has reported on it. In either case, attention is needed.

2 An experiment with "pink noise," sometimes used to mask conversation in offices, proved it to be counterproductive in hospital settings because it reduced the ability of staff to determine the direction of monitor alarms: see Mazer 2002.

3 If you are a listener who simply wants to improve the variety of your playlists, branching into new experiences can benefit you as well. While familiarity is a major determinant of musical taste – we tend to like what we have heard before and often – it isn't a limitation. Try what's new more than once to give your nervous system a chance to stop being surprised and listen more deeply. Consider the music an antidote to noise, whether the source of noise is ambient or internal. Consider it a portal to the experience of conscious listening.

4 How many? There is no statistical record of the number; it would be necessary to survey individual hospitals.

CHAPTER SIXTEEN

1 If this sounds intriguing, consult a textbook on music history for numerous examples of historical warnings given by government and religious officials against new musical styles as sources of corruption.

2 Essentially, there's no accounting for individual taste.

3 Since research has not yet been done in this area, no conclusions can be drawn. The question is raised as a plea for it to be started.

4 The documentary film *Alive Inside* gives a moving demonstration of this principle. It's available on YouTube and Netflix.

5 "How Opening Our Ears Can Open Our Minds: Hildegard Westerkamp," *Ideas*, CBC Radio, 31 August 2017, http://www.cbc.ca/radio/ideas/ecology-of-sound-hildegard-westerkamp-1.3962163.

6 Andra McCartney, "While Walking," posted 15 May 2018, on Soundwalking Interactions, https://soundwalkinginteractions.wordpress.com.

7 For literature about soundwalking practices and purposes, see the website of Hildegard Westerkamp, http://www.sfu.ca/~westerka/writings.html.

CHAPTER SEVENTEEN

1 Internet founder and Google vice-president Vint Cerf, quoted in *TechRepublic* magazine (Patterson 2015).

2 See Tangermann (2018), who also points out that the variety of "beeps and chirps" currently in use will confuse listeners: standardization is needed.

3 Perhaps low-income neighbourhoods will also become a thing of the past, the eventual result of more humane systems for income distribution and housing policies.

4 See, for example, Gidlöf-Gunnarsson and Öhrström (2007) and Jackson (2003). Nilsson and Berglund (2006) specifically emphasize the need to keep daytime traffic noise in parks and suburban green spaces below 50 dBA.

5 Whether this attitude results from the civic-mindedness of Chinese culture is not discussed. In any case, the Huangzhou survey and its results provide a model for similar tests of public support for urban renewal plans.

6 For examples drawn from several cultures, see Gouk (2000).

7 In other cultural traditions of music – many of which are not based on harmonic structures – the satisfaction can come from well-turned melodic phrases, the expression of an emotion or image by the music, compelling rhythms, and/or passionate performance. My point here is not to regard Western harmony as definitive, but to present the concept of harmony as a symbol for universally beneficial interaction.

8 Brief portions of this section have previously appeared in my article "Soundscaping Health: Resonant Speculations" (Epstein 2016).

References

Abram, David. 1996. *The Spell of the Sensuous: Perception and Language in a More-than-Human World*. New York: Pantheon.

Ackerman, Diane. 1990. *A Natural History of the Senses*. New York: Random House.

Agrawal, Yuri, Elizabeth A. Platz, and John K. Niparko. 2008. "Risk Factors for Hearing Loss in US Adults: Data from the National Health and Nutrition Examination Survey, 1999 to 2002." *Otological Neurotology* 30, no. 2: 139–45. doi:10.1097/MAO.0b013e318192483c.

Ahissar, Merav, Athanassios Protopapas, Miriam Reid, and Michael M. Merzenich. 2000. "Auditory Processing Parallels Reading Abilities in Adults." *Proceedings of the National Academy of Sciences* 97, no. 12: 6832–7. doi:10.1073/pnas.97.12.6832.

Akansel, Neriman, and Şenay Kaymakçi. 2008. "Effects of Intensive Care Unit Noise on Patients: A Study on Coronary Artery Bypass Graft Surgery Patients." *Journal of Clinical Nursing* 17, no. 12: 1581–90. doi:10.1111/j.1365-2702.2007.02144.x.

Akiyama, Mitchell. 2015. "Unsettling the World Soundscape Project: Sound-scapes of Canada and the Politics of Self-Recognition." *Sounding Out!* 20 August. https://soundstudiesblog.com/2015/08/20/unsettling-the-world-soundscape-project-soundscapes-of-canada-and-the-politics-of-self-recognition/.

Alink, Arjen, Felix Euler, Elena Galeano, Alexandra Krugliak, Wolf Singer, and Axel Kohler. 2012. "Auditory Motion Capturing Ambiguous Visual Motion." *Frontiers in Psychology* 2: 391. doi:10.3389/fpsyg.2011.00391.

Alleyne, Brian C., Ronald M. Dufresne, Nasim Kanji, and Michael R. Reesal. 1989. "Costs of Workers' Compensation Claims for Hearing Loss." *Journal of Occupational and Environmental Medicine* 31, no. 2: 134–8.

https://journals.lww.com/joem/Abstract/1989/02000/Costs_of_Workers
    _Compensation_Claims_for_Hearing.14.aspx.

Al-Mannai, Haya, and John Everatt. 2005. "Phonological Processing Skills as Pre-
    dictors of Literacy amongst Arabic Speaking Bahraini Children." *Dyslexia* 11,
    no. 4: 269–91. doi:10.1002/dys.303.

Al-Mutairi, Nayef Z., M.A. Al-Attar, and Fahad S. Al-Rukaibi. 2011. "Traffic-Gen-
    erated Noise Pollution: Exposure of Road Users and Populations in Metropol-
    itan Kuwait." *Environmental Monitoring Assessments* 183, nos. 1–4: 65–75.
    doi:10.1007/s10661-011-1906-0.

Altmann, Jürgen. 2001. "Acoustic Weapons: A Prospective Assessment." *Science
    and Global Security* 9, no. 3: 165–234. doi:10.1080/08929880108426495.

Alves-Pereira, Mariana, and Nuno Castelo Branco. 2007. "Vibroacoustic Disease:
    Biological Effects of Infrasound and Low-Frequency Noise Explained by
    Mechanotransduction Cellular Signalling." *Progress in Biophysics and Molecular
    Biology* 93, no. 1–3: 256–79. doi:10.1016/j.pbiomolbio.2006.07.011.

Amengual, Julià L., Nuria Rojo, Misericordia V. de las Heras, Josep Marco-
    Pallarés, Jennifer Grau-Sánchez, Sabine Schneider, Lucía Vaquero, et al. 2013.
    "Sensorimotor Plasticity after Music-Supported Therapy in Chronic Stroke
    Patients Revealed by Transcranial Magnetic Stimulation." *PLOS One* 8, no. 4:
    e61883. doi:10.1371/journal.pone.0061883.

American Academy of Pediatrics. 1997. "Noise: A Hazard for the Fetus and New-
    born." *Pediatrics* 100, no. 4: 724–7. https://pediatrics.aappublications.org
    /content/100/4/724.long.

American Society for Testing and Materials. 2003. "Standardization News."
    Accessed 29 December 2018. http://www.astm.org/SNEWS/MARCH_2003
    /lawrence_mar03table1.html.

American Speech-Language-Hearing Association. 2010. "Wind Turbine Design
    and Noise." *ASHA Leader* 15, no. 11: 5. doi:10.1044/leader.NIB.15112010.5.

American Tinnitus Association. 2018. "Audiometric Evaluations for Hearing
    Loss and Tinnitus." Accessed 29 December. http://www.ata.org/understanding-
    facts/measuring-tinnitus.

Amitay, Sygal, Meray Ahissar, and Israel Nelken. 2002. "Auditory Processing
    Deficits in Reading Disabled Adults." *Journal of the Association for Research in
    Otolaryngology* 3, no. 3: 302–20. doi:10.1007/s101620010093.

Andersson, Gerhard, Linda Jüris, Viktor Kaldo, David M. Baguley, Hans C.
    Larsen, and Lisa Ekselius. 2005. "Hyperacusis: An Unexplored Field. Cognitive
    Behavior Therapy Can Relieve Problems in Auditory Intolerance, a Condi-
    tion with Many Questions." English abstract of article in Swedish. *Läkartidnin-
    gen* 102, no. 44: 3210–12. https://www.ncbi.nlm.nih.gov/pubmed/16329450.

Ando, Yoichi, and H. Hattori. 1970. "Effects of Intense Noise during Fetal Life upon Postnatal Adaptability (Statistical Study of the Reactions of Babies to Aircraft Noise)." *Journal of the Acoustical Society of America* 47, no. 4B: 1128–30. doi:10.1121/1.1912014.

– 1973. "Statistical Studies on the Effects of Intense Noise during Human Fetal Life." *Journal of Sound and Vibration* 27, no. 1: 101–10. doi:10.1016/0022-460X(73)90038-2.

Arias, Pablo, and Javier Cudeiro. 2010. "Effect of Rhythmic Auditory Stimulation on Gait in Parkinsonian Patients with and without Freezing of Gait." *PLOS One* 5, no. 3: e9675. doi:10.1371/journal.pone.0009675.

Arnold, Paul, and David Canning. 1999. "Does Classroom Amplification Aid Comprehension?" *British Journal of Audiology* 33, no. 3: 171–8. doi:10.3109 /03005369909090096.

Arsel, Zeynep, and Jonathan Bean. 2013."Taste Regimes and Market-Mediated Practice." *Journal of Consumer Research* 39, no. 5: 899–917. doi:10.1086/666595.

Attali, Jacques. 1985. *Noise: The Political Economy of Music.* Translated by Brian Massumi. Minneapolis: University of Minnesota Press.

Babisch, Wolfgang. 2005. "Noise and Health." *Environmental Health Perspectives* 113, no. 1: A14–15. doi:10.1289/ehp.113-a14.

Babisch, Wolfgang, Bernd Beule, Marianne Schust, Norbert Kersten, and Hartmut Ising. 2005. "Traffic Noise and Risk of Myocardial Infarction." *Epidemiology* 16, no. 1: 33–40. doi:10.1097/01.ede.0000147104.84424.24.

Baerwald, Erin F., Genevieve H. D'Amours, Brandon J. Klug, and Robert M.R. Barclay. 2008. "Barotrauma Is a Significant Cause of Bat Fatalities at Wind Turbines." *Current Biology* 18, no. 16: R695–6. doi:10.1016/j.cub.2008.06.029.

Bailey, Peter. [1998] 2004. "Breaking the Sound Barrier: A Historian Listens to Noise." In *Hearing History: A Reader*, edited by Mark Smith, 23–35. Athens: University of Georgia Press.

Bakalar, Nicholas. 2006. "Hazards: New York Subway Clatter Exceeds Safe Levels." *New York Times*, 17 October. https://www.nytimes.com/2006/10/17 /health/17nois.html.

– 2008. "Hazards: Sleeping through Noise, but Still Feeling Its Effects." *New York Times,* 4 March. https://www.nytimes.com/2008/03/04/health/research /04haza.html.

Baldwin, Ann L. 2007. "Effects of Noise on Rodent Physiology." *International Journal of Comparative Psychology* 20, no. 2: 134–44. https://escholarship.org /uc/item/04m5m3h1.

Baldwin, Roberto. 2014. "What Is the LRAD Sound Cannon?" 14 August. http://gizmodo.com/what-is-the-lrad-sound-cannon-5860592. Accessed 29 December 2018.

Baliatsas, Christos, Irene van Kamp, John Bolte, Maarten Schipper, Joris Yzermans, and Erik Lebret. 2012. "Non-Specific Physical Symptoms and Electromagnetic Field Exposure in the General Population: Can We Get More Specific? A Systematic Review." *Environment International* 41: 15–28. doi:10.1016/j.envint.2011.12.002.

Baliatsas, Christos, Irene van Kamp, Mariette Hooiveld, Joris Yzermans, and Erik Lebret. 2014. "Comparing Non-Specific Physical Symptoms in Environmentally Sensitive Patients: Prevalence, Duration, Functional Status and Illness Behavior." *Journal of Psychosomatic Research* 76, no. 5: 405–13. doi:10.1016/j.jpsychores.2014.02.008.

Banbury, Simon, and Dianne C. Berry. 1998. "Disruption of Office-Related Tasks by Speech and Office Noise." *British Journal of Psychology* 89, no. 3: 499–517. doi:10.1111/j.2044-8295.1998.tb02699.x.

– 2005. "Office Noise and Employee Concentration: Identifying Causes of Disruption and Potential Improvements." *Ergonomics* 48, no. 1: 25–37. doi:10.1080/00140130412331311390.

Bangjun, Zhang, Shi Lili, and Di Guoqing. 2003. "The Influence of the Visibility of the Source on the Subjective Annoyance Due to Noise." *Applied Acoustics* 64, no. 12: 1205–15. doi:10.1016/S0003-682X(03)00074-4.

Barclay, Robert M.R., Erin F. Baerwald, and Jeffrey C. Gruver. 2007. "Variation in Bat and Bird Fatalities at Wind Energy Facilities: Assessing the Effects of Rotor Size and Tower Height." *Canadian Journal of Zoology* 85, no. 3: 381–7. doi:10.1139/Z07-011.

Barlow, Christopher. 2010. "Potential Hazard of Hearing Damage to Students in Undergraduate Popular Music Courses." *Medical Problems of Performing Artists* 25, no. 4: 175–82. https://www.ncbi.nlm.nih.gov/pubmed/21170480.

Barlow, Christopher, and Francisco Castilla-Sanchez. 2012. "Occupational Noise Exposure and Regulatory Adherence in Music Venues in the United Kingdom." *Noise and Health* 14, no. 57: 86–90. http://www.noiseandhealth.org/text.asp?2012/14/57/86/95137.

Barone, J.A., J.M. Peters, D.H. Garabrant, L. Bernstein, and R. Krebsbach. 1987. "Smoking as a Risk Factor in Noise-Induced Hearing Loss." *Occupational Medicine* 29, no. 9: 741–5. https://europepmc.org/abstract/med/3681506.

Bartosińska, Maria, and Jan Ejsmont. 2002. "Health Condition of Employees Exposed to Noise: Extra Auditory Health Effects." English abstract of article in Polish. *Wiadomości Lekarskie* 55, no. 1: 20–5. https://www.ncbi.nlm.nih.gov/pubmed/15002214.

Bartz, Andrea. 2017. "How City Noise Is Slowly Killing You." *Harper's Bazaar*. 25 July. https://www.harpersbazaar.com/culture/features/a10295155/noise-detox/.

Battaner-Moro, Juan, Chris Barlow, and Paul Wright. 2010. "Social Deprivation and Accessibility to Quiet Areas in Southampton." Paper presented at the

Noise in the Built Environment 2010 Conference, Ghent, BE, 29–30 April. http://ssudl.solent.ac.uk/2413/.

BBC. 2006. "The Sound That Repels Troublemakers." 4 April. http://www.bbc.co .uk/wiltshire/content/articles/2006/04/04/mosquito_sound_wave_feature .shtml.

Beach, Elizabeth F., Megan Gilliver, and Warwick Williams. 2013. "Leisure Noise Exposure: Participation Trends, Symptoms of Hearing Damage, and Perception of Risk." *International Journal of Audiology* 52, no. S1: S20–5. doi:10.3109 /14992027.2012.743050.

Beach, Elizabeth F., and Valerie Nie. 2014. "Noise Levels in Fitness Classes Are Still Too High: Evidence from 1997–1998 and 2009–2011." *Archives of Environmental and Occupational Health* 69, no. 4: 223–30. doi:10.1080/19338244.2013 .771248.

Beer, David. 2007. "Tune Out: Music, Soundscapes and the Urban Mise-en-Scène." *Information, Communication and Society* 10, no. 6: 846–66. doi:10.1080 /13691180701751031.

Beijing Municipal Science Research Institute on Labor Protection. n.d. "Noise Map of Beijing." (Article in Mandarin; unpublished translation by Naiwen Liu). Accessed 2014. http://www.hudong.com/wiki.

Belachew, Ayele, and Yemane Berhane. 1999. "Noise-Induced Hearing Loss among Textile Workers." *Ethiopian Journal of Health Development* 13, no. 2: 69–75.

Belin, Pascal, Shirley Fecteau, Ian Charest, Nicholas Nicastro, Marc D. Hauser, and Jorge L. Armony. 2007. "Human Cerebral Response to Animal Affective Vocalizations." *Proceedings of the Royal Society B: Biological Sciences* 275, no. 1634: 473–81. doi:10.1098/rspb.2007.1460.

Belojevic, Goran, Branko Jakovljevic, and Vesna Slepcevic. 2003. "Noise and Mental Performance: Personality Attributes and Noise Sensitivity." *Noise and Health* 6, no. 21: 77–89. http://www.noiseandhealth.org/text.asp?2003/6 /21/77/31680.

Belojevic, Goran, Branko Jakovljevic, Vesna Stojanov, Katarina Paunovic, and Jelena Ilic. 2008. "Urban Road-Traffic Noise and Blood Pressure and Heart Rate in Preschool Children." *Environment International* 34, no. 2: 226–31. doi:10.1016/j.envint.2007.08.003.

Bendor, Daniel, and Xiaoqin Wang. 2005. "The Neuronal Representation of Pitch in Primate Auditory Cortex." *Nature* 436: 1161–5. doi:10.1038/nature 03867.

Bengtsson, Johanna. 2004. "Is a 'Pleasant' Low-Frequency Noise also Less Annoying?" *Journal of Sound and Vibration* 277, no 3: 535–7. doi:10.1016/j.jsv.2004 .03.014.

Bergdahl, Jan, Göran Anneroth, and Evert Stenman. 1994. "Description of Persons with Symptoms Presumed to Be Caused by Electricity or Visual Display Units: Oral Aspects." *Scandinavian Journal of Dental Research* 102, no. 1: 41–5. doi:10.1111/j.1600-0722.1994.tb01150.x.

Berglund, Birgitta, Peter Hassmén, and Anna Preis. 2002. "Annoyance and Spectral Contrast Are Cues for Similarity and Preference of Sounds." *Journal of Sound and Vibration* 250, no. 1: 53–64. doi:10.1006/jsvi.2001.3889.

Berglund, Birgitta, Peter Hassmén, and Raymond F. Soames Job. 1996. "Sources and Effects of Low-Frequency Noise." *Journal of the Acoustical Society of America* 99, no. 5: 2985–3002. doi:10.1121/1.414863.

Berglund, Birgitta, Thomas Lindvall, and Dietrich H. Schwela. 1999. *Guidelines for Community Noise*. Geneva: World Health Organization. Accessed 30 December 2018.

Bernardi, Luciano, Cesare Porta, and Peter Sleight. 2006. "Cardiovascular, Cerebrovascular, and Respiratory Changes Induced by Different Types of Music in Musicians and Non-Musicians: The Importance of Silence." *Heart* 92, no. 4: 445–52. doi:10.1136/hrt.2005.064600.

Bess, Fred H., and Larry E. Humes. 2008. *Audiology: The Fundamentals*. 4th ed. Philadelphia: Lippincott Williams & Wilkins.

Bijsterveld, Karin. 2001. "The Diabolical Symphony of the Mechanical Age: Technology and Symbolism of Sound in European and North American Noise Abatement Campaigns, 1900–40." *Social Studies of Science* 31, no. 1: 37–70. doi:10.1177/030631201031001003.

– 2008. *Mechanical Sound: Technology, Culture, and Public Problems of Noise in the Twentieth Century*. Cambridge, MA: MIT Press.

Blackwell, Debra L., Jacqueline W. Lucas, and Tainya C. Clarke. 2014. "Summary Health Statistics for U.S. Adults: National Health Interview Survey, 2012." *Vital and Health Statistics* 10, no. 260: 1–161. https://www.cdc.gov/nchs /data/series/sr_10/sr10_260.pdf.

Blesser, Barry. 2007. "The Seductive (Yet Destructive) Appeal of Loud Music." *eContact!* 9, no. 4. http://cec.sonus.ca/econtact/9_4/blesser.html.

Blesser, Barry, and Linda-Ruth Salter. 2006. *Spaces Speak, Are You Listening? Experiencing Aural Architecture*. Cambridge, MA: MIT Press.

– 2008. "The Unexamined Rewards for Excessive Loudness." In *Proceedings of the 9th International Congress on Noise as a Public Health Problem*, 254–61. Foxwoods, CT: International Commission on Biological Effects of Noise. http://citeseerx.ist.psu.edu/viewdoc/summary?doi=10.1.1.501.2230.

Block, Melissa. 2005. "For a Non-Lethal Weapon, Israel Uses Sound." National Public Radio, 13 June. https://www.npr.org/templates/story/story.php ?storyId=4701588.

Blomberg, Leslie D. 2014. "Noise Pollution in the 21st Century." *Journal of the Acoustical Society of America* 135, no. 4: 2271. doi:10.1121/1.4877441.

Blood, Anne J., and Robert J. Zatorre. 2001. "Intensely Pleasurable Responses to Music Correlate with Activity in Brain Regions Implicated in Reward and Emotion." *Proceedings of the National Academy of Sciences of the United States of America* 98, no. 20: 11818–23. doi:10.1073/pnas.191355898.

Bluhm, Gösta, Emma Nordling, and Niklas Berglind. 2004. "Road Traffic Noise and Annoyance: An Increasing Environmental Health Problem." *Noise and Health* 6, no. 24: 43–9. http://www.noiseandhealth.org/text.asp?2004/6/24/43/31656.

Bodin, Theo, Maria Albin, Jonas Ardö, Emilie Stroh, Per-Olof Östergren, and Jonas Björk. 2009. "Road Traffic Noise and Hypertension: Results from a Cross-Sectional Public Health Survey in Southern Sweden." *Environmental Health* 8, no. 38. doi:10.1186/1476-069X-8-38.

Boehm, Heidi, and Stacy Morast. 2009. "Quiet Time: A Daily Period without Distractions Benefits Both Patients and Nurses." *American Journal of Nursing* 109, no. 11: 29–32. doi:10.1097/01.NAJ.0000362015.55128.54.

Boets, Bart, Pol Ghesquière, Astrid van Wieringen, and Jan Wouters. 2007. "Speech Perception in Preschoolers at Family Risk for Dyslexia: Relations with Low-Level Auditory Processing and Phonological Ability." *Brain and Language* 101, no. 1: 19–30. doi:10.1016/j.bandl.2006.06.009.

Boets, Bart, Jan Wouters, Astrid van Wieringen, and Pol Ghesquière. 2007. "Auditory Processing, Speech Perception and Phonological Ability in Pre-School Children at High-Risk for Dyslexia: A Longitudinal Study of the Auditory Temporal Processing Theory." *Neuropsychologia* 45, no. 8: 1608–20. doi:10.1016/j.neuropsychologia.2007.01.009.

Bogoch, Isaac I., Ronald A. House, and Irena Kudla. 2005. "Perceptions about Hearing Protection and Noise-Induced Hearing Loss of Attendees of Rock Concerts." *Canadian Journal of Public Health* 96, no. 1: 69–72. http://journal.cpha.ca/index.php/cjph/article/view/583/583.

Bohlin, Margareta C., and Soly I. Erlandsson. 2007. "Risk Behaviour and Noise Exposure among Adolescents." *Noise and Health* 9, no. 36: 55–63. http://www.noiseandhealth.org/text.asp?2007/9/36/55/36981.

Borchgrevink, Hans M. 2003. "Does Health Promotion Work in Relation to Noise?" *Noise and Health* 5, no. 18: 25–30. http://www.noiseandhealth.org/text.asp?2003/5/18/25/31821.

Bosmans, Annick M.M., and Luk Warlop. 2005. "How Vulnerable Are Consumers to Blatant Persuasion Attempts?" *Social Science Research Network*. 15 December. doi:10.2139/ssrn.869978.

Bostick, James H. 2013. "Man-Portable Non-Lethal Pressure Shield." US Patent

US20110235467A1, filed 31 December 2010, and issued 26 March 2013. https://patents.google.com/patent/US20110235467A1/en.

Bradshaw, David H., Gary W. Donaldson, Robert C. Jacobson, Yoshio Nakamura, and C. Richard Chapman. 2011. "Individual Differences in the Effects of Music Engagement on Responses to Painful Stimulation." *Journal of Pain* 12, no. 12: 1262–73. doi:10.1016/j.jpain.2011.08.010.

Bradt, Joke, Cheryl Dileo, Denise Grocke, Lucanne Magill, and Aaron Teague. 2016. "Music Interventions for Improving Psychological and Physical Outcomes in Cancer Patients." *Cochrane Database of Systematic Reviews* 2016, no. 8: CD006911. doi:10.1002/14651858.CD006911.pub3.

Bradt, Joke, Minjung Shim, and Sherry W. Goodill. 2015. "Dance/Movement Therapy for Improving Psychological and Physical Outcomes in Cancer Patients." *Cochrane Database of Systematic Reviews* 2015, no. 1. doi:10.1002/14651858.CD007103.pub3.

Brammer, Anthony J., and Chantal Laroche. 2012. "Noise and Communication: A Three-Year Update." *Noise and Health* 14, no. 61: 281–6. doi:10.4103/1463-1741.104894.

Britten, Nick. "Aerospace Firm Bans iPods over Hearing Concerns." *Telegraph*, 1 July 2009. http://www.telegraph.co.uk/technology/apple/5711455/Aerospace-firm-bans-iPods-over-hearing-concerns.html.

Bronzaft, Arline L. n.d. "A Quieter School: An Enriched Learning Environment." Accessed July 2019. https://www.musicmotion.com/images/quieter_school.pdf.

– 1988. "Noise and Health: A Warning to Adolescents." *Children's Environments Quarterly* 5, no. 2: 40–5. https://www-jstor-org.ezproxy.lib.ucalgary.ca/stable/pdf/41514672.pdf?refreqid=excelsior%3A495f0f96bbaaf6babcf9fc7a017c2f5d.

Bronzaft, Arline, and Dennis P. McCarthy. 1975. "The Effect of Elevated Train Noise on Reading Ability." *Environment and Behavior* 7, no. 4: 517–28. doi:10.1177/001391657500700406.

Brookhouser, Patrick E., Don W. Worthington, and William J. Kelly. 1992. "Noise-Induced Hearing Loss in Children." *Laryngoscope* 102, no. 6: 645–55. doi:10.1288/00005537-199206000-00010.

Brown and Associates Planning Group and Russ Gerrish Consulting. 2003. "Spy Hill Lands Development Project: Phase 1 Report." Prepared for Alberta Transportation and Alberta Infrastructure. February. Accessed 3 January 2019. http://www.transportation.alberta.ca/Content/docType182/Production/Phase1Rept.pdf.

Brown, Gemma. 2009. "NICU Noise and the Preterm Infant." *Neonatal Network* 28, no. 3: 165–73. doi:10.1891/0730-0832.28.3.165.

Bruck, Margaret. 1987. "The Adult Outcomes of Children with Learning Disabilities." *Annals of Dyslexia* 37, no. 1: 252–63. doi:10.1007/BF02648071.

Bull, Michael. 2004. "Thinking about Sound, Proximity, and Distance in Western Experience: The Case of Odysseus's Walkman." In *Hearing Cultures: Essays on Sound, Listening and Modernity*, edited by Veit Erlmann, 173–90. Oxford: Berg Publishers.

– 2007. *Sound Moves: iPod Culture and Urban Experience*. London: Routledge.

Bull, Michael, and Les Back, eds. 2003. *The Auditory Culture Reader*. Oxford: Berg.

Burnett, Charles, Michael Fend, and Penelope Gouk, eds. 1991. *The Second Sense: Studies in Hearing and Musical Judgement from Antiquity to the Seventeenth Century*. London: Warburg Institute, University of London.

Burnham, Denis, Kaoru Sekiyama, and Doğu Erdener. 2008. "Cross-Language Auditory-Visual Speech Perception Development." *Journal of the Acoustical Society of America* 123, no. 5. doi:10.1121/1.2935787.

Busch-Vishniac, Ilene J., James E. West, Colin Barnhill, Tyrone Hunter, Douglas Orellana, and Ram Chivukula. 2005. "Noise Levels in Johns Hopkins Hospital." *Journal of the Acoustical Society of America* 118, no 6: 3629–45. doi:10.1121/1.2118327.

Buss, Claudia, Elysia Poggi Davis, Quetzal A. Class, Matt Gierczak, Carol Pattillo, Laura M. Glynn, and Curt A. Sandman. 2009. "Maturation of the Human Fetal Startle Response: Evidence for Sex-Specific Maturation of the Human Fetus." *Early Human Development* 85, no. 10: 633–8. doi:10.1016/j.earlhumdev.2009.08.001.

Bütikofer, R. 2012. "Airport Noise." In *Noise Mapping in the EU: Models and Procedures*, edited by Gaetano Licitra, 129–58. Boca Raton, FL: CRC Press.

Byers, Robert, and Bill Gordon. 2011. "Wearing Earphones at Work?" *Ask the Workplace Doctors*, 30 October. Accessed 9 January 2019. http://workplacedr .comm.kent.edu/wordpress/wearing-earphones-at-work/.

Cacioppo, John T., and Richard E. Petty. 1989. "Effects of Message Repetition on Argument Processing, Recall, and Persuasion." *Basic and Applied Social Psychology* 10, no. 1: 3–12. doi:10.1207/s15324834basp1001_2.

Cadigan, Mary E., Nancy A. Caruso, Sioban M. Haldeman, Mary Ellen McNamara, Dorothy A. Noyes, M. Ann Spadafora, and Diane L. Carroll. 2001. "The Effects of Music on Cardiac Patients on Bed Rest." *Progress in Cardiovascular Nursing* 16, no. 1: 5–13. doi:10.1111/j.0889-7204.2001.00797.x.

Cain, Rebecca, Paul Jennings, and John Poxon. 2013. "The Development and Application of the Emotional Dimensions of a Soundscape." *Applied Acoustics* 74, no. 2: 232–9. doi:10.1016/j.apacoust.2011.11.006.

Camara, Jorge G., Joseph M. Ruszkowski, and Sandra R. Worak. 2008. "The Effect of Live Classical Piano Music on the Vital Signs of Patients Undergoing Ophthalmic Surgery." *Medscape Journal of Medicine* 10, no. 6: 149. https://www.ncbi.nlm.nih.gov/pmc/articles/PMC2491669/.

Carmichael, Stephanie. 2017. "3 Free Classroom Noise Level Monitors." *Classcraft Blog*, 11 September. https://www.classcraft.com/blog/features/classroom-noise-level-monitors/.

Carroll, Terry. 2019. "Benching: An Important Piece of Today's Workplace." Accessed 21 February. https://www.kimball.com/getattachment/2855f215-4f0b-4751-871b-1fdbd6f89753/Benching_Important_Piece_Dec2018.pdf.

Carter, Paul. 2004. "Ambiguous Traces, Mishearing, and Auditory Space." In *Hearing Cultures: Essays on Sound, Listening and Modernity*, edited by Veit Erlmann, 43–63. Oxford: Berg Publishers.

Carter, Sarah. 2010. "Vuvuzela-Haters Rush for World Cup Ear Plugs." *CBS News*, 15 June. http://www.cbsnews.com/8301-31751_162-20007720-10391697.html.

Cassano, E., P. Bavaro, Ingrid Aloise, Elena Bobbio, and Margareth Renna. 2008. "Music through Earphones: An Underestimated Risk." English abstract of article in Italian. *La Medicina del Lavoro* 99, no. 5: 362–65. https://europepmc.org/abstract/MED/18828535.

Cate, Anthony D., Timothy J. Herron, E. William Yund, G. Christopher Stecker, Teemu Rinne, Xiaojian Kang, Christopher I. Petkov, Elizabeth A. Disbrow, and David L. Woods. 2009. "Auditory Attention Activates Peripheral Visual Cortex." *PLoS One* 4, no. 2: e4645. doi:10.1371/journal.pone.0004645.

Cepeda, M. Soledad, Daniel B. Carr, Joseph Lau, and Hernando Alvarez. 2006. "Music for Pain Relief." *Cochrane Database of Systematic Reviews* 2. doi:10.1002/14651858.CD004843.pub2.

Čeponienė, Rita, Elisabet Service, Sanna Kurjenluoma, Marie Cheour, and Risto Näätänen. 1999. "Children's Performance on Pseudoword Repetition Depends on Auditory Trace Quality: Evidence from Event-Related Potentials." *Developmental Psychology* 35, no. 3: 709–20. doi:10.1037/0012-1649.35.3.709.

Chait, Maria, Guinevere Eden, David Poeppel, Jonathan Z. Simon, Deborah F. Hill, and D. Lynn Flowers. 2007. "Delayed Detection of Tonal Targets in Background Noise in Dyslexia." *Brain and Language* 102, no. 1: 80–90. doi:10.1016/j.bandl.2006.07.001.

Chang, Charis. 2016. "Long Range Acoustic Devices Terrorising Protesters." *News.com.au*, 22 December. https://www.news.com.au/technology/innovation/long-range-acoustic-devices-terrorising-protesters/news-story/83a47dbd6e64095549f09a89ef020f94.

Chang, E.F., and M.M. Merzenich. 2003. "Environmental Noise Retards Auditory Cortical Development." *Science* 300 (5618): 498–502. doi:10.1037/0012-1649.35.3.709.

Chapman, Simon, Alexis St George, Karen Waller, and Vince Cakic. 2013. "The Pattern of Complaints about Australian Wind Farms Does Not Match the Establishment and Distribution of Turbines: Support for the Psychogenic, 'Communicated Disease' Hypothesis." *PLoS One* 8, no. 10: e76584. doi:10.1371/journal.pone.0076584.

Charbonneau, Danielle, and Catherine Goldschmidt, with Richard Larocque. 2004. "Safety of Noisy Toys: A Current Assessment." Office of Consumer Affairs, Industry Canada. https://option-consommateurs.org/wp-content/uploads/2017/08/oc-noisy-toys-eng1004.pdf.

Chauhan, Avnish. 2008. "Study of Noise Pollution Level in Different Places of Haridwar and Dehradun City (India)." *Environment Conservation Journal* 9, no. 3: 21–5. https://www.cabdirect.org/cabdirect/abstract/20093085007.

Chavalitsakulchi, Pranee, Tsuyoshi Kawakami, Udomsak Kongmuang, Pongkasew Vivatjestsadawut, and Winit Leongsrisook. 1989. "Noise Exposure and Permanent Hearing Loss of Textile Workers in Thailand." *Industrial Health* 27, no. 4: 165–73. https://doi.org/10.2486/indhealth.27.165.

Chen, Bo, Zhiyi Bao, and Zhujun Zhu. 2006. "Assessing the Willingness of the Public to Pay to Conserve Urban Green Space: The Hangzhou City, China, Case." *Journal of Environmental Health* 69, no. 5: 26–30. https://www.jstor.org/stable/44532793.

Chepesiuk, Ron. 2005. "Decibel Hell: The Effects of Living in a Noisy World." *Environmental Health Perspectives* 113, no. 1: A34–41. doi:10.1289/ehp.113-a34.

Chou, Peter T. 2010. "Attention Drainage Effect: How Background Music Effects Concentration in Taiwanese College Students." *Journal of the Scholarship of Teaching and Learning* 10, no. 1: 36–46. https://scholarworks.iu.edu/journals/index.php/josotl/article/view/1733.

Clark, Charlotte, Rocio Martin, Elise van Kempen, Tamuno Alfred, Jenny Head, Hugh W. Davies, Mary M. Haines, Isabel L. Barrio, Mark Matheson, and Stephen A. Stansfield. 2006. "Exposure-Effect Relations between Aircraft and Road Traffic Noise Exposure at School and Reading Comprehension: The RANCH Project." *American Journal of Epidemiology* 163, no. 1: 27–37. doi:10.1093/aje/kwj001.

Clark, David. 2007. "From Margins to Centre: A Review of the History of Palliative Care in Cancer." *Lancet Oncology* 8, no. 5: 430–8. doi:10.1016/S1470-2045(07)70138-9.

Coates, Peter A. 2005. "The Strange Stillness of the Past: Toward an Environmental History of Sound and Noise." *Environmental History* 10, no. 4: 636–65. doi:10.1093/envhis/10.4.636.

Coghaln, Andy. 2007. "Plant Genes Switched on by Sound Waves." *New Scientist*, 29 August. http://www.newscientist.com/article/mg19526196.100-plant-genes-switched-on-by-sound-waves.html.

Cohen, S. Robin, Patricia Boston, Balfour M. Mount, and Pat Porterfield. 2001. "Changes in Quality of Life Following Admission to Palliative Care Units." *Palliative Medicine* 15, no. 5: 363–71. doi:10.1191/026921601680419401.

Cohen, Sheldon, and Shirlynn Spacapan. 1984. "The Social Psychology of Noise." In *Noise and Society*, edited by Dylan M. Jones and Antony J. Chapman, 221–45. Chichester, UK: Wiley.

Cohen, Sheldon, and Neil Weinstein. 1981. "Nonauditory Effects of Noise on Behavior and Health." *Journal of Social Issues* 37, no. 1: 36–70. doi:10.1111/j.1540-4560.1981.tb01057.x.

Columbia Hearing Center. 2017. "More Teens Showing Signs of Hearing Loss." Accessed 9 January 2017. http://columbiamohearingcenter.com/2014/10/more-teens-showing-signs-of-hearing-loss/.

Connor, Alison, and Elizabeth Ortiz. 2009. "Staff Solutions for Noise Reduction in the Workplace." *Permanente Journal* 13, no. 4: 23–7. doi:10.7812/TPP/09-057.

Connor, Steven. 2004. "Edison's Teeth: Touching Hearing." In *Hearing Cultures: Essays on Sound, Listening and Modernity*, edited by Veit Erlmann, 153–72. Oxford: Berg Publishers.

Cordova, Alfredo C., Kartik Logishetty, James A. Fauerbach, Leigh A. Price, B. Robert Gibson, and Stephen M. Milner. 2013. "Noise Levels in a Burn Intensive Care Unit." *Burns* 39, no. 1: 44–8. doi:10.1016/j.burns.2012.02.033.

"Court Rules That Law Restricting Car Stereo Volume Is Unconstitutional." 2011. *Tampa Bay Times*. 11 May.

Cowan, James P. 2008. "The Results of Hum Studies in the United States." In *Proceedings of the 9th International Congress on Noise as a Public Health Problem*, 626–30. Foxwoods, CT: International Commission on Biological Effects of Noise. http://www.icben.org/2008/PDFs/Cowan.pdf.

Crandell, Carl C., and Joseph J. Smaldino. 1999. "Improving Classroom Acoustics: Utilizing Hearing-Assistive Technology and Communication Strategies in the Educational Setting." *Volta Review* 101, no. 5: 47–62. https://eric.ed.gov/?id=EJ627953.

Crichton, Fiona, George Dodd, Gian Schmid, and Keith J. Petrie. 2015. "Framing Sound: Using Expectations to Reduce Environmental Noise Annoyance." *Environmental Research* 142: 609–14. doi:10.1016/j.envres.2015.08.016.

Cruickshanks, Karen J., Ronald Klein, Barbara E.K. Klein, Terry L. Wiley, David M. Nondahl, and Ted S. Tweed. 1998. "Cigarette Smoking and Hearing Loss:

The Epidemiology of Hearing Loss Study." *Journal of the American Medical Association* 279, no. 21: 1715–19. doi:10.1001/jama.279.21.1715.

Cusick, Suzanne G. 2006. "Music as Torture / Music as Weapon." *Revista Transcultural de Música / Transcultural Music Review* 10. https://www.sibetrans.com /trans/articulo/152/music-as-torture-music-as-weapon.

Cutshall, Susanne M., Patricia G. Anderson, Sharon K. Prinsen, Laura J. Wentworth, Tammy L. Olney, Penny K. Messner, Karen M. Brekke, et al. 2011. "Effect of the Combination of Music and Nature Sounds on Pain and Anxiety in Cardiac Surgical Patients: A Randomized Study." *Alternative Therapies in Health and Medicine* 17, no. 4: 16–23. https://mayoclinic.pure.elsevier.com /en/publications/effect-of-the-combination-of-music-and-nature-sounds-on-pain-and-.

Danhauer, Jeffrey L., Carole E. Johnson, Anne Byrd, Laura DeGood, Caitlin Meuel, Angela Pecile, and Lindsey L. Koch. 2009. "Survey of College Students on iPod Use and Hearing Health." *Journal of the American Academy of Audiology* 20, no. 1: 5–27. doi:10.3766/jaaa.20.1.2.

Danhauer, Jeffrey L., Carole E. Johnson, Aislinn F. Dunne, Matthew D. Young, Suzanne N. Rotan, Tasha A. Snelson, Jennifer S. Stockwell, and Michelle J. McLain. 2012. "Survey of High School Students' Perceptions about Their iPod Use, Knowledge of Hearing Health, and Need for Education." *Language, Speech, and Hearing Services in Schools* 43, no. 1: 14–35. doi:10.1044/0161-1461(2011/10-0088).

Daniel, Eileen. 2007. "Noise and Hearing Loss: A Review." *Journal of School Health* 77, no. 5: 225–31. doi:10.1111/j.1746-1561.2007.00197.x.

Daniell, William E., Deborah Fulton-Kehoe, Martin Cohen, Susan S. Swan, and Gary M. Franklin. 2002. "Increased Reporting of Occupational Hearing Loss: Workers' Compensation in Washington State, 1984–1998." *American Journal of Industrial Medicine* 42, no. 6: 502–10. doi:10.1002/ajim.10146.

Daniell, William E., Susan S. Swan, Mary M. McDaniel, Janice E. Camp, Michael A. Cohen, and John G. Stebbins. 2006. "Noise Exposure and Hearing Loss Prevention Programmes after 20 Years of Regulations in the United States." *Occupational and Environmental Medicine* 63, no. 5: 343–51. doi:10.1136/oem.2005 .024588. Correction, doi:10.1136/oem.2005.024588corr1.

Daniell, William E., Susan S. Swan, Mary M. McDaniel, John G. Stebbins, Noah S. Seixas, and Michael S. Morgan. 2002. "Noise Exposure and Hearing Conservation Practices in an Industry with High Incidence of Workers' Compensation Claims for Hearing Loss." *American Journal of Industrial Medicine* 42, no. 4: 309–17. doi:10.1002/ajim.10124.

Darcy, Ashley E., Lauren E. Hancock, and Emily J. Ware. 2008. "A Descriptive Study of Noise in the Neonatal Intensive Care Unit Ambient Levels and Perceptions of Contributing Factors." *Advances in Neonatal Care* 8, no. 3: 165–75. doi:10.1097/01.ANC.0000324341.24841.6e.

Darroch, Manpreet. 2010. "Tune into Traffic: Earphone Advert." *Injury Prevention* 16, no. 1: A286. doi:10.1136/ip.2010.029215.1025.

Dassa, Ayelet, and Dorit Amir. 2014. "The Role of Singing Familiar Songs in Encouraging Conversation among People with Middle to Late Stage Alzheimer's Disease." *Journal of Music Therapy* 51, no. 2: 131–53. doi:10.1093/jmt/thu007.

Davies, Hugh W., Kay Teschke, Susan M. Kennedy, Murray R. Hodgson, and Paul A. Demers. 2008. "Occupational Noise Exposure and Hearing Protector Use in Canadian Lumber Mills." *Journal of Occupational and Environmental Hygiene* 6, no. 1: 32–41. doi:10.1080/15459620802548940.

Dean, Roger T., Freya Bailes, and Emery Schubert. 2011. "Acoustic Intensity Causes Perceived Changes in Arousal Levels in Music: An Experimental Investigation." *PLOS One* 6, no. 4: e18591. doi:10.1371/journal.pone.0018591.

De Croon, Einar M., Judith K. Sluiter, P. Paul Kuijer, and Monique H. Frings-Dresen. 2005. "The Effect of Office Concepts on Worker Health and Performance: A Systematic Review of the Literature." *Ergonomics* 48, no. 2: 119–34. doi:10.1080/00140130512331319409.

Degeest, Sofie, Paul Corthals, Bart Vinck, and Hannah Keppler. 2014. "Prevalence and Characteristics of Tinnitus after Leisure Noise Exposure in Young Adults." *Noise and Health* 16, no. 68: 26–33. doi:10.4103/1463-1741.127850.

de Jong, Ronald G., Tammo Houtgast, Ellis A.M. Franssen, and Winni F. Hofman, eds. 2003. *Proceedings of the 8th International Congress on Noise as a Public Health Problem*. Schiedam, NL: International Commission on Biological Effects of Noise. Accessed 2 February 2019. http://www.gbv.de/dms/tib-ub-hannover/367330717.pdf.

Delaney, Lori J., Marian J. Currie, Hsin-Chia C. Huang, Violeta Lopez, Edward Litton, and Frank Van Haren. 2017. "The Nocturnal Acoustic Intensity of the Intensive Care Environment: An Observational Study." *Journal of Intensive Care* 5, no. 41. doi:10.1186/s40560-017-0237-9.

DeLoach, Alana G., Jeff P. Carter, and Jonas Braasch. 2015. "Tuning the Cognitive Environment: Sound Masking with 'Natural' Sounds in Open-Plan Offices." *Journal of the Acoustical Society of America* 137, no. 4: 2291. doi:10.1121/1.4920363.

Dennis, Christina M., Robert Lee, Elizabeth K. Woodard, Jeffrey J. Szalaj, and Catrice A. Walker. 2010. "Benefits of Quiet Time for Neuro-Intensive Care Patients." *Journal of Neuroscience Nursing* 42, no. 4: 217–24. doi:10.1097/JNN .0b013e3181e26c20.

Desourdie, Todd. "Municipal Corporations: By-Laws, Permits and Legal Proceedings." In *The Canadian Encyclopedic Digest Western*. 4th ed. Vol. 41A, Title 107.1. Toronto: Carswell, 2010.

Dettoni, Josilene L., Fernanda M. Consolim-Colombo, Luciano F. Drager, Marcelo C. Rubira, Silvia B.P.C. de Souza, Maria C. Irigoyen, Cristiano Mostarda, et al. 2012. "Cardiovascular Effects of Partial Sleep Deprivation in Healthy Volunteers." *Journal of Applied Physiology* 113, no. 2: 232–6. doi:10.1152/japplhysiol.01604.2011.

de Vos, P. 2012."Railway Noise." In *Noise Mapping in the EU: Models and Procedures*, edited by Gaetano Licitra, 81–108. Boca Raton, FL: CRC Press.

Dickson, Neil. 2014. "Aircraft Noise Technology and International Noise Standards." ICAO Environment Committee on Aviation Environmental Protection. https://www.icao.int/Meetings/EnvironmentalWorkshops/Documents /Env-Seminars-Lima-Mexico/Lima/10_ICAO_NoiseTech-Standards.pdf.

Divon, Michael Y., Lawrence D. Platt, Cathy J. Cantrell, Carl V. Smith, Sze-Ya Yeh, and Richard H. Paul. 1985. "Evoked Fetal Startle Response: A Possible Intrauterine Neurological Examination." *American Journal of Obstetrics and Gynecology* 153, no. 4: 454–6. doi:10.1016/0002-9378(85)90086-9.

Dobnik, Verena. 2007. "African Drummers Face 'New Harlemites.'" Associated Press, 12 August: http://nigeriavillagesquare.com/forum/lounge/11949-african-drummers-face-new-harlemites.html.

Drobnick, Jim, ed. 2006. *The Smell Culture Reader*. Oxford: Berg.

Drouot, Xavier, Belen Cabello, Marie-Pia d'Ortho, and Laurent Brochard. 2008. "Sleep in the Intensive Care Unit." *Sleep Medicine Reviews* 12, no 5: 391–403. doi:10.1016/j.smrv.2007.11.004.

Dubbs, Dana. 2004. "Sound Effects: Design and Operations Solutions to Hospital Noise." *Health Facilities Management* 17, no. 9: 14–18. https://www.ncbi .nlm.nih.gov/pubmed/15478713.

Duck, Francis A. 2007. "Medical and Non-Medical Protection Standards for Ultrasound and Infrasound." *Progress in Biophysics and Molecular Biology* 93, no. 1: 176–91. doi:10.1016/j.pbiomolbio.2006.07.008.

Dzhambov, Angel, and Donka Dimitrova. 2014. "Neighborhood Noise Pollution as a Determinant of Displaced Aggression: A Pilot Study." *Noise and Health* 16: 95–101. http://www.noiseandhealth.org/text.asp?2014/16 /69/95/132090.

Ecob, Russell, Graham Sutton, Alicja Rudnicka, Pauline Smith, Chris Power, David Strachan, and Adrian Davis. 2008. "Is the Relation of Social Class to Change in Hearing Threshold Levels from Childhood to Middle Age Explained by Noise, Smoking, and Drinking Behaviour?" *International Journal of Audiology* 47, no. 3: 100–8. doi:10.1080/14992020701647942.

Edelstein, Miren, David Brang, Romke Ruow, and Vilayanur S. Ramachandran.

2013. "Misophonia: Physiological Investigations and Case Descriptions." *Frontiers in Human Neuroscience* 7: 296. doi:10.3389/fnhum.2013.00296.

Eggemann, C., M. Koester, and Patrick Zorowka. 2002. "Hearing Loss Due to Leisure Time Noise Is on the Rise. The Ear also Needs a Rest Period." English abstract of article in German. *MMW Fortschritte der Medezin* 144, no. 49: 30–3. https://www.ncbi.nlm.nih.gov/pubmed/12577736.

Eggermont, Jos J. 2013. "Hearing Loss, Hyperacusis, or Tinnitus: What Is Modeled in Animal Research?" *Hearing Research* 295: 140–9. doi:10.1016/j.heares.2012.01.005.

Eggertson, Laura. 2012. "Hospital Noise." *Canadian Nurse* 108, no. 4: 28–31. https://www.canadian-nurse.com/articles/issues/2012/april-2012/hospital-noise.

Eivazzadeh, Mehran, Sahra Vahedi, Mohammad A. Jafarabadi, and Akbar Gholampour. 2017. "Noise Pollution as a Public Health Concern in Pediatric Hospitals: A Case Study in Tabriz, Iran." *Journal of Air Pollution and Health* 2, no. 2: 87–94. Accessed 6 January 2019. http://japh.tums.ac.ir/index.php/japh/article/view/104.

Ekirch, A. Roger. 2006. *At Day's Close: Night in Times Past*. New York: W.W. Norton.

El Dib, Regina P., Edina M. Silva, José F. Morais, and Virgínia F. Trevisani. 2008. "Prevalence of High Frequency Hearing Loss Consistent with Noise Exposure among People Working with Sound Systems and General Population in Brazil: A Cross-Sectional Study." *BioMedCentral Public Health* 8: 151. doi:10.1186/1471-2458-8-151.

Ellenbogen, Jeffrey, Sheryl Grace, Wendy J. Heiger-Bernays, James F. Manwell, Dora A. Mills, Kimberly A. Sullivan, and Marc G. Weisskopf. 2012. "Wind Turbine Health Impact Study: Report of Independent Expert Panel." January. Prepared for the Massachusetts Department of Environmental Protection and the Massachusetts Department of Public Health. Accessed 7 January 2019. http://www.scotianwindfields.ca/sites/default/files/publications/massachusetts_turbine_impact_study_2012.pdf.

Elliott, Robert, and Leslie S. Greenberg. 1997. "Multiple Voices in Process-Experiential Therapy: Dialogues between Aspects of the Self." *Journal of Psychotherapy Integration* 7, no. 3: 225–39. doi:10.1037/h0101127.

Elliott, Rosalind M., Sharon M. McKinley, and David Eager. 2010. "A Pilot Study of Sound Levels in an Australian Adult General Intensive Care Unit." *Noise and Health* 12, no. 46: 26–36. doi:10.4103/1463-1741.59997.

Eltiti, Stacy, Denise Wallace, Anna Ridgewell, Konstantina Zougkou, Ricardo Russo, Francisco Sepulveda, Dariush Mirshekar-Syahkal, et al. 2007. "Does Short-Term Exposure to Mobile Phone Base Station Signals Increase Symptoms in Individuals Who Report Sensitivity to Electromagnetic Fields? A Double-Blind Randomized Provocation Study." *Environmental Health Perspectives* 115, no. 11: 1603–8. doi:10.1289/ehp.10286.

Emmerich, Edeltraut, Lars Rudel, and Frank Richter. 2008. "Is the Audiologic Status of Professional Musicians a Reflection of the Noise Exposure in Classical Orchestral Music?" *European Archives of Oto-Rhino-Laryngology* 265, no. 7: 753–8. doi:10.1007/s00405-007-0538-z.

Enbom, Håkan, and Inga M. Enbom. 2013. "Infrasound from Wind Turbines: An Overlooked Health Hazard." English abstract of article in Swedish. *Läkartidningen* 110 (2013): 1388–9. Accessed 7 January 2019. https://waubrafoundation.org.au/wp-content/uploads/2013/08/Enbom-H-I.-Infrasound-an-Overlooked-Health-Hazard_.pdf.

Epstein, Marcia Jenneth. 1995. "Accessing Trauma through the Voice." In *Caring for Your Voice: Teachers and Coaches*, edited by E. Lisbeth Donaldson, 113–32. Toronto: Detselig.

– 2003. "Growing an Interdisciplinary Hybrid: The Case of Acoustic Ecology." *History of Intellectual Culture* 3, no. 1. https://www.ucalgary.ca/hic/issues/vol3/9.

– 2004a. "The Acoustics of Health." In *Creating Connections between Nursing Care and the Creative Arts Therapies: Expanding the Concept of Holistic Care*, edited by Carole-Lynne Le Navenec and Laurel Bridges, 187–97. Springfield, IL: Charles C. Thomas.

– 2004b. "Teaching a 'Humanistic' Science: Reflections on Interdisciplinary Course Design at the Post-Secondary Level." *Current Issues in Education* 7, no. 3. https://cie.asu.edu/ojs/index.php/cieatasu/article/view/822/248.

– 2016. "Soundscaping Health: Resonant Speculations." *SoundEffects: An Interdisciplinary Journal of Sound and Sound Experience* 6, no. 1: 105–20. https://www.soundeffects.dk/article/view/24916.

Erlmann, Veit, ed. 2004. *Hearing Cultures: Essays on Sound, Listening and Modernity*. Oxford: Berg Publishers,

Ernst, Liz. 2012. "Noise Related Hearing Loss in Industrialized Nations: New Discoveries through Electron Tomography." *Acoustiblok*. 4 May. Accessed 7 January 2019. http://info.acoustiblok.com/blog/author/liz-ernst/page/5.

European Commission. 2002. "The Environmental Noise Directive (2002/49/EC)." Accessed 7 January 2019. http://ec.europa.eu/environment/archives/noise/directive.htm.

European Environment Agency. 2014. *Noise in Europe*. European Environment Agency Report no. 10. Luxembourg: Publications Office of the European Union. Accessed 7 January 2019. https://www.eea.europa.eu/publications/noise-in-europe-2014.

Evans, David. 2002. "The Effectiveness of Music as an Intervention for Hospital Patients." *Journal of Advanced Nursing* 37, no. 1: 8–18. doi:10.1046/j.1365-2648.2002.02052.x.

Evers, Marco. 2005. "The Weapon of Sound: Sonic Cannon Gives Pirates an Earful." *Spiegel International*, 15 November. https://www.spiegel.de/international

/spiegel/the-weapon-of-sound-sonic-canon-gives-pirates-an-earful-a-385048.html.

Feder, Katya Polena, David Michaud, James McNamee, Elizabeth Fitzpatrick, Pamela Ramage-Morin, and Yves Beauregard. 2016. "Prevalence of Hearing Loss among a Representative Sample of Canadian Children and Adolescents, 3 to 19 Years of Age." *Ear and Hearing* 38, no. 1: 7–20. https://dx-doi-org.ezproxy.lib.ucalgary.ca/10.1097%2FAUD.000000000000345.

Fei, Chen, Zhu Yazhou, Yang Jun, and Xu Gang. 2012. "Field Performance Assessment of Porous Asphalt Pavement under Heavy Traffic on Yan-Tong Expressway in China." *Journal of Testing and Evaluation* 40, no. 7: 1238–43. doi:10.1520/JTE20120123.

Ferrite, Silvia, and Vilma Santana. 2005. "Joint Effects of Smoking, Noise Exposure and Age on Hearing Loss." *Occupational Medicine* 55, no. 1: 48–53. doi:10.1093/occmed/kqi002.

Filmer, Paul. 2003. "Songtime: Sound Culture, Rhythm and Sociality." In *The Auditory Culture Reader*, edited by Michael Bull and Les Back, 91–112. Oxford: Berg.

Firestone, Robert W. 1986. "The 'Inner Voice' and Suicide." *Psychotherapy: Theory, Research, Practice, Training* 23, no. 3: 439–47. doi:10.1037/h0085636.

Firestone, Robert W., and Lisa Firestone. 1998. "Voices in Suicide: The Relationship between Self-Destructive Thought Processes, Maladaptive Behavior, and Self-Destructive Manifestations." *Death Studies* 22, no 5: 411–43. doi:10.1080/074811898201443.

Fischlin, Daniel, Ajay Heble, and George Lipsitz. 2013. *The Fierce Urgency of Now: Improvisation, Rights, and the Ethics of Cocreation*. Durham, NC: Duke University Press.

Fiumara, Gemma C. 1990. *The Other Side of Language: A Philosophy of Listening*. Translated by Charles Lambert. London: Routledge.

Fleischer, G., E. Hoffmann, R. Lang, and R. Müller. 1999. "Documentation of the Effects of Child Cap Pistols." English abstract of article in German. *Hals-Nasen-Ohrenheilkunde* 47, no. 6: 535–40. Accessed 8 January 2019. https://study libde.com/doc/6899422/dokumentation-der-auswirkungen-von—hearing.

Fleming, Neil, and David Baume. 2006. "Learning Styles Again: VARKing Up the Right Tree!" *Educational Developments* 7, no. 4: 4–7. Accessed 8 January 2019. https://semcme.org/wp-content/uploads/Flora-Educational-Developments.pdf.

Fligor, Brian, J., and Lakeisha C. Cox. 2004. "Output Levels of Commercially Available Portable Compact Disc Players and the Potential Risk to Hearing." *Ear and Hearing* 25, no. 6: 513–27. doi:10.1097/00003446-200412000-00001.

Fogari, Roberto, Annalisa Zoppi, Luca Corradi, Gianluigi Marasi, Alessandro Vanasia, and Alberto Zanchetti. 2001. "Transient but Not Sustained Blood

Pressure Increments by Occupational Noise. An Ambulatory Blood Pressure Measurement Study." *Journal of Hypertension* 19, no. 6: 1021–7. doi:10.1097 /00004872-200106000-00005.

Folmer, Robert L. 2002. "Long-Term Reductions in Tinnitus Severity." *BioMed-Central Ear, Nose and Throat Disorders* 2, no. 3. doi:10.1186/1472-6815-2-3.

– 2006. "Noise-induced Hearing Loss in Young People." *Pediatrics* 117, no. 1: 248–9. doi:10.1542/peds.2005-1702.

Folmer, Robert L., and Susan E. Griest. 2000. "Tinnitus and Insomnia." *American Journal of Otolaryngology* 21, no. 5 (2000): 287–93. doi:10.1053/ajot .2000.9871.

Forbes, Daniel. 2003. "NYC Used City Vehicles, Extreme Noise against Peaceful Protesters." *BBS News*, 7 March. http://www.bbsnews.net/2003-03-07b.html.

Forquer, LeAnne M., and C. Merle Johnson. 2005. "Continuous White Noise to Reduce Resistance Going to Sleep and Night Wakings in Toddlers." *Child and Family Behavior Therapy* 27, no. 2: 1–10. doi:10.1300/J019v27n02_01.

Foy, George M. 2010. *Zero Decibels: The Quest for Absolute Silence.* New York: Scribner.

Franklin, Ursula. 2000. "Silence and the Notion of the Commons." *Soundscape* 1, no. 2. http://ecoear.proscenia.net/wfaelibrary/library/articles/franklin _commons.pdf.

Franklin, Ursula. 2006. *The Ursula Franklin Reader: Pacifism as a Map.* Toronto: Between the Lines.

Fransen, Erik, Vedat Topsakal, Jan-Jaap Hendrickx, Lut Van Laer, Jeroen R. Huyghe, Els Van Eyken, Nele Lemkens, et al. 2008. "Occupational Noise, Smoking, and a High Body Mass Index Are Risk Factors for Age-Related Hearing Impairment and Moderate Alcohol Consumption Is Protective: A European Population-Based Multicenter Study." *Journal of the Association for Research in Otolaryngology* 9, no. 3: 264–76. doi:10.1007/s10162-008-0123-1.

Freedman, Neil S., Joost Gazendam, Lachelle Levan, Allan I. Pack, and Richard J. Schwab. 2001. "Abnormal Sleep/Wake Cycles and the Effect of Environmental Noise on Sleep Disruption in the Intensive Care Unit." *American Journal of Respiratory and Critical Care Medicine* 163, no. 2: 451–7. doi:10.1164/ajrccm.163.2.9912128.

Frey, Allan H. 1962. "Human Auditory System Response to Modulated Electromagnetic Energy." *Journal of Applied Physiology* 17, no. 4: 689–92. doi:10.1152/jappl.1962.17.4.689.

Frey, Barbara J., and Peter J. Hadden. 2007. "Noise Radiation from Wind Turbines Installed Near Homes: Effects on Health." Accessed 8 January 2019. https://docs.wind-watch.org/wtnoisehealth.pdf.

Fritsch, Michael H., Chris E. Chacko, and Emily B. Patterson. 2010. "Operating

Room Sound Level Hazards for Patients and Physicians." *Otology and Neurotology* 31, no. 5: 715–21. doi:10.1097/MAO.0b013e3181d8d717.

Frosch, Franz G. 2013. "Hum and Otoacoustic Emissions May Arise Out of the Same Mechanisms." *Journal of Scientific Exploration* 27, no. 4: 603–24. https://www.scientificexploration.org/docs/27/jse_27_4_Frosch.pdf.

– 2017. "Possible Joint Involvement of the Cochlea and Semicircular Canals in the Perception of Low-Frequency Tinnitus, Also Called 'The Hum' or 'Taos Hum." *International Tinnitus Journal* 21, no. 1: 63–7. doi: 10.5935/0946-5448.20170012.

Fuente, Adrian, and Louise Hickson. 2011. "Noise-Induced Hearing Loss in Asia." *International Journal of Audiology* 50, no. 1: S3–10. doi:10.3109/14992027.2010.540584.

Furnham, Adrian, and Lisa Strbac. 2002. "Music Is as Distracting as Noise: The Differential Distraction of Background Music and Noise on the Cognitive Test Performance of Introverts and Extraverts." *Egonomics* 45, no. 3: 203–17. doi:10.1080/00140130210121932.

Gabor, Jonathan Y., Andrew B. Cooper, Shelley A. Crombach, Bert Lee, Nisha Kadikar, Harald E. Bettger, and Patrick J. Hanly. 2003. "Contribution of the Intensive Care Unit Environment to Sleep Disruption in Mechanically Ventilated Patients and Healthy Subjects." *American Journal of Respiratory and Critical Care Medicine* 167, no. 5: 708–15. doi:10.1164/rccm.2201090.

Gabor, Jonathan Y., Andrew B. Cooper, and Patrick J. Hanly. 2001. "Sleep Disruption in the Intensive Care Unit." *Current Opinion in Critical Care* 7, no. 1: 21–7. https://journals.lww.com/co-criticalcare/Abstract/2001/02000/Sleep_disruption_in_the_intensive_care_unit.4.aspx.

Gangi, Shabnam, and O. Johansson. 2000. "A Theoretical Model Based upon Mast Cells and Histamine to Explain the Recently Proclaimed Sensitivity to Electric and/or Magnetic Fields in Humans." *Medical Hypotheses* 54, no. 4: 663–671. doi:10.1054/mehy.1999.0923.

Gardner, Glenn, Christine Collins, Sonya Osborne, Amanda Henderson, and Misha Eastwood. 2009. "Creating a Therapeutic Environment: A Non-Randomised Controlled Trial of a Quiet Time Intervention for Patients in Acute Care." *International Journal of Nursing Studies* 46, no. 6: 778–86. doi:10.1016/j.ijnurstu.2008.12.009.

Gaser, Christian, Igor Nenadic, Hans-Peter Volz, Christian Büchel, and Heinrich Sauer. 2004. "Neuroanatomy of 'Hearing Voices': A Frontotemporal Brain Structural Abnormality Associated with Auditory Hallucinations in Schizophrenia." *Cerebral Cortex* 14, no. 1: 91–6. doi:10.1093/cercor/bhg107.

Gates, George A., and John H. Mills. 2005. "Presbycusis." *Lancet* 366, no. 9491: 1111–20. doi:10.1016/S0140-6736(05)67423-5.

Gehring, Ulrike, Lillian Tamburic, Hind Sbihi, Hugh W. Davies, and Michael Brauer. 2014. "Impact of Noise and Air Pollution on Pregnancy Outcomes." *Epidemiology* 25, no. 3: 351–8. doi:10.1097/EDE.0000000000000073.

Gendron, Renée. 2011. "The Meanings of Silence during Conflict." *Journal of Conflictology* 2, no. 1: 1–7. http://journal-of-conflictology.uoc.edu/joc/en/index.php/journal-of-conflictology/article/download/vol2iss1-gendron/1091-2001-1-PB.pdf.

Genuis, Stephen J., and Christopher T. Lipp. 2012. "Electromagnetic Hypersensitivity: Fact or Fiction?" *Science of the Total Environment* 414: 103–12. doi:10.1016/j.scitotenv.2011.11.008.

Gershon, Robyn R.M., Richard Neitzel, Marissa A. Barrera, and Muhammad Akram. 2006. "Pilot Survey of Subway and Bus Stop Noise Levels." *Journal of Urban Health* 83, no. 5: 802–12. doi:10.1007/s11524-006-9080-3.

Gibney, Alex, director. *Taxi to the Dark Side*. New York: THINKFilm, 2007.

Gidlöf-Gunnarsson, Anita, and Evy Öhrström. 2007. "Noise and Well-Being in Urban Residential Environments: The Potential Role of Perceived Availability to Nearby Green Areas." *Landscape and Urban Planning* 83, nos. 2–3: 115–26. doi:10.1016/j.landurbplan.2007.03.003.

Gill, Alexandra. 2008. "I Really Hope You Said Fork." *Globe and Mail*, 13 February. https://www.theglobeandmail.com/life/i-really-hope-you-said-fork/article1051265/.

Gilles, Annick, Dirk De Ridder, Guido Van Hal, Kristien Wouters, Andrea K. Punte, and Paul Van de Heyning. 2012. "Prevalence of Leisure Noise-Induced Tinnitus and the Attitude Toward Noise in University Students." *Otology and Neurology* 33, no. 6: 899–906. doi:10.1097/MAO.0b013e31825d640a.

Gilles, Annick, Guido Van Hal, Dirk De Ridder, Kristien Wouters, and Paul Van de Heyning. 2013. "Epidemiology of Noise-Induced Tinnitus and the Attitudes and Beliefs towards Noise and Hearing Protection in Adolescents." *PLOS One* 8, no. 7: e70297. doi:10.1371/journal.pone.0070297.

Gjestland, Truls. 2002. "Current Research Topics and Problems: The Role of ICBEN." *Journal of Sound and Vibration* 250, no. 1: 5–8. doi:10.1006/jsvi.2001.3883.

Glass, Elisabeth, Steffi Sachse, and Waldemar von Suchodoletz. 2008. "Development of Auditory Sensory Memory from 2 to 6 Years: An MMN Study." *Journal of Neural Transmission* 115, no. 8: 1221–9. doi:10.1007/s00702-008-0088-6.

Glover, Paul, Jonathon Hykin, Penny Gowland, Jeff Wright, Ian Johnson, and Peter Mansfield. 1995. "An Assessment of the Intrauterine Sound Intensity Level during Obstetric Echo-Planar Magnetic Resonance Imaging." *British Journal of Radiology* 68, no. 814: 1090–4. doi:10.1259/0007-1285-68-814-1090.

Goines, Lisa. 2008. "The Importance of Quiet in the Home: Teaching Noise

Awareness to Parents before the Infant Is Discharged from the NICU." *Neonatal Network* 27, no. 3: 171–6. doi:10.1891/0730-0832.27.3.171.

Goines, Lisa, and Louis Hagler. 2007. "Noise Pollution: A Modern Plague." *Southern Medical Journal* 100, no. 3: 287–94. doi:10.1097/SMJ.0b013e 3180318be5.

Goldsmith, Mike. 2012. *Discord: The Story of Noise.* Oxford,: Oxford University Press.

Goodman, Steve. 2009. *Sonic Warfare: Sound, Affect, and the Ecology of Fear.* Cambridge, MA: MIT Press.

Görges, Matthias, Boaz A. Markewitz, and Dwayne R. Westenskow. 2009. "Improving Alarm Performance in the Medical Intensive Care Unit Using Delays and Clinical Context." *Anesthesia and Analgesia* 108, no. 5: 1546–52. doi:10.1213/ane.0b013e31819bdfbb.

Goto, Kyoichi, and Tetsuya Kaneko. 2002. "Distribution of Blood Pressure Data from People Living Near an Airport." *Journal of Sound and Vibration* 250, no. 1: 145–9. doi:10.1006/jsvi.2001.3895.

Gouk, Penelope, ed. 2000. *Musical Healing in Cultural Contexts.* Aldershot, UK: Ashgate.

Government of Canada. 2019. "Personal Stereo Systems and the Risk of Hearing Loss." https://www.canada.ca/en/health-canada/services/healthy-living/your-health/lifestyles/personal-stereo-systems-risk-hearing-loss.html#a5.

Government of Nova Scotia. 2011. "Wind Turbine Effects." Accessed February. https://energy.novascotia.ca/sites/default/files/Wind%20Turbine %20Effects.pdf.

Government of Rhode Island. 2016. "Rhode Island Land-Based Wind Siting Guidelines." Rhode Island Office of Energy Resources. Accessed January 2017. http://www.energy.ri.gov/documents/landwind/WindSitingDoc_2016-1-6_FINALforPublicReview.pdf.

Government of the United Kingdom. Ministry of Housing, Communities and Local Government. 2019. "Guidance: Noise." Accessed 3 February 2019. https://www.gov.uk/guidance/noise-2.

Goycoolea, Marcos V., Ismael Mena, Sonia G. Neubauer, Raquel G. Levy, Margarita Fernández Grez, and Claudia G. Berger. 2007. "Musical Brains: A Study of Spontaneous and Evoked Musical Sensations without External Auditory Stimuli." *Acta Oto-Laryngologica* 127, no. 7: 711–21. doi:10.1080/00016480601053057.

Graeven, David B. 1975. "Necessity, Control and Predictability of Noise as Determinants of Noise Annoyance." *Journal of Social Psychology* 95, no. 1: 85–90. doi:10.1080/00224545.1975.9923237.

Graham, Kelly C., and Maria Cvach. 2010. "Monitor Alarm Fatigue: Standardizing Use of Physiological Monitors and Decreasing Nuisance Alarms." *American Journal of Critical Care* 19, no. 1: 28–34. doi:10.4037/ajcc2010651.

Grahn, Jessica A., and Matthew Brett. 2007. "Rhythm and Beat Perception in Motor Areas of the Brain." *Journal of Cognitive Neuroscience* 19, no. 5: 893–906. doi:10.1162/jocn.2007.19.5.893.

Grahn, Jessica A., and James B. Rowe. 2013. "Finding and Feeling the Musical Beat: Striatal Dissociations between Detection and Prediction of Regularity." *Cerebral Cortex* 23, no, 4: 913–21. doi:10.1093/cercor/bhs083.

Granier-Deferre, Carolyn, Sophie Bassereau, Aurélie Ribeiro, Anne-Yvonne Jacquet, and Anthony J. DeCasper. 2011. "A Melodic Contour Repeatedly Experienced by Human Near-Term Fetuses Elicits a Profound Cardiac Reaction One Month after Birth." *PLOS One* 6, no. 2: e17304. doi:10.1371/journal.pone.0017304.

Granlund-Lind, Rigmor, and John Lind. 2004. *Black on White: Voices and Witnesses about Electro-Hypersensitivity*. Translated by Jeffrey Ganellen. Stockholm: Mimers Brunn Kunskapsförlaget. Accessed 10 January 2019. https://www.avaate.org/IMG/pdf/blackonwhite-complete-book.pdf.

Griefahn, Barbara, Peter Bröde, Anke Marks, and Mathias Basner. 2008. "Autonomic Arousals Related to Traffic Noise during Sleep." *Sleep* 31, no. 4: 569–77. doi:10.1093/sleep/31.4.569.

Griffiths, Scott K., Linda L. Pierson, Kenneth J. Gerhardt, Robert M. Abrams, and Aemil J.M. Peters. 1994. "Noise Induced Hearing Loss in Fetal Sheep." *Hearing Research* 74, no. 1–2: 221–30. doi:10.1016/0378-5955(94)90190-2.

Guéguen, Nicolas, C. Jacob, H. Le Guellec, T. Morineau, M. Lourel. 2008. "Sound Level of Environmental Music and Drinking Behavior: A Field Experiment with Beer Drinkers." *Alcohol Clinical Experimental Research* 32, no. 10: 1795–8. doi: 10.1111/j.1530-0277.2008.00764.x.

Gunaratnam, Yasmin. 2009. "Auditory Space, Ethics and Hospitality: 'Noise,' Alterity and Care at the End of Life." *Body and Society* 15, no. 4: 1–19. doi:10.1177/1357034X09337781.

Gunderson, Erik, Jacqueline Moline, and Peter Catalano. 1997. "Risks of Developing Noise-Induced Hearing Loss in Employees of Urban Music Clubs." *American Journal of Industrial Medicine* 31, no. 1: 75–9. doi:10.1002/(SICI)1097-0274(199701)31:1<75::AID-AJIM11>3.0.CO;2-4.

Gupta, Deepak, and S.K. Vishwakarma. 1989. "Toy Weapons and Firecrackers: A Source of Hearing Loss." *Laryngoscope* 99, no. 3: 330–4. doi:10.1288/00005537-198903000-00018.

Hackney, Madeleine E., and Gammon M. Earhart. 2010. "Effects of Dance on Balance and Gait in Severe Parkinson Disease: A Case Study." *Disability and Rehabilitation* 32, no. 8: 679–84. doi:10.3109/09638280903247905.

Hackney, Madeleine E., Svetlana Kantorovich, Rebecca Levin, and Gammon M. Earhart. 2007. "Effects of Tango on Functional Mobility in Parkinson's Disease: A Preliminary Study." *Journal of Neurologic Physical Therapy* 31, no. 4: 173–9. doi:10.1097/NPT.0b013e31815ce78b.

Haines, Mary, and Stephen Stansfeld. 2003. "Ambient Neighbourhood Noise and Children's Mental Health." *Occupational and Environmental Medicine* 60: 146. doi:10.1136/oem.60.2.146.

Haines, Mary, Stephen Stansfeld, Raymond F. Soames Job, Birgitta Berglund, and Jenny Head. 2001. "Chronic Aircraft Noise Exposure, Stress Responses, Mental Health and Cognitive Performance in School Children." *Psychological Medicine* 31, no. 2: 265–77. doi:10.1017/S0033291701003282.

Hall, Deborah A., Haúla Haider, Dimitris Kikidis, Marzena Mielczarek, Birgit Mazurek, Agnieszka J. Szczepek, and Christopher R. Cederroth. 2015. "Toward a Global Consensus on Outcome Measures for Clinical Trials in Tinnitus: Report from the First International Meeting of the COMiT Initiative, 14 November 2014, Amsterdam, The Netherlands." *Trends in Hearing* 19. doi:10.1177/2331216515580272.

Halonen, Jaana I., Anna L. Hansell, John Gulliver, David Morley, Marta Blangiardo, Daniela Fecht, Mireille B. Toledano, et al. 2015. "Road Traffic Noise Is Associated with Increased Cardiovascular Morbidity and Mortality and All-Cause Mortality in London." *European Heart Journal* 36, no. 39: 2653–61. doi:10.1093/eurheartj/ehv216.

Halpern, David. 1995. *Mental Health and the Built Environment: More Than Bricks and Mortar?* London: Taylor and Francis.

Hambling, David. 2008. "US 'Sonic Blasters' Sold to China." *Wired*, 15 May. https://www.wired.com/2008/05/us-sonic-blaste/

– 2010. "A Sonic Blaster So Loud, It Could Be Deadly." *Wired*, 18 January. http://www.wired.com/dangerroom/2010/01/a-sonic-blaster-so-loud-it-could-be-deadly/.

Hamdan, Abdul-Latif, Reem Deeb, Abla Sibai, Charbel Rameh, Hani Rifai, and John Fayyad. 2009. "Vocal Characteristics in Children with Attention Deficit Hyperactivity Disorder." *Journal of Voice* 23, no. 2: 190–4. doi:10.1016/j.jvoice.2007.09.004.

Hammer, Monica S., Tracy K. Swinburn, and Richard L. Neitzel. 2014. "Environmental Noise Pollution in the United States: Developing an Effective Public Health Response." *Environmental Health Perspectives* 122, no. 2: 115–19. doi:10.1289/ehp.1307272.

Hansell, Anna L., Marta Blangiardo, Lea Fortunato, Sarah Floud, Kees de Hoogh, Daniela Fecht, Rebecca E. Ghosh, et al. 2013. "Aircraft Noise and Cardiovascular Disease Near Heathrow Airport in London: Small Area Study." *BMJ* 347, no. 7928: f5432. doi:10.1136/bmj.f5432.

Haralabidis, Alexandros S., Konstantina Dimakopoulou, Federica Vigna-Taglianti, Matteo Giampaolo, Alessandro Borgini, Marie-Louise Dudley, Göran Pershagen, et al. 2008. "Acute Effects of Night-Time Noise Exposure on Blood Pressure in Populations Living Near Airports." *European Heart Journal* 29, no. 5: 658–64. doi:10.1093/eurheartj/ehn013.

Harden, Michael. 2008. "It's Your Shout." *The Age*. 22 April. http://theage.com.au /news/epicure/its-your-shout/2008/04/21/1208629783592.html

Harding, Anne-Helen, Gillian A. Frost, Emma Tan, Aki Tsuchiya, and Howard M. Mason. 2013. "The Cost of Hypertension-Related Ill-Health Attributable to Environmental Noise." *Noise and Health* 15, no, 67: 437–45. doi:10.4103/1463-1741.121253.

Hasegawa, Takehiro, Ken-Ichi Matsuki, Takashi Ueno, Yasuhiro Maeda, Yoshi-hiko Matsue, Yukuo Konishi, and Norihiro Sadato. 2004. "Learned Audio-Visual Cross-Modal Associations in Observed Piano Playing Activate the Left Planum Temporale. An fMRI Study." *Cognitive Brain Research* 20, no. 3: 510–18. doi:10.1016/j.cogbrainres.2004.04.005.

Hatfield, Julie, Raymond F. Soames Job, Andrew J. Hede, Norman L. Carter, Peter Peploe, Richard Taylor, and Stephen Morrell. 2002. "Human Response to Environmental Noise: The Role of Perceived Control." *International Journal of Behavioral Medicine* 9, no. 4: 341–59. doi:10.1207/S15327558IJBM0904_04.

Hayne, M., L.J.H. Schulze, and L. Quintana. 1997. "Personal and Car Stereo Volume Levels: Hazards of Leisure Listening Activities." In *Advances in Occupational Ergonomics and Safety: Proceedings of the Annual International Occupational Ergonomics and Safety Conference*, edited by Biman Das, Waldemar Karwowski, 581–4. Amsterdam: IOS Press.

He, Qinxian. 2010. "Development of an Income-Based Hedonic Monetization Model for the Assessment of Aviation-Related Noise Impacts." MSc thesis, Massachusetts Institute of Technology. Accessed 12 January 2019. http://hdl .handle.net/1721.1/59559.

Health Canada. 2014. "Wind Turbine Noise and Health Study: Summary of Results." 30 October. Accessed 12 January 2019. http://www.hc-sc.gc.ca/ewh-semt/noise-bruit/turbine-eoliennes/summary-resume-eng.php.

Hearing Health Foundation. 2019. "Hearing Health and Tinnitus Statistics." Accessed 10 May 2020. https://hearinghealthfoundation.org/hearing-loss-tinnitus-statistics.

Hébert, Sylvie, Renée Béland, Odrée Dionne-Fournelle, Martine Crête, and Sonia J. Lupien. 2005. "Physiological Stress Response to Video-Game Playing: The Contribution of Built-in Music." *Life Sciences* 76, no. 20: 2371–80. doi:10.1016/j.lfs.2004.11.011.

Hellstrom, Per Anders, Harold A. Dengerink, and Alf Axelsson. 1992. "Noise Levels from Toys and Recreational Articles for Children and Teenagers."

British Journal of Audiology 26, no. 5: 267–70. doi:10.3109/0300536920 9076646.

Helmkamp, James C., Evelyn O. Talbott, and Helene Margolis. 1984. "Occupational Noise Exposure and Hearing Loss Characteristics of a Blue-Collar Population." *Journal of Occupational Medicine* 26, no. 12: 885–91. https://www.osti .gov/biblio/6096420-occupational-noise-exposure-hearing-loss-characteristics-blue-collar-population.

Henderson, Elizabeth, Marcia A. Testa, and Christopher Hartnick. 2011. "Prevalence of Noise-Induced Hearing-Threshold Shifts and Hearing Loss among US Youths." *Pediatrics* 127, no. 1: 39–46. http://pediatrics.aappublications .org/content/127/1/e39.short.

Herbinger, Ilka, Sarah Papworth, Christophe Boesch, and Klaus Zuberbühler. 2009. "Vocal, Gestural and Locomotor Responses of Wild Chimpanzees to Familiar and Unfamiliar Intruders: A Playback Study." *Animal Behaviour* 78, no. 6: 1389–96. doi:10.1016/j.anbehav.2009.09.010.

Ho, Judy Woon Yee. 2006. "Functional Complementarity between Two Languages in ICQ." *International Journal of Bilingualism* 10, no. 4: 429–51. doi:10.1177/13670069060100040301.

Hodgetts, William, and Richard Liu. 2006. "Can Hockey Playoffs Harm Your Hearning?" *Canadian Medical Association Journal* 175, no. 12: 1541–2. https://doi.org/10.1503/cmaj.060789.

Hodgetts, William, Ryan Szarko, and Jana Rieger. 2009. "What Is the Influence of Background Noise and Exercise on the Listening Levels of iPod Users?" *International Journal of Audiology* 48, no. 12: 825–32. doi:10.3109/14992020903082104.

Hohmann, Cynthia, Linus Grabenhenrich, Yvonne de Kluizenaar, Christina Tischer, Joachim Heinrich, Chih-Mei Chen, Carel Thijs, Mark Nieuwenhuijsen, and Thomas Keil. 2013. "Health Effects of Chronic Noise Exposure in Pregnancy and Childhood: A Systematic Review Initiated by ENRIECO." *International Journal of Hygiene and Environmental Health* 216, no. 3: 217–29. doi:10.1016/j.ijheh.2012.06.001.

Holmberg, Sharon K., and Sharon Coon. 1999. "Ambient Sound Levels in a State Psychiatric Hospital." *Archives of Psychiatric Nursing* 13, no. 3: 117–26. doi:10.1016/S0883-9417(99)80042-9.

Holmes, Alice E., Stephen E. Widén, Soly Erlandsson, Courtney L. Carver, and Lori L. White. 2007. "Perceived Hearing Status and Attitudes toward Noise in Young Adults." *American Journal of Audiology* 16, no. 2: S182–9. doi:10.1044/1059-0889(2007/022).

Holt, Avril G., David Bissig, Najab Mirza, Gary Rajah, and Bruce Berkowitz. 2010. "Evidence of Key Tinnitus-Related Brain Regions Documented by a

Unique Combination of Manganese-Enhanced MRI and Acoustic Startle Reflex Testing." *PLOS One* 5, no. 12: e14260. doi:10.1371/journal.pone .0014260.

Honos-Webb, Lara, and William B. Stiles. 1998. "Reformulation of Assimilation Analysis in Terms of Voices." *Psychotherapy: Theory, Research, Practice, Training* 35, no. 1: 23–33. doi:10.1037/h0087682.

Horden, Peregrine, ed. 2000. *Music as Medicine: The History of Music Therapy since Antiquity*. Aldershot, UK: Ashgate.

Horvath, Janet. 2015. "A Musician Afraid of Sound." *Atlantic*. October. http://www.theatlantic.com/health/archive/2015/10/a-musician-afraid-of-sound/411367/

Huang, Xinyan, Kenneth J. Gerhardt, Robert M. Abrams, and Patrick J. Antonelli. 1997. "Temporary Threshold Shifts Induced by Low-Pass and High-Pass Filtered Noises in Fetal Sheep in Utero." *Hearing Research* 113, no. 1–2: 173–81. doi:10.1016/S0378-5955(97)00139-1.

Hubbard, Timothy L. 2010. "Auditory Imagery: Empirical Findings." *Psychological Bulletin* 136, no. 2: 302–29. doi:10.1037/a0018436.

Hugh, Sarah C., Nikolaus E. Wolter, Evan J. Propst, Karen A. Gordon, Sharon L. Cushing, and Blake C. Papsin. 2014. "Infant Sleep Machines and Hazardous Sound Pressure Levels." *Pediatrics* 133, no. 4: e67347. http://pediatrics .aappublications.org/content/133/4/677.full.

Hughes, Robert, and Dylan M. Jones. 2001. "The Intrusiveness of Sound: Laboratory Findings and Their Implications for Noise Abatement." *Noise and Health* 4, no. 13: 51–70. http://www.noiseandhealth.org/text.asp?2001/4/13/51/31802.

Humes, Larry E. 2005. "Do 'Auditory Processing' Tests Measure Auditory Processing in the Elderly?" *Ear and Hearing* 26, no. 2: 109–19. doi:10.1097/00003446-200504000-00001.

Huylebrouck, Dirk. 1996. "Puzzles, Patterns Drums: The Dawn of Mathematics in Rwanda and Burundi." *Humanistic Mathematics Network Journal* 14, no. 6: 9–22. doi:10.5642/hmnj.199601.14.06.

Hviid, Steen. 2010. "Electromagnetic Hypersensitivity." *Townsend Letter: The Examiner of Alternative Medicine*, January. http://www.townsendletter.com /Jan2010/electromag0110.html.

Hyperacusis Network. 2019. "4 Types of Sound Sensitivity." Accessed 17 February. http://www.hyperacusis.net/what-is-it/4-types-of-sound-sensitivity/.

Ihde, Don. *Listening and Voice: Phenomenologies of Sound*, 2nd ed. Albany: State University of New York Press, 2007.

Inam-ul-Haq, Taqmeem Hussain, Hashim Farooq, and Muhammad Raza Ahmad. 2014. "Evaluation of the Traffic Noise Pollution at Some Busiest Sites of Faisalabad City, Pakistan." *Academic Research International* 5, no. 2: 23–6.

https://www.academia.edu/10144056/Evaluation_of_the_Traffic_Noise
_Pollution_at_Some_Busiest_Sites_of_Faisalabad_City_Pakistan.

Institute of Medicine. 2006. *Noise and Military Service: Implications for Hearing Loss and Tinnitus*. Washington, DC: National Academies Press. doi:10.17226/11443.

Ising, Hartmut, Wolfgang Babisch, J. Hanee, and Barbara Kruppa. 1997. "Loud Music and Hearing Risk." *Journal of Audiological Medicine* 6, no. 3: 123–33. https://www.researchgate.net/publication/283944919_Loud_music_and _hearing_risk.

Ising, Hartmut, and Ran Michalak. 2004. "Stress Effects of Noise in a Field Experiment in Comparison to Reactions to Short Term Noise Exposure in the Laboratory." *Noise and Health* 6, no. 24: 1–7. http://www.noiseandhealth .org/text.asp?2004/6/24/1/31660.

Iwanaga, Akoto, and Takako Ito. 2002. "Disturbance Effect of Music on Processing of Verbal and Spatial Memories." *Perceptual and Motor Skills* 94, no. 3: 1251–8. doi:10.2466/pms.2002.94.3c.1251.

Jackson, Laura E. 2003. "The Relationship of Urban Design to Human Health and Condition." *Landscape and Urban Planning* 64, no. 4: 191–200. doi:10.1016/S0169-2046(02)00230-X.

Jacobsen, Jörn-Henrik, Johannes Stelzer, Thomas Hans Fritz, Gael Chételat, Renaud La Joie, and Robert Turner. 2015. "Why Musical Memory Can Be Preserved in Advanced Alzheimer's Disease." *Brain* 138, no. 8: 2438–50. doi:10.1093/brain/awv135.

Jahncke, Helena. 2012. "Cognitive Performance and Restoration in Open-Plan Office Noise." PhD diss., Luleå University.

Jahncke, Helena, Valtteri Hongisto, and Petra Virjonen. 2013. "Cognitive Performance during Irrelevant Speech: Effects of Speech Intelligibility and Office-Task Characteristics." *Applied Acoustics* 74, no. 3: 307–16. doi:10.1016/j .apacoust.2012.08.007.

Jahncke, Helena, Staffan Hygge, Niklas Halin, Anne Marie Green, and Kenth Dimberg. 2011. "Open-Plan Office Noise: Cognitive Performance and Restoration." *Journal of Environmental Psychology* 31, no. 4: 373–82. doi:10.1016/j.jenvp.2011.07.002.

Jamieson, Donald G., Garry Kranjc, Karen Yu, and William E. Hodgetts. 2004. "Speech Intelligibility of Young School-Aged Children in the Presence of Real-Life Classroom Noise." *Journal of the American Academy of Audiology* 15, no. 7: 508–17. doi:10.3766/jaaa.15.7.5.

Jardin, Xeni. 2005. "Focused Sound 'Laser' for Crowd Control." National Public Radio, 21 September. https://www.npr.org/templates/story/story.php ?storyId=4857417.

Jarup, Lars, Wolfgang Babisch, Danny Houthuijs, Göran Pershagen, Klea Katsouyanni, Ennio Cadum, Marie-Louise Dudley, et al. 2008. "Hypertension and Exposure to Noise Near Airports: The HYENA Study." *Environmental Health Perspectives* 116, no. 3: 329–33. doi:10.1289/ehp.10775. Correction at doi:10.1289/ehp.10775-erratum.

Jenkins, Linda, Alex Tarnopolsky, and David Hand. 1981. "Psychiatric Admissions and Aircraft Noise from London Airport: Four-Year, Three-Hospitals' Study." *Psychological Medicine* 11, no. 4: 765–82. doi:10.1017/S0033291700041271.

Jeon, Jin Yong, Pyoung Jik Lee, Jin You, and Jian Kang. 2010. "Perceptual Assessment of Quality of Urban Soundscapes with Combined Noise Sources and Water Sounds." *Journal of the Acoustical Society of America* 127, no. 3: 1357–66. doi:10.1121/1.3298437.

Jiang, Wen, Fei Zhao, Nicola Guderley, and Vinaya Manchaiah. 2016. "Daily Music Exposure Dose and Hearing Problems Using Personal Listening Devices in Adolescents and Young Adults: A Systematic Review." *International Journal of Audiology* 55, no. 4: 197–205. doi:10.3109/14992027.2015.1122237.

Johansson, Lotta, Ingegerd Bergbom, and Berit Lindahl. 2012. "Meanings of Being Critically Ill in a Sound-Intensive ICU Patient Room - A Phenomenological Hermeneutical Study." *Open Nursing Journal* 6: 108–16. doi:10.2174/1874434601206010108.

Johansson, Olle, Shabnam Gangi, Yong Liang, Ken Yoshimura, Chen Jing, and Peng-Yue Liu. 2001. "Cutaneous Mast Cells Are Altered in Normal Healthy Volunteers Sitting in Front of Ordinary TVs/PCs - Results from Open-Field Provocation Experiments." *Journal of Cutaneous Pathology* 28, no. 10: 513–19. doi:10.1034/j.1600-0560.2001.281004.x.

John, Ayugi, Loyal Poonamjeet, Mugwe Peter, and Nyandusi Musa. 2015. "Demographic Patterns of Acoustic Shock Syndrome as Seen in a Large Call Centre." *Occupational Medicine and Health Affairs* 3, no. 4: 212. doi:10.4172/2329-6879.1000212.

Jokitulppo, Jaana S., Erkki A. Björki, and Eero Akaan-Penttiä. 1997. "Estimated Leisure Noise Exposure and Hearing Symptoms in Finnish Teenagers." *Scandinavian Audiology* 26, no. 4: 257–62. doi:10.3109/01050399709048017.

Jones, Dylan M., and Antony J. Chapman, eds. 1984. *Noise and Society.* Chichester, UK: John Wiley and Sons.

Jones, Simon C., and Thomas G. Schumacher. 1992. "Muzak: On Functional Music and Power." *Critical Studies in Mass Communication* 9, no. 2: 156–69. doi:10.1080/15295039209366822.

Kaarlela-Tuomaala, A., Riikka Helenius, Esko Keskinen, and Valtteri Hongisto. 2009. "Effects of Acoustic Environment on Work in Private Office Rooms and Open-Plan Offices: Longitudinal Study during Relocation." *Ergonomics* 52, no. 11: 1423–44. doi:10.1080/00140130903154579.

Kang, Jian, Francesco Aletta, Truls T. Gjestland, Lex A. Brown, Dick Bottel-dooren, Brigitte Schulte-Fortkamp, Peter Lercher, et al. 2016. "Ten Questions on the Soundscapes of the Built Environment." *Building and Environment* 108: 284–94. doi:10.1016/j.buildenv.2016.08.011.

Karlsson, Henrik. 2000. "The Acoustic Environment as a Public Domain." *Soundscape* 1, no. 2. https://www.wfae.net/journal.html.

Katz, Jonathan. 2014. "Noise in the Operating Room." *Anesthesiology* 121, no. 4: 894–8. doi:10.1097/ALN.0000000000000319.

Keizer, Garrett. 2010. *The Unwanted Sound of Everything We Want: A Book about Noise.* Philadelphia: PublicAffairs.

Keppler, Hannah, Ingeborg Dhooge, Leen Maes, Wendy D'haenens, Annelies Bockstael, Birgit Philips, Freya Swinnen, and Bart Vinck. 2010. "Short-Term Auditory Effects of Listening to an MP3 Player." *Archives of Otolaryngology: Head and Neck Surgery* 136, no. 6: 538–48. doi:10.1001/archoto.2010.84.

Keppler, Hannah, Ingeborg Dhooge, and Bart Vinck. 2015. "Hearing in Young Adults. Part II: The Effects of Recreational Noise Exposure." *Noise and Health* 17, no. 78: 245–52. doi:10.4103/1463-1741.165026.

Keuroghlian, Alex S., and Eric I. Knudsen. 2007. "Adaptive Auditory Plasticity in Developing and Adult Animals." *Progress in Neurobiology* 82, no. 3: 109–21. doi:10.1016/j.pneurobio.2007.03.005.

Kight, Caitlin R., and John P. Swaddle. 2011. "How and Why Environmental Noise Impacts Animals: An Integrative, Mechanistic Review." *Ecology Letters* 14, no. 10: 1052–61. doi:10.1111/j.1461-0248.2011.01664.x.

Kim, Hong, Myoung-Hwa Lee, Hyun-Kyung Chang, Taeck-Hyun Lee, Hee-Hyuk Lee, Min-Chul Shin, Mal-Soon Shin, Ran Won, Hye-Sook Shin, and Chang-Ju Kim. 2006. "Influence of Prenatal Noise and Music on the Spatial Memory and Neurogenesis in the Hippocampus of Developing Rats." *Brain and Development* 28, no. 2: 109–14. doi:10.1016/j.braindev.2005.05.008.

Kim, Jungsoo, and Richard de Dear. 2013. "Workspace Satisfaction: The Privacy-Communication Trade-Off in Open-Plan Offices." *Journal of Environmental Psychology* 36: 18–26. doi:10.1016/j.jenvp.2013.06.007.

Kim, Soo J. 2010. "Music Therapy Protocol Development to Enhance Swallowing Training for Stroke Patients with Dysphagia." *Journal of Music Therapy* 47, no. 2: 102–19. doi:10.1093/jmt/47.2.102.

Klatte, Maria, Thomas Lachmann, and Markus Meis. 2010. "Effects of Noise and Reverberation on Speech Perception and Listening Comprehension of Children and Adults in a Classroom-Like Setting." *Noise and Health* 12, no. 49: 270–82. doi:10.4103/1463-1741.70506.

Kleebauer, Alistair. 2015. "Music Played during Operations Can Impair Team Communication." *Nursing Standard* 29, no. 50: 11. doi:10.7748/ns.29.50.11.s12.

Knobel, Keila A., and Maria C. Lima. 2012. "Knowledge, Habits, Preferences, and Protective Behavior in Relation to Loud Sound Exposures among Brazilian Children." *International Journal of Audiology* 51, sup. 1: S12–19. doi:10.3109 /14992027.2011.637307.

Kocisky, Katherine A. 2016. "Evaluating Residents' Willingness to Pay and Volunteer for Urban Street Trees on Logan Square's Historic Green Boulevards in Chicago." MA thesis, Northeastern Illinois University.

Koelsch, Stefan. 2009. "A Neuroscientific Perspective on Music Therapy." *Annals of the New York Academy of Sciences* 1169, no. 1: 374–84. doi:10.1111/j.1749-6632.2009.04592.x.

Konkani, Avinash, and Barbara Oakley. 2012. "Noise in Hospital Intensive Care Units: A Critical Review of a Critical Topic." *Journal of Critical Care* 27, no. 5: 522.e1–9. doi:10.1016/j.jcrc.2011.09.003.

Konnikova, Maria. 2014. "The Open-Office Trap." *New Yorker*, 7 January. https://www.newyorker.com/business/currency/the-open-office-trap.

Konrad, John. 2019. "Wind Turbine Designs: The 11 Most Interesting." Accessed 18 January. https://gcaptain.com/the-most-interesting-wind-turbine-designs/.

Kraus, Kari S., and Barbara Canlon. 2012. "Neuronal Connectivity and Interactions between the Auditory and Limbic Systems. Effects of Noise and Tinnitus." *Hearing Research* 288, no. 1–2: 34–46. doi:10.1016/j.heares.2012.02.009.

Kraus, Ute, Alexandra Schneider, Susanne Breitner, Regina Hampel, Regina Rückerl, Mike Pitz, Uta Geruschkat, Petra Belcredi, Katja Radon, and Annette Peters. 2013. "Individual Daytime Noise Exposure during Routine Activities and Heart Rate Variability in Adults: A Repeated Measures Study." *Environmental Health Perspectives* 121, no. 5: 607–12. doi:10.1289/ehp.1205606.

Krause, Bernie. 2002. *Wild Soundscapes: Discovering the Voice of the Natural World.* Berkeley: Wilderness Press.

Krishnamurti, Sridhar, and Peter W. Grandjean. 2003. "Effects of Simultaneous Exercise and Loud Music on Hearing Acuity and Auditory Function." *Journal of Strength Conditioning Research* 17, no. 2: 307–13. https://www.academia.edu /7212361/Effects_of_Simultaneous_Exercise_and_Loud_Music_on_Hearing _Acuity_and_Auditory_Function.

Krogh, Carmen. 2012. "Re: Wind Turbine Health Impact Study: Report of Independent Expert Panel." National Wind Watch. 29 January. Accessed 18 January 2019. https://www.wind-watch.org/documents/re-wind-turbine-health-impact-study-report-of-independent-expert-panel/.

Kshettry, Vibhu R., Linda F. Carole, Susan J. Henly, Sue Sendelbach, and Barbara Kummer. 2006. "Complementary Alternative Medical Therapies for Heart Surgery Patients: Feasibility, Safety, and Impact." *Annals of Thoracic Surgery* 81, no. 1: 201–5. doi:10.1016/j.athoracsur.2005.06.016.

Kujawa, Sharon G., and M. Charles Liberman. 2006. "Acceleration of Age-Related Hearing Loss by Early Noise Exposure: Evidence of a Misspent Youth." *Journal of Neuroscience* 26, no. 7: 2115–23. doi:10.1523/JNEUROSCI.4985-05.2006.

– 2009. "Adding Insult to Injury: Cochlear Nerve Degeneration after 'Temporary' Noise-Induced Hearing Loss." *Journal of Neuroscience* 29, no. 45: 14077–85. doi:10.1523/JNEUROSCI.2845-09.2009.

Kumar, Vivek, Tapas Chandra Nag, Uma Sharma, Sujeet Mewar, Naranamangalam R. Jagannathan, and Shashi Wadhwa. 2014. "High Resolution ¹H NMR-Based Metabonomic Study of the Auditory Cortex Analogue of Developing Chick (*Gallus Gallus Domesticus*) Following Prenatal Chronic Loud Music and Noise Exposure." *Neurochemistry International* 76: 99–108. doi:10.1016/j.neuint.2014.07.002.

Kurmis, Andrew P., and Stacey A. Apps. 2007. "Occupationally Acquired Noise-Induced Hearing Loss: A Senseless Workplace Hazard." *International Journal of Occupational Medicine and Environmental Health* 20, no. 2: 127–36. doi:10.2478/v10001-007-0016-2.

Labelle, Brandon. 2010. *Acoustic Territories: Sound Culture and Everyday Life.* New York: Continuum.

Lammers, H. Bruce. 2003. "An Oceanside Field Experiment on Background Music Effects on the Restaurant Tab." *Perceptual and Motor Skills* 96, no. 3: 1025–6. doi:10.2466/pms.2003.96.3.1025.

Landälv, Daniel, Lennart Malmström, and Stephen E. Widén. 2013. "Adolescents' Reported Hearing Symptoms and Attitudes toward Loud Music." *Noise and Health* 15, no. 66: 347–54. doi:10.4103/1463-1741.116584.

Larsson, Britt-Marie. 2011. "Noise during Pregnancy: Occupational Noise Exposure during Pregnancy – Effects on Hearing and Birth Weight." Karolinska Institutet. https://web.archive.org/web/20120907212615/http://ki.se/ki/jsp/polopoly.jsp?d=28275&a=94538&l=en.

Lawrence, Catherine, Jason Jones, and Myra Cooper. 2010. "Hearing Voices in a Non-Psychiatric Population." *Behavioural and Cognitive Psychotherapy* 38, no. 3: 363–73. doi:10.1017/S1352465810000172.

Leaver, Amber M., Laurent Renier, Mark A. Chevillet, Susan Morgan, Hung J. Kim, and Josef P. Rauschecker. 2011. "Dysregulation of Limbic and Auditory Networks in Tinnitus." *Neuron* 69, no. 1: 33–43. doi:10.1016/j.neuron.2010.12.002.

Leaver, Amber M., Anna Seydell-Greenwald, Ted K. Turesky, Susan Morgan, Hung J. Kim, and Josef P. Rauschecker. 2012. "Cortico-Limbic Morphology Separates Tinnitus from Tinnitus Distress." *Frontiers in Systems Neuroscience* 6: 21. doi:10.3389/fnsys.2012.00021.

Lee, Ji H., Weechang Kang, Seung R. Yaang, Nari Choy, and Choong R. Lee.

2009. "Cohort Study for the Effect of Chronic Noise Exposure on Blood Pressure among Male Workers in Busan, Korea." *American Journal of Industrial Medicine* 52, no. 6: 509–17. doi:10.1002/ajim.20692.

Lee, L.T. 1999. "A Study of the Noise Hazard to Employees in Local Discotheques." *Singapore Medical Journal* 40, no. 9: 571–4. http://www.smj.org.sg /sites/default/files/4009/4009a2.pdf.

Lehnert, Kevin, Brian D. Till, and Brad D. Carlson. 2013. "Advertising Creativity and Repetition: Recall, Wearout and Wearin Effects." *International Journal of Advertising* 32, no. 2: 211–31. doi:10.2501/IJA-32-2-211-231.

Leighton, T.G. 2016. "Are Some People Suffering as a Result of Increasing Mass Exposure of the Public to Ultrasound in Air?" In *Proceedings of the Royal Society A: Mathematical, Physical and Engineering Sciences* 472. doi:10.1098/rspa .2015.0624.

– 2017. "Comment on 'Are Some People Suffering as a Result of Increasing Mass Exposure of the Public to Ultrasound in Air?'" *Proceedings of the Royal Society* 473. doi:10.1098/rspa.2016.0828.

Lena, Jennifer C., and Richard A. Peterson. 2008. "Classification as Culture: Types and Trajectories of Music Genres." *American Sociological Review* 73, no. 5: 697–718. doi:10.1177/000312240807300501.

Lercher, Peter, Mark Brink, Johannes Rüdisser, Timothy Van Renterghem, Dick Botteldooren, Michel Baulac, and Jérôme Defrance. 2010. "The Effects of Railway Noise on Sleep Medication Intake: Results from the ALPNAP-Study." *Noise and Health* 12, no. 47: 110–19. doi:10.4103/1463-1741.63211.

Lercher, Peter, Gary W. Evans, M. Meis, and Walter Kofler. 2002. "Ambient Neighbourhood Noise and Children's Mental Health." *Occupational and Environmental Medicine* 59: 380–6. doi:10.1136/oem.59.6.380.

Lesiuk, Teresa. 2000. "The Effect of Music Listening on a Computer Programming Task." *Journal of Computer Information Systems* 40, no. 3: 50–7. doi:10.1080/08874417.2000.11647454.

– 2005. "The Effect of Music Listening on Work Performance." *Psychology of Music* 33, no. 2: 173–91. doi:10.1177/0305735605050650.

Levin, Stephen R., Thomas V. Petros, and Florence W. Petrella. 1982. "Pre-Schoolers' Awareness of Television Advertising." *Child Development* 53, no. 4: 933–7. doi:10.2307/1129131.

Leus, Maria. 2011. "The Soundscape of Cities: A New Layer in City Renewal." *Sustainable Development and Planning* 5: 355–70.

Levitin, Daniel J. 2006. *This Is Your Brain on Music.* New York: Plume Books.

Levitin, Daniel J., and Anna K. Tirovolas. 2009. "Current Advances in the Cognitive Neuroscience of Music." *Annals of the New York Academy of Sciences* 1156, no. 1: 211–31. doi:10.1111/j.1749-6632.2009.04417.x.

Li, Hui-Juan, Wen-Bo Yu, Jing-Qiao Lu, Lin Zeng, Nan Li, and Yi-Ming Zhao. 2008. "Investigation of Road-Traffic Noise and Annoyance in Beijing: A Cross-Sectional Study of 4th Ring Road." *Archives of Environmental & Occupational Health* 63, no. 1: 27–33. doi:10.3200/AEOH.63.1.27-33.

Li, Shu-Yen, Tsae-Jyy Wang, Shu F. Vivienne Wu, Shu-Yuan Liang, and Heng-Hsin Tung. 2011. "Efficacy of Controlling Night-Time Noise and Activities to Improve Patients' Sleep Quality in a Surgical Intensive Care Unit." *Journal of Clinical Nursing* 20, nos. 3–4: 396–407. doi:10.1111/j.1365-2702.2010.03507.x.

Liberman, M. Charles, Michael J. Epstein, Sandra S. Cleveland, Haobing Wang, and Stéphane F. Maison. 2016. "Toward a Differential Diagnosis of Hidden Hearing Loss in Humans." *PLOS One* 11, no. 9. doi:10.1371/journal.pone .0162726.

Lichenstein, Richard, Daniel C. Smith, Jordan L. Ambrose, and Laurel A. Moody. 2012. "Headphone Use and Pedestrian Injury and Death in the United States: 2004–2011." *Injury Prevention* 18: 287–90. doi:10.1136/injuryprev-2011-040161.

Lim, Changwoo, Jaehwan Kim, Jiyoung Hong, and Soogab Lee. 2008. "Effect of Background Noise Levels on Community Annoyance from Aircraft Noise." *Journal of the Acoustical Society of America* 123, no. 2: 766–71. doi:10.1121/1.2821985.

Lin, F.R., L. Ferrucci, Y. An, J.O. Goh, Jimit Doshi, E.J. Metter, C. Davatzikos, M.A. Kraut, and S.M. Resnick. 2014. "Association of Hearing Impairment with Brain Volume Changes in Older Adults." *NeuroImage* 90, 84–92. doi:10.1016/j.neuroimage.2013.12.059.

Linares, Cristina, Julio Díaz, A. Tobías, J.M. De Miguel, and Angel Otero. 2006. "Impact of Urban Air Pollutants and Noise Levels over Daily Hospital Admissions in Children in Madrid: A Time Series Analysis." *International Archives of Occupational and Environmental Health* 79, no. 2: 143–52. doi:10.1007/s00420-005-0032-0.

Lissek, Shmuel, Johanna M.P. Baas, Daniel S. Pine, Kaebah Orme, Sharone Dvir, Monique Nugent, Emily Rosenberger, Elizabeth Rawson, and Christian Grillon. 2005. "Airpuff Startle Probes: An Efficacious and Less Aversive Alternative to White-Noise." *Biological Psychology* 68, no. 3: 283–97. doi:10.1016/j.biopsycho.2004.07.007.

Liu, Jiang, Jian Kang, Holger Behm, and Tao Luo. 2014. "Effects of Landscape on Soundscape Perception: Soundwalks in City Parks." *Landscape and Urban Planning* 123: 30–40. doi:10.1016/j.landurbplan.2013.12.003.

Liu, Yuewei, Haijiao Wang, Shaofan Weng, Wenjin Su, Xin Wang, Yanfei Guo, Dan Yu, et al. 2015. "Occupational Hearing Loss among Chinese Municipal Solid Waste Landfill Workers: A Cross-Sectional Study." *PLOS ONE* 10 (6): e0128719. https://doi.org/10.1371/journal.pone.0128719.

Llano, Samuel. 2018. "Mapping Street Sounds in the Nineteenth-Century City: A Listener's Guide to Social Engineering." *Sound Studies: An Interdisciplinary Journal* 4, no. 2: 143–61. doi:10.1080/20551940.2018.1476305.

Locke, Christine L., and Diana S. Pope. 2017. "Assessment of Medical-Surgical Patients' Perception of Hospital Noises and Reported Ability to Rest." *Clinical Nurse Specialist* 31, no. 5: 261–7. doi:10.1097/NUR.0000000000000321.

Locker, George, and Arline Bronzaft. 1996. "New York City's Noise Laws: Do They Work?" *New York Law Journal*, 24 December. Reprinted on Tenants Services website. Accessed 20 January 2019. http://www.tenantsservices.com/art_nyc_noise.php.

Lockwood, Alan H., David S. Wack, Robert F. Burkard, Mary L. Coad, Samuel A. Reyes, Shanna A. Arnold, and Richard J. Salvi. 2001. "The Functional Anatomy of Gaze-Evoked Tinnitus and Sustained Lateral Gaze." *Neurology* 56, no. 4: 472–80. doi:10.1212/WNL.56.4.472.

Loewen, Laura J., and Peter Suedfeld. 1992. "Cognitive and Arousal Effects of Masking Office Noise." *Environment and Behavior* 24, no. 3: 381–95. doi:10.1177/0013916592243006.

Longcamp, Marieke, Marie-Thérèse Zerbato-Poudou, and Jean-Luc Velay. 2005. "The Influence of Writing Practice on Letter Recognition in Preschool Children: A Comparison between Handwriting and Typing." *Acta Psychologica* 119, no. 1: 67–79. doi:10.1016/j.actpsy.2004.10.019.

López, Hassan H., Adam S. Bracha, and H. Stefan Bracha. 2002. "Evidence Based Complementary Intervention for Insomnia." *Hawaii Medical Journal* 61, no. 9: 192, 213. https://core.ac.uk/download/pdf/77123142.pdf.

Lower, Judith S., Carrie Bonsack, and Julie Guion. 2003. "Peace and Quiet." *Nursing Management* 34, no. 4: 40A–40D. doi:10.1097/00006247-200304000-00017.

Lowther, Jill E. 2012. "Understanding the Subjective Norms Surrounding Noise Exposure and Hearing Conservation in Children." MSc thesis, University of Western Ontario. Accessed 20 January 2019. http://ir.lib.uwo.ca/etd/506.

Luckyj, Christina. 2002. *"A Moving Rhetoricke": Gender and Silence in Early Modern England*. Manchester: Manchester University Press.

Lujan, Heidi L., and Stephen E. DiCarlo. 2006. "First-Year Medical Students Prefer Multiple Learning Styles." *Advances in Physiology Education* 30, no. 1: 13–16. doi:10.1152/advan.00045.2005.

Lundsteen, Sarah. 1971. *Listening: Its Impact on Reading and the Other Language Arts*. Urbana: National Council of Teachers of English, Educational Resources Information Center, Clearinghouse on the Teaching of English.

Ma, Guoxia, Yujun Tian, Tianzhen Ju, and Zhengwu Ren. 2006. "Assessment of Traffic Noise Pollution from 1989 to 2003 in Lanzhou City." *Environmental Mon-itoring and Assessment* 123, no. 1–3: 413–30. doi:10.1007/s10661-006-1494-6.

Ma, Hui, Xianrong Ji, and Takashi Yano. 2008. "Analysis of Community Response to Noise in Chinese City." *Acta Acustica-Peking* 33, no. 3: 275.

Ma, Hui, and Takashi Yano. 2004. "An Experiment on Auditory and Non-Auditory Disturbances Caused by Railway and Road Traffic Noises in Outdoor Conditions." *Journal of Sound and Vibration* 277, no. 3: 501–9. doi:10.1016/j.jsv.2004 .03.011.

Maassen, Marcus M., W. Babisch, K.D. Bachmann, Hartmut Ising, G. Lehnert, Peter Plath, Peter K. Plinkert, Ekkehard Rebentisch, G. Schuschke, M. Spreng, et al. 2001. "Ear Damage Caused by Leisure Noise." *Noise and Health* 4, no. 13: 1–16. http://www.noiseandhealth.org/text.asp?2001/4/13/1/31806.

MacArthur, John D. "Low-Level Noise Is an Insidious Stressor." Accessed January 2004. http://www.noiselaw.org/article18.html.

Mackrill, Jamie B., Rebecca Cain, and Paul A. Jennings. 2013. "Experiencing the Hospital Ward Soundscape: Towards a Model." *Journal of Environmental Psychology* 36: 1–8. doi:10.1016/j.jenvp.2013.06.004.

Mackrill, Jamie B., Rebecca Cain, Paul A. Jennings, and Michelle England. 2013. "Sound Source Information to Improve Cardiothoracic Patients' Comfort." *British Journal of Nursing* 22, no. 7: 387–93. doi:10.12968/bjon.2013.22.7.387.

Mackrill, Jamie B., Paul A. Jennings, and Rebecca Cain. 2013. "Improving the Hospital 'Soundscape': A Framework to Measure Individual Perceptual Response to Hospital Sounds." *Ergonomics* 56, no. 11: 1687–97. doi:10.1080 /00140139.2013.835873.

– 2014. "Exploring Positive Hospital Ward Soundscape Interventions." *Applied Ergonomics* 45, no. 6: 1454–60. doi:10.1016/j.apergo.2014.04.005.

MacPherson, Glen. 2012. "The World Hum Map and Database Project." Accessed 3 February 2019. http://www.thehum.info.

MacSweeney, Mairéad, Dafydd Waters, Michael J. Brammer, Bencie Woll, and Usha Goswami. 2008. "Phonological Processing in Deaf Signers and the Impact of Age of First Language Acquisition." *NeuroImage* 40, no. 3: 1369–79. doi:10.1016/j.neuroimage.2007.12.047.

Magee, Wendy L., Imogen Clark, Jeanette Tamplin, and Joke Bradt. 2017. "Music Interventions for Acquired Brain Injury." *Cochrane Database of Systematic Reviews* 2017, no. 1: CD006787. doi:10.1002/14651858.CD006787.pub3.

Mahendra Prashanth, K.V., and V. Sridhar. 2008. "The Relationship between Noise Frequency Components and Physical, Physiological and Psychological Effects of Industrial Workers." *Noise and Health* 10, no. 40: 90–8. http://www.noiseandhealth.org/text.asp?2008/10/40/90/44347.

Mahoney, Colin J., Jonathan D. Rohrer, Johanna C. Goll, Nick C. Fox, Martin N. Rossor, and Jason D. Warren. 2011. "Structural Neuroanatomy of Tinnitus and Hyperacusis in Semantic Dementia." *Journal of Neurology, Neurosurgery and Psychiatry* 82, no. 11: 1274–8. doi:10.1136/jnnp.2010.235473.

Mai, François M.M. 2006. "Beethoven's Terminal Illness and Death." *Journal of the Royal College of Physicians of Edinburgh* 36, no. 3: 258–63. http://www.rcpe.ac.uk/sites/default/files/t_100506_a_mai.pdf.

Maitland, Sara. *A Book of Silence*. Bodmin, UK: Granta, 2008.

Mak, Cheuk M., and Y.P. Lui. 2012. "The Effect of Sound on Office Productivity." *Building Services Engineering Research and Technology* 33, no. 3: 339–45. doi:10.1177/0143624411412253.

Mallozzi, Vincent M. 2011. "On Train, a Fight between Silent and Merely Quiet." *New York Times*, 9 January. http://www.nytimes.com/2011/01/10/nyregion/10quiet.html.

Marcus, Amy D. 2015. "Army Tests Hearing Drug at the Rifle Range." *Wall Street Journal*, 21 August. https://www.wsj.com/articles/army-tests-hearing-drug-at-the-rifle-range-1440182197.

Marshall, Lynne, John F. Brandt, and Larry E. Marston. 1975. "Anticipatory Middle-Ear Reflex Activity from Noisy Toys." *Journal of Speech and Hearing Disorders* 40, no. 3: 320–6. doi:10.1044/jshd.4003.320. Erratum *Speech and Hearing Disorders* 40, no. 4: 549. doi:10.1044/jshd.4004.549b.

Martinez, Carlos, Christopher Wallenhorst, Don McFerran, and Deborah A. Hall. 2015. "Incidence Rates of Clinically Significant Tinnitus: 10-Year Trend from a Cohort Study in England." *Ear and Hearing* 36, no. 3: e69–75. doi:10.1097/AUD.0000000000000121.

Martinez-Devesa, Pablo, Rafael Perera, Megan Theodoulou, and Angus Waddell. 2010. "Cognitive Behavioural Therapy for Tinnitus." *Cochrane Database of Systematic Reviews* 9. doi:10.1002/14651858.CD005233.pub3.

Mathews, Kenneth E., and Lance K. Canon. 1975. "Environmental Noise Level as a Determinant of Helping Behavior." *Journal of Personality and Social Psychology* 32, no. 4: 571–7. doi:10.1037/0022-3514.32.4.571.

Matsui, Toshihito, Stephen A. Stansfeld, Michael Haines, and Jenny J. Head. 2004. "Children's Cognition and Aircraft Noise Exposure at Home: The West London Schools Study." *Noise and Health* 7, no. 25: 49–57. http://www.noiseandhealth.org/text.asp?2004/7/25/49/31647.

Matsui, Toshihito, T. Uehara, Takashi Miyakita, K. Hiramatsu, Y. Osada, and T. Yamamoto. 2004. "The Okinawa Study: Effects of Chronic Aircraft Noise on Blood Pressure and Some Other Physiological Indices." *Journal of Sound and Vibration* 277, no. 3: 469–70. doi:10.1016/j.jsv.2004.03.007.

Maxwell, Lorraine E., and Gary W. Evans. 2000. "The Effects of Noise on Pre-School Children's Pre-Reading Skills." *Journal of Environmental Psychology* 20, no. 1: 91–7. doi:10.1006/jevp.1999.0144.

May, Cindi. 2014. "A Learning Secret: Don't Take Notes with a Laptop." *Scientific American*, 3 June. https://www.scientificamerican.com/article/a-learning-secret-don-t-take-notes-with-a-laptop/.

Mayr, Albert. 2007. "Above and Below Acoustic Ecology." Accessed 24 January 2019. http://ecoear.proscenia.net/wfaelibrary/library/articles/mayer_above_below.pdf.

Mazer, Susan E. 2002. "Sound Advice: Seven Steps for Abating Hospital Noise Problems." *Health Facilities Management* 15, no. 5: 24–6. https://www.research gate.net/publication/11339621_Sound_advice_Seven_steps_for_abating _hospital_noise_problems.

McAllister, Anita M., Svante Granqvist, Peta Sjölander, and Johan Sundberg. 2009. "Child Voice and Noise: A Pilot Study of Noise in Day Cares and the Effects on 10 Children's Voice Quality According to Perceptual Evaluation." *Journal of Voice* 23, no. 5: 587–93. doi:10.1016/j.jvoice.2007.10.017.

McBride, David, Frank Gill, David Proops, Malcolm Harrington, Kerry Gardiner, and Cynthia Attwell. 1992. "Noise and the Classical Musician." *British Medical Journal* 305, no. 6868: 1561–3. doi:10.1136/bmj.305.6868.1561.

McCullough, Paul, and J. Oliver Hetherington. 2005. "A Practical Evaluation of Objective Noise Criteria Used for the Assessment of Disturbance Due to Entertainment Music." *Journal of Environmental Health Research* 4, no. 2: 69–74. https://pdfs.semanticscholar.org/f212/1e3fb07a637c832ca875cb29 a0e2cfb779c9.pdf.

McFerran, Don J., and D.M. Baguley. 2007. "Acoustic Shock." *Journal of Laryngology and Otology* 121, no. 4: 301–5. doi:10.1017/S0022215107006111.

McFerran, Katerina Skewes. 2014. "Depending on Music to Feel Better: Being Conscious of Responsibility When Appropriating the Power of Music." *Arts in Psychotherapy* 41, no. 1: 89–97. doi:10.1016/j.aip.2013.11.007.

McGregor, Jena. 2014. "9 Things You Didn't Know About the Office Cubicle." *Washington Post*, 8 April. https://www.washingtonpost.com/news/on-leader-ship/wp/2014/04/18/9-things-you-didnt-know-about-the-office-cubicle/.

McLean, E.K., and A. Tarnopolsky. 1977. "Noise, Discomfort and Mental Health: A Review of the Socio-Medical Implications of Disturbance by Noise." *Psychological Medicine* 7, no. 1: 19–62. doi:10.1017/S0033291700023138.

Meister, Edward A., and Rebecca J. Donatelle. 2000. "The Impact of Commercial-Aircraft Noise on Human Health: A Neighborhood Study in Metropolitan Minnesota." *Journal of Environmental Health* 63, no. 4: 9–15. https://www .jstor.org/stable/44528021.

Melamed, Samuel, Yitzhak Fried, and Paul Froom. 2004. "The Joint Effect of Noise Exposure and Job Complexity on Distress and Injury Risk among Men and Women: The Cardiovascular Occupational Risk Factors Determination in Israel Study." *Journal of Occupational and Environmental Medicine* 46, no. 10: 1023–32. doi:10.1097/01.jom.0000141661.66655.a5.

Melamed, Samuel, Paul Froom, Estela Kristal-Boneh, Dafna Gofer, and Joseph Ribak. 1997. "Industrial Noise Exposure, Noise Annoyance, and Serum Lipid

Levels in Blue-Collar Workers: The CORDIS Study." *Archives of Environmental Health: An International Journal* 52, no. 4: 292–8. doi:10.1080/00039899 709602201.

Mercier, Vlasta, and Beat W. Hohmann. 2002. "Is Electronically Amplified Music Too Loud? What Do Young People Think?" *Noise and Health* 4, no. 16: 47–55. http://www.noiseandhealth.org/text.asp?2002/4/16/47/31828.

Meriläinen, Merja, Helvi Kyngäs, and Tero Ala-Kokko. 2010. "24-Hour Intensive Care: An Observational Study of an Environment and Events." *Intensive and Critical Care Nursing* 26, no. 5: 246–53. doi:10.1016/j.iccn.2010.06.003.

Merluzzi, F., Enrico Pira, and Luciano Riboldi. 2006. "The Italian Decree 195/2006 on the Protection of Workers against Risks Arising from Noise." English abstract of article in Italian. *Giornale Italiano di Medicina del Lavoro ed Ergonomia* 28, no. 3: 245–7. https://www.researchgate.net/publication/6655438 _The_Italian_decree_1952006_on_the_protection_of_workers_against_risks _arising_from_noise.

Metternich, Frank U., and Tilman Brusis. 1999. "Acute Hearing Loss and Tinnitus Caused by Amplified Recreational Music." English abstract of article in German. *Laryngo-Rhino-Otologie* 78, no. 11: 614–19. doi:10.1055/s-1999-8763.

Meyer-Bisch, Christian. 1996. "Epidemiological Evaluation of Hearing Damage Related to Strongly Amplified Music (Personal Cassette Players, Discotheques, Rock Concerts): High-Definition Audiometric Survey on 1364 Subjects." *Audiology* 35, no. 3: 121–42. doi:10.3109/00206099609071936.

Michaud, David S., Stephen E. Keith, and Dale McMurchy. 2008. "Annoyance and Disturbance of Daily Activities from Road Traffic Noise in Canada." *Journal of the Acoustical Society of America* 123, no. 2: 784–92. doi:10.1121/1 .2821984.

Miller, Wreford. "Silence in the Contemporary Soundscape." MA thesis, Simon Fraser University, 1993. https://core.ac.uk/download/pdf/56369830.pdf.

Mitchell, Lisa. 2008. "It's Your Shout." *Age*, 22 April. http://theage.com.au/news /epicure/its-your-shout/2008/04/21/1208629783592.html.

Mok, Esther, and Kwai-Yiu Wong. 2003. "Effects of Music on Patient Anxiety." *AORN Journal* 77, no. 2: 396–410. doi:10.1016/S0001-2092(06)61207-6.

Molholm, Sophie, Walter Ritter, Micah M. Murray, Daniel C. Javitt, Charles E. Schroeder, and John J. Foxe. 2002. "Multisensory Auditory-Visual Interactions during Early Sensory Processing in Humans: A High-Density Electrical Mapping Study." *Cognitive Brain Research* 14, no. 1: 115–28. doi:10.1016/S0926-6410(02)00066-6.

Morais, Dario, José I. Benito, and Ana Almaraz. 2007. "Acoustic Trauma in Classical Music Players." English abstract of article in Spanish. *Acta Otorrinolaringologica Española* 58, no. 9: 401–7. doi:10.1016/S2173-5735(07)70378-2.

Morat, Daniel, ed. 2014. *Sounds of Modern History: Auditory Cultures in 19th- and 20th-Century Europe*. New York: Berghahn.

Morata, Thaís C., Ann-Christin Johnson, Per Nylen, Eva B. Svensson, Jun Cheng, Edward F. Krieg, Ann-Cathrine Lindblad, Lena Ernstgård, and John Franks. 2002. "Audiometric Findings in Workers Exposed to Low Levels of Styrene and Noise." *Journal of Occupational and Environmental Medicine* 44, no. 9: 806–14. doi:10.1097/00043764-200209000-00002.

Morata, Thaís C., and Mark B. Little. 2002. "Suggested Guidelines for Studying the Combined Effects of Occupational Exposure to Noise and Chemicals on Hearing." *Noise and Health* 4, no. 14: 73–87. http://www.noiseandhealth.org /text.asp?2002/4/14/73/31807.

Morgan, Charles A. III, Christian Grillon, Steven M. Southwick, Michael Davis, and Dennis S. Charney. 1996. "Exaggerated Acoustic Startle Reflex in Gulf War Veterans with Posttraumatic Stress Disorder." *American Journal of Psychiatry* 153, no. 1: 64–8. doi:10.1176/ajp.153.1.64.

Moudon, Anne Vernez. 2009. "Real Noise from the Urban Environment: How Ambient Community Noise Affects Health and What Can Be Done about It." *American Journal of Preventive Medicine* 37, no. 2: 167–71. doi:10.1016/j.amepre .2009.03.019.

Muchnik, Chava, Noam Amir, Ester Shabtai, and Ricky Kaplan-Neeman. 2012. "Preferred Listening Levels of Personal Listening Devices in Young Teenagers: Self Reports and Physical Measurements." *International Journal of Audiology* 51, no. 4: 287–93. doi:10.3109/14992027.2011.631590.

Mueller, Pam A., and Daniel M. Oppenheimer. 2014. "The Pen Is Mightier than the Keyboard: Advantages of Longhand over Laptop Note Taking." *Psychological Science* 25, no. 6: 1159–68. doi:10.1177/0956797614524581.

Mullins, Joe H., and James P. Kelly. 1995. "The Mystery of the Taos Hum." *Echoes: The Newsletter of the Acoustical Society of America* 5, no. 3: 1, 4, 6. https://acousticalsociety.org/wp-content/uploads/2018/02/v5n3.pdf.

Multi-Municipal Wind Turbine Working Group. 2019. "Designing a Credible, Effective Study Based on Indisputable Science." Health Canada Wind Turbine Noise and Health Study. Accessed 29 January. http://www.na-paw.org/hc /hcdoc/multi-municipal-wind-turbine-working-group.pdf.

Münzel, Thomas, Tommaso Gori, Wolfgang Babisch, and Mathias Basner. 2014. "Cardiovascular Effects of Environmental Noise Exposure." *European Heart Journal* 35, no. 13: 829–36. doi:10.1093/eurheartj/ehu030.

Murphy, Gina, Anissa Bernardo, and Joanne Dalton. 2013. "Quiet at Night: Implementing a Nightingale Principle." *American Journal of Nursing* 113, no. 12: 43–51. doi:10.1097/01.NAJ.0000438871.60154.a8.

Murray, Christopher, Donald E. Goldstein, Stephen Nourse, and Eugene Edgar.

2000. "The Postsecondary School Attendance and Completion Rates of High School Graduates with Learning Disabilities." *Learning Disabilities Research and Practice* 15, no. 3: 119–27. doi:10.1207/SLDRP1503_1.

Na, Hyun Joo, and Soo Yang. 2009. "Effects of Listening to Music on Auditory Hallucination and Psychiatric Symptoms in People with Schizophrenia." English abstract of article in Korean. *Journal of the Korean Academy of Nursing* 39, no. 1: 62–71. doi:10.4040/ikan.2009.39.1.62.

Ndrepepa, Ana, and Dorothée Twardella. 2011. "Relationship between Noise Annoyance from Road Traffic Noise and Cardiovascular Diseases: A Meta-Analysis." *Noise and Health* 13, no. 52: 251–9. http://www.noiseandhealth.org/text.asp?2011/13/52/251/80163.

Neider, Mark B., Jason S. McCarley, James A. Crowell, Henry Kaczmarski, and Arthur F. Kramer. 2010. "Pedestrians, Vehicles, and Cell Phones." *Accident Analysis and Prevention* 42, no. 2: 589–94. doi:10.1016/j.aap.2009.10.004.

Neitzel, Richard, Robyn Gershon, Marina Zeltser, Allison Canton, and Muhammad Akram. 2009. "Noise Levels Associated with New York City's Mass Transit Systems." *American Journal of Public Health* 99, no. 8: 1393–9. doi:10.2105/AJPH.2008.138297.

Nelsen, Aaron. 2012. "Could the Wind Turbines of Chile Harm Blue Whales?" *Time*, 7 February. http://content.time.com/time/world/article/0,8599,2106064,00.html.

Nelson, Deborah I., Robert Y. Nelson, Marisol Concha-Barrientos, and Marilyn Fingerhut. 2005. "The Global Burden of Occupational Noise-Induced Hearing Loss." *American Journal of Industrial Medicine* 48, no. 6: 446–58. doi:10.1002/ajim.20223.

Nelson, Peggy, Kathryn Kohnert, Sabina Sabur, and Daniel Shaw. 2005. "Classroom Noise and Children Learning through a Second Language: Double Jeopardy?" *Language, Speech, and Hearing Services in Schools* 36, no. 3: 219–29. doi:10.1044/0161-1461(2005/022).

Nevill, A.M., N.J. Balmer, and A. Mark Williams. 2002. "The Influence of Crowdnoise and Experience upon Refereeing Decisions in Football." *Psychology of Sport and Exercise* 3, no. 4: 261–72. doi:10.1016/S1469-0292(01)00033-4.

Newman, Lily Hay. 2014. "This Is the Sound Cannon Used against Protesters in Ferguson." *Slate*, 14 August. http://www.slate.com/blogs/future_tense/2014/08/14/lrad_long_range_acoustic_device_sound_cannons_were_used_for_crowd_control.html.

Nguyen, An Luong, The Cong Nguyen, Trinh Van Le, Minh Hien Hoang, Sy Nguyen, Hiroshi Jonai, Maria Beatriz Villanueva, Shinya Matsuda, Midori Sotoyama, and Ayako Sudo. 1998. "Noise Levels and Hearing Ability in

Female Workers in a Textile Factory Vietnam." *Industrial Health* 36, no.1: 61–5. doi: 10/2486/indhealth.36.61.

Nielsen-Bohlman, Lynn, Robert T. Knight, David L. Woods, and Kelly Woodward. 1991. "Differential Auditory Processing Takes Place during Sleep." *Electroencephalography and Clinical Neuroscience* 79, no. 4: 281–90. doi:10.1016 /0013-4694(91)90124-M.

Niemann, Hildegard, X. Bonnefoy, Matthias Braubach, K. Hecht, Christian Maschke, C. Rodrigues, and N. Robbel. 2006. "Noise-Induced Annoyance and Morbidity Results from the Pan-European LARES Study." *Noise and Health* 8, no. 31: 63–79. http://www.noiseandhealth.org/text.asp?2006/8/31/63/33537.

Niemann, Hildegard, and Christian Maschke. 2004. "WHO LARES Final Report: Noise Effects and Morbidity." World Health Organization / Berlin Center of Public Health. http://www.euro.who.int/__data/assets/pdf_file/0015 /105144/WHO_Lares.pdf.

Nilsson, Mats E., and Birgitta Berglund. 2006. "Noise Annoyance and Activity Disturbance before and after the Erection of a Roadside Noise Barrier." *Journal of the Acoustical Society of America* 119, no. 4: 2178–88. doi:10.1121/1 .2169906.

"NIOSH Releases Materials on Call Center Noise Hazards." 2011. *Occupational Health and Safety*. 16 October. Accessed 3 February 2019. http://ohsonline .com/articles/2011/10/16/niosh-releases-materials-on-call-center-noise-hazards.aspx.

Niskar, Amanda S., Stephanie M. Kieszak, Alice E. Holmes, Emilio Esteban, Carol Rubin, and Debra J. Brody. 2001. "Estimated Prevalence of Noise-Induced Hearing Threshold Shifts among Children 6 to 19 Years of Age: The Third National Health and Nutrition Examination Survey, 1988–1994, United States." *Pediatrics* 108, no. 1: 40–3. doi:10.1542/peds.108.1.40.

"Noise Regulation." 2019. *Wikipedia*. Accessed 11 February. https://en.wikipedia .org/wiki/Noise_regulation.

"The Noise of Summer." 2019. *Character Counts!* Accessed 4 January. https://charactercounts.org/the-noise-of-summer/.

Nomura, Kyoko, Mutsuhiro Nakao, and Eiji Yano. 2005. "Hearing Loss Associated with Smoking and Occupational Noise Exposure in a Japanese Metal Working Company." *International Archive of Occupational and Environmental Health* 78, no. 3: 178–84. doi:10.1007/s00420-005-0604-z.

"N[orth] W[est] Calgary Asphalt Plant: A Study in Noise Pollution." 2007. Unpublished document and poster by students of the "Introduction to Acoustic Ecology/Acoustic Communications" course, University of Calgary.

Nozza, Robert J., Reva N.F. Rossman, Linda C. Bond, and Sandra L. Miller. 1990. "Infant Speech-Sound Discrimination in Noise." *Journal of the Acoustical Society of America* 87, no. 1: 339–50. doi:10.1044/jshr.3403.643.

Nurminen, Tuula. 1995. "Female Noise Exposure, Shift Work, and Reproduction." *Journal of Occupational and Environmental Medicine* 37, no. 8: 945–51. doi:10.1097/00043764-199508000-00010.

Nurminen, Tuula, and Kari Olavi Kurppa. 1989. "Occupational Noise Exposure and Course of Pregnancy." *Scandinavian Journal of Work, Environment and Health* 15, no. 2: 117–24. doi:10.5271/sjweh.1873.

O'Brien, Ian, Wayne James Wilson, and Andrew P. Bradley. 2008. "Nature of Orchestral Noise." *Journal of the Acoustical Society of America* 124, no. 2: 926–39. doi:110.1121/1.2940589.

Obelenis, Vytautis, and Vilija Malinauskienė. 2007. "The Influence of Occupational Environment and Professional Factors on the Risk of Cardiovascular Disease." English abstract of article in Lithuanian. *Medicina* 43, no. 2: 96–102. doi:10.3390/medicina43020011.

Oishi, Naoki, and Jochen Schacht. 2011. "Emerging Treatments for Noise-Induced Hearing Loss." *Expert Opinion: Emerging Drugs* 16, no. 2: 235–45. doi:10.1517/14728214.2011.552427.

Okamoto, Hidehiko, Henning Teismann, Ryusuke Kakigi, and Christo Pantev. 2011. "Broadened Population-Level Frequency Tuning in Human Auditory Cortex of Portable Music Player Users." *PLoS One* 6, no. 3: e17022. doi:10.1371/journal.pone.0017022.

Oliveira, Maria João R., Mariana P. Monteiro, Andreia M. Ribeiro, Duarte Pignatelli, Artur P. Águas. 2009. "Chronic Exposure of Rats to Occupational Textile Noise Causes Cytological Changes in Adrenal Cortex." *Noise and Health* 11, no. 43: 118–23. http://www.noiseandhealth.org/text.asp?2009/11/43/118/50697.

Oliveira, Maria João R., A.S. Pereira, N.A.A. Castelo Branco, N.R. Grande, and Artur P. Águas. 2001. "In Utero and Postnatal Exposure of Wistar Rats to Low Frequency/High Intensity Noise Depletes the Tracheal Epithelium of Ciliated Cells." *Lung* 179, no. 4: 225–32. doi:10.1007/s004080000063.

Ong, Walter. 1982. *Orality and Literacy: The Technologizing of the Word*. London: Routledge.

Orr, Scott P., Natasha B. Lasko, Arieh Y. Shalev, and Roger K. Pitman. 1995. "Physiologic Responses to Loud Tones in Vietnam Veterans with Posttraumatic Stress Disorder." *Journal of Abnormal Psychology* 104, no. 1: 75–82. doi:10.1037/0021-843X.104.1.75.

Orr, Scott P., Linda J. Metzger, Natasha B. Lasko, Michael L. Macklin, Frank B. Hu, Arieh Y. Shalev, Roger K. Pitman, Seth A. Eisen, Mark W. Gilbertson, Gregory M. Gillette, et al. 2003. "Physiologic Responses to Sudden, Loud Tones in Monozygotic Twins Discordant for Combat Exposure: Association with Posttraumatic Stress Disorder." *Archives of General Psychiatry* 60, no. 3: 283–8. doi:10.1001/archpsyc.60.3.283.

Osatuke, Katerine, Carol L. Humphreys, Meredith J. Glick, Robin L. Graff-Reed, LaTasha McKenzie Mack, and William B. Stiles. 2005. "Vocal Manifestations of Internal Multiplicity: Mary's Voices." *Psychology and Psychotherapy: Theory, Research and Practice* 78, no. 1: 21–44. doi:10.1348/147608304X22364.

Osatuke, Katerine, James K. Mosher, Jacob Z. Goldsmith, William B. Stiles, David A. Shapiro, Gillian E. Hardy, and Michael Barkham. 2007. "Submissive Voices Dominate in Depression: Assimilation Analysis of a Helpful Session." *Journal of Clinical Psychology* 63, no. 2: 153–64. doi:10.1002/jclp.20338.

Packwood, William T. 1974. "Loudness as a Variable in Persuasion." *Journal of Counseling Psychology* 21, no. 1: 1–2. doi:10.1037/h0036065.

Padnani, Amisha. 2012. "The Power of Music, Tapped in a Cubicle." *New York Times*, 11 August. http://www.nytimes.com/2012/08/12/jobs/how-music-can-improve-worker-productivity-workstation.html.

Page, Richard A. 1977. "Noise and Helping Behavior." *Environment and Behavior* 9, no. 3: 311–34. doi:10.1177/001391657700900302.

Palisson, J., C. Roussel-Baclet, D. Maillet, C. Belin, J. Ankri, and P. Narme. 2015. "Music Enhances Verbal Episodic Memory in Alzheimer's Disease." *Journal of Clinical and Experimental Neuropsychology* 37, no. 5: 503–17. doi:10.1080/13803395.2015.1026802.

Pallesen, Karen J., Elvira Brattico, Christopher J. Bailey, Antti Korvenoja, Juha Koivisto, Albert Gjedde, and Synnöve Carlson. 2010. "Cognitive Control in Auditory Working Memory Is Enhanced in Musicians." *PLOS One* 5, no. 6: e11120. doi:10.1371/journal.pone.0011120.

Palmer, Jason. 2012. "Science Decodes 'Internal Voices.'" *BBC News*, 31 January. http://www.bbc.co.uk/news/science-environment-16811042.

Palmer, K., M.J. Griffin, H.E. Syddall, and D. Coggon. 2004. "Cigarette Smoking, Occupational Exposure to Noise, and Self Reported Hearing Difficulties." *Occupational and Environmental Medicine* 61: 340–4. doi:10.1136/oem.2003.009183.

Paquette, David. 2004. "Describing the Contemporary Sound Environment: An Analysis of Three Approaches, Their Synthesis, and a Case Study of Commercial Drive, Vancouver, BC." MA thesis, Simon Fraser University. https://core.ac.uk/download/pdf/56372892.pdf.

Parga, Joanna, Robert Daland, Kalpashri Kesavan, Paul Macey, Lonnie Zeltzer, and Ronald Harper. 2018. "A Description of Externally Recorded Womb Sounds in Human Subjects during Gestation." *PLOS One* 13, no. 5: e0197045. doi:10.1371/journal.pone.0197045.

Partanen, Eino, Teija Kujala, Mari Tervaniemi, and Minna Huotilainen. 2013. "Prenatal Music Exposure Induces Long-Term Neural Effects." *PloS One* 8, no. 10: e78946. doi:10.1371/journal.pone.0078946.

Pashler, H., M. McDaniel, D. Rohrer, and R. Bjork. 2008. "Learning Styles: Concepts and Evidence." *Psychological Science in the Public Interest* 9, no. 3: 105–19. doi:10.1111/j.1539-6053.2009.01038.x.

Pasley, Brian N., Stephen V. David, Nima Mesgarani, Adeen Flinker, Shihab A. Shamma, Nathan E. Crone, Robert T. Knight, and Edward F. Chang. 2012. "Reconstructing Speech from Human Auditory Cortex." *PLOS Biology* 10, no. 1: e1001251. doi:10.1371/journal.pbio.1001251.

Patel, Aniruddh. 2008. *Music, Language, and the Brain*. Oxford: Oxford University Press.

Patel, Jacqueline A., and Keith Broughton. 2002. "Assessment of the Noise Exposure of Call Centre Operators." *Annals of Occupational Hygiene* 46, no. 8: 653–61. doi:10.1093/annhyg/mef091.

Patterson, Dan. 2015. "Cities First to Benefit from Internet of Things, If We Can Write Better Software." *TechRepublic*, 5 November. https://www.techrepublic.com/article/vint-cerf-cities-first-to-benefit-from-internet-of-things-if-we-can-write-better-software/.

Paul, Stephan, Isabel C. Kuniyoshi, Flávio André M. de Araújo, and Lucinara Camargo. 2013. "Noise Pollution in Urban Settings of the Western Amazonia and an Approach to Cope with." *Journal of the Acoustical Society of America* 133, no. 5: 3276. doi:10.1121/1.4805334.

Pawlaczyk-Łuszczyńska, Małgorzata, Adam Dudarewicz, and Mariola Śliwińska-Kowalska. 2007. "Sources of Occupational Exposure to Ultrasonic Noise." *Medycyna Pracy* 58, no. 2: 105–16. https://europepmc.org/abstract/med/17926499.

Pawlaczyk-Łuszczyńska, Małgorzata, Adam Dudarewicz, Małgorzata Waszkowska, Wiesła Szymczak, and Mariola Śliwińska-Kowalska. 2005. "The Impact of Low-Frequency Noise on Human Mental Performance." *International Journal of Occupational Medicine and Environmental Health*, 18, no. 2: 185–98. https://www.ncbi.nlm.nih.gov/pubmed/16201210.

Pawlaczyk-Łuszczyńska, Małgorzata, Małgorzata Zamojska-Daniszewska, Kamil Zaborowski, and Adam Dudarewicz. 2019. "Evaluation of Noise Exposure and Hearing Threshold Levels among Call Centre Operators." *Archives of Acoustics* 44, no. 4: 747–59. doi:10.24425/aoa.2019.129730.

Peng, J.H., Z.Z. Tao, and Z.W. Huang. 2007. "Risk of Damage to Hearing from Personal Listening Devices in Young Adults." *Journal of Otolaryngology* 36, no. 3: 181–5. https://www.ncbi.nlm.nih.gov/pubmed/17711774.

Pérez-Lloret, Santiago, Joaquín Diez, María Natalia Domé, Andrea Alvarez Delvenne, Nestor Braidot, Daniel P. Cardinali, and Daniel Eduardo Vigo. 2014. "Effects of Different 'Relaxing' Music Styles on the Autonomic Nervous System." *Noise and Health* 16, no. 72: 279–84. http://www.noiseandhealth.org/text.asp?2014/16/72/279/140507.

Perham, Nick, and Martinne Sykora. 2012. "Disliked Music Can Be Better for

Performance than Liked Music." *Applied Cognitive Psychology* 26, no. 4: 550–5. doi:10.1002/acp.2826.

Perrin, Fabien, Luis García-Larrea, François Mauguière, and Hélène Bastuji. 1999. "A Differential Brain Response to the Subject's Own Name Persists during Sleep." *Clinical Neurophysiology* 110, no. 12: 2153–64. doi:10.1016/S1388-2457(99)00177-7.

Perrine, Ruby, Anne Caclin, Sabrina Boulet, Claude Delpuech, and Dominique Morlet. 2008. "Odd Sound Processing in the Sleeping Brain." *Journal of Cognitive Neuroscience* 20, no. 2: 296–311. doi:10.1162/jocn.2008.20023.

Persson Waye, Kerstin, and Erica Ryberg. 2013. "Achieving a Healthy Sound Environment in Hospitals." Keynote paper at InterNoise conference, Innsbruck, Austria. September.

Persson Waye, Kerstin, Johanna Bengtsson, Anders Kjellberg, and Stephen Benton. 2001. "Low Frequency Noise 'Pollution' Interferes with Performance." *Noise and Health* 4, no. 13: 33–49. http://www.noiseandhealth.org/text.asp ?2001/4/13/33/31803.

Persson Waye, Kerstin, Johanna Bengtsson, Ragnar Rylander, Frank Hucklebridge, Phil Evans, and Angela Clow. 2002. "Low Frequency Noise Enhances Cortisol among Noise Sensitive Subjects during Work Performance." *Life Sciences* 70, no. 7: 745–58. doi:10.1016/S0024-3205(01)01450-3.

Persson Waye, Kerstin, Martin Björkman, and Ragnar Rylander. 1985. "An Experimental Evaluation of Annoyance Due to Low-Frequency Noise." *Journal of Low-Frequency Noise, Vibration and Active Control* 4, no. 4: 145–53. doi:10.1177/026309238500400401.

Persson Waye, Kerstin, and Erica Ryherd. 2013. "Achieving a Healthy Sound Environement in Hospitals." 42nd International Congress and Exposition on Noise Control Engineering. *Inter-Noise* 1: 38–45. https://experts.nebraska .edu/en/publications/achieving-a-healthy-sound-environment-in-hospitals.

Persson Waye, Kerstin, Ragnar Rylander, Stephen Benton, and H.G. Leventhal. 1997. "Effects on Performance and Work Quality Due to Low Frequency Ventilation Noise." *Journal of Sound and Vibration* 205, no. 4: 467–74. doi:10.1006/jsvi.1997.1013.

Philbin, M. Kathleen, Alex Robertson, and James W. Hall III. 1997. "Recommended Permissible Noise Criteria for Occupied, Newly Constructed or Renovated Hospital Nurseries." *Journal of Perinatology* 19, no. 8, part 1: 559–63. doi:10.1038/sj.jp.7200279.

Phillips, Carl V. 2011. "Properly Interpreting the Epidemiologic Evidence about the Health Effects of Industrial Wind Turbines on Nearby Residents." *Bulletin of Science, Technology and Society* 31, no. 4: 303–15. doi:10.1177/027046761 1412554.

Picker, John. 2003. *The Victorian Soundscape*. Oxford: Oxford University Press.

Picard, Michel, Serge André Girard, Marilène Courteau, Tony Leroux, et al. 2008. "Could Driving Safety Be Compromised by Noise Exposure at Work and Noise-Induced Hearing Loss?" *Traffic Injury Prevention* 9, no. 5: 489–99. doi:10.1080/15389580802271478.

Pimperton, Hannah, and Kate Nation. 2010. "Understanding Words, Understanding Numbers: An Exploration of the Mathematical Profiles of Poor Comprehenders." *British Journal of Educational Psychology* 80, no. 2: 255–68. doi:10.1348/000709909X477251.

Pinch, Trevor, and Karin Bijsterveld. 2004. "Sound Studies: New Technologies and Music." *Social Studies of Science* 34/5: 635–48. doi:10.1177/0306312704 047615.

Pope, D. 2010. "Decibel Levels and Noise Generators on Four Medical/Surgical Nursing Units." *Journal of Clinical Nursing* 19, nos. 17–18: 2463–70. doi:10.1111/j.1365-2702.2010.03263.x.

Pouryaghoub, Gholamreza, Ramin Mehrdad, and Saber Mohammadi. 2007. "Interaction of Smoking and Occupational Noise Exposure on Hearing Loss: A Cross-Sectional Study." *BioMedClinical (BMC) Public Health* 7, no. 137. doi:10.1186/1471-2458-7-137.

Prado Saldivar, Reina Alejandra. 2013. "Sonic Brownface: Representations of Mexicanness in an Era of Discontent." *Sounding Out!* 10 June. https://sound studiesblog.com/2013/06/10/sonic-brownface-representations-of-mexicanness-in-an-era-of-discontent/.

Prashanth, K.V. Mahendra, and V. Sridhar. 2008. "The Relationship between Noise Frequency Components and Physical, Physiological and Psychological Effects of Industrial Workers." *Noise and Health* 10, no. 40: 90–8. doi:10.4103/1463-1741.44347.

Prochnik, George. 2010. *In Pursuit of Silence: Listening for Meaning in a World of Noise*. New York: Doubleday.

Purper-Ouakil, D., M. Wohl, G. Michel, M.C. Mouren, and P. Gorwood. 2004. "Symptom Variations in ADHD: Importance of Context, Development and Comorbidity." English abstract of article in French. *Encephale* 30, no. 6: 533–9. https://www.ncbi.nlm.nih.gov/pubmed/15738855.

Rabat, Arnaud. 2007. "Extra-Auditory Effects of Noise in Laboratory Animals: The Relationship between Noise and Sleep." *Journal of the American Association for Laboratory Animal Science* 46, no. 1: 35–41. https://www.researchgate.net/publication/6597032_Extra-auditory_effects_of _noise_in_laboratory_animals_The_relationship_between_noise_and_sleep.

Rabinowitz, Peter M., Deron Galusha, Christine Dixon-Ernst, Martin D. Slade, and Mark R. Cullen. 2007. "Do Ambient Noise Exposure Levels Predict Hearing Loss in a Modern Industrial Cohort?" *Occupational and Environmental Medicine* 64, no. 1: 53–9. doi:10.1136/oem.2005.025924.

Radicchi, Antonella. 2017. "Beyond the Noise: Open Source Soundscapes – A Mixed Methodology to Analyse, Evaluate and Plan 'Everyday' Quiet Areas." *Proceedings of Meetings on Acoustics* 30, no. 1. doi:10.1121/2.0000565.

Radicchi, Antonella, Dietrich Henckel, and Martin Memmel. 2017. "Citizens as Smart, Active Sensors for a Quiet and Just City: The Case of the 'Open Source Soundscapes' Approach to Identify, Assess and Plan 'Everyday Quiet Areas' in Cities." *Noise Mapping* 4, no. 1: 104–23. https://www.degruyter.com/view/j /noise.2017.4.issue-1/noise-2017-0008/noise-2017-0008.xml.

Ramage-Morin, Pamela L., and Marc Gosselin. 2018. "Canadians Vulnerable to Workplace Noise." Statistics Canada *Health Reports* 29, no. 8: 9–17. https://www150.statcan.gc.ca/n1/pub/82-003-x/2018008/article/00002-eng.htm.

Ramesh, A., P.N. Suman Rao, G. Sandeep, M. Nagapoornima, V. Srilakshmi, M. Dominic, and Swarnarekha. 2009. "Efficacy of a Low Cost Protocol in Reducing Noise Levels in the Neonatal Intensive Care Unit." *Indian Journal of Pediatrics* 76, no. 5: 475–8. doi:10.1007/s12098-009-0066-5.

Rauscher, Frances H., K. Desix Robinson, and Jason J. Jens. 1998. "Improved Maze Learning through Early Music Exposure in Rats." *Neurological Research* 20, no. 5: 427–32. doi:10.1080/01616412.1998.11740543.

Rawnsley, Adam. 2011. "'The Scream': Israel Blasts Protesters with Sonic Gun." *Wired*, 23 September. http://www.wired.com/2011/09/the-scream-israel-blasts-rioters-with-sonic-gun/.

Redström, Johan. 2007. "Is Acoustic Ecology about Ecology?" Accessed 18 February 2019. http://ciufo.org/classes/ae_sp14/reading/redstrom_aeecology.pdf.

Reid, Marylou, and Katerine Osatuke. 2006. "Acknowledging Problematic Voices: Processes Occurring at Early Stages of Conflict Assimilation in Patients with Functional Somatic Disorder." *Psychology and Psychotherapy: Theory, Research and Practice* 79, no. 4: 539–55. doi:10.1348/147608305X90467.

Reilly, Mary Jo, Kenneth D. Rosenman, and Douglas J. Kalinowski. 1998. "Occupational Noise-Induced Hearing Loss Surveillance in Michigan." *Journal of Occupational and Environmental Medicine* 40, no. 8: 667–74. doi:10.1097 /00043764-199808000-00002.

Reissland, Nadja, Brian Francis, Louisa Buttanshaw, Joe M. Austen, and Vincent Reid. 2016. "Do Fetuses Move Their Lips to the Sound That They Hear? An Observational Feasibility Study on Auditory Stimulation in the Womb." *Pilot and Feasibility Studies* 2, no. 14: 1–7. doi:10.1186/s40814-016-0053-3.

Ressel, Wolfram, C.D. Eisenbach, Stefan Alber, B. Bergk, and Frederik Wurst. 2007. "Enduring Traffic Noise Reduction by Improved Porous Asphalt." Paper presented at 4th International SIIV Congress, Palermo, IT, 12–14 September.

http://www.siiv.net/site/sites/default/files/Documenti/palermo/63_2848
_20080107214247.pdf.

Revill, Jo. 2006. "1,000 Call Centre Workers Suffer from Noise Shock." *Guardian*,
19 November. http://www.guardian.co.uk/money/2006/nov/19/medicineand
health.workandcareers.

Reynolds, James, Alastair McClelland, and Adrian Furnham. 2013. "An Investiga-
tion of Cognitive Test Performance across Conditions of Silence, Background
Noise and Music as a Function of Neuroticism." *Anxiety, Stress and Coping: An
International Journal* 27, no. 4. doi:10.1080/10615806.2013.864388.

Richards, Douglas S., Barbara Frentzen, Kenneth J. Gerhardt, Mary E. McCann,
and Robert M. Abrams. 1992. "Sound Levels in the Human Uterus." *Obstetrics
and Gynecology* 80, no. 2: 187–90. https://journals.lww.com/greenjournal
/Abstract/1992/08000/Sound_Levels_in_the_Human_Uterus.6.aspx.

Ristovska, Gordana, Dragan Gjorgjev, and Nada P. Jordanova. 2004. "Psychosocial Effects of Community Noise: Cross Sectional Study of School Children
in Urban Center of Skopje, Macedonia." *Croatian Medical Journal* 45, no. 4:
473–6. http://europepmc.org/abstract/med/15311422.

Ristovska, Gordana, Helga E. Laszlo, and Anna L. Hansell. 2014. "Reproductive
Outcomes Associated with Noise Exposure: A Systematic Review of the Literature." *International Journal of Environmental Research and Public Health* 11, no.
8: 7931–52. doi:10.3390/ijerph110807931.

Rochester, Lynn, Victoria Hetherington, Diana Jones, Alice Nieuwboer, Anne-
Marie Willems, Gert Kwakkel, and Erwin Van Wegen. 2005. "The Effect of
External Rhythmic Cues (Auditory and Visual) on Walking during a
Functional Task in Homes of People with Parkinson's Disease." *Archives of Physi-
cal Medicine and Rehabilitation* 86, no. 5: 999–1006. doi:10.1016/j.apm
.2004.10.040.

Roelofsen, Paul. 2008. "Performance Loss in Open-Plan Offices Due to Noise by
Speech." *Journal of Facilities Management* 6, no. 3: 202–11. doi:10.1108/1472596
0810885970.

Romei, Vincenzo, Micah M. Murray, Lotfi B. Merabet, and Gregor Thut. 2007.
"Occipital Transcranial Magnetic Stimulation Has Opposing Effects on Visual
and Auditory Stimulus Detection: Implications for Multisensory Interactions."
Journal of Neuroscience 27, no. 43: 11465–72. doi:10.1523/JNEUROSCI.2827-
07.2007.

Rosanowski, Frank, Ulrich Eysholdt, and Ulrich Hoppe. 2006. "Influence of
Leisure-Time Noise on Outer Hair Cell Activity in Medical Students." *International Archives of Occupational and Environmental Health* 80, no. 1: 25–31.
doi:10.1007/s00420-006-0090-y.

Roy, Kenneth, Amy Costello, and Anita Snader. 2008. "Case Study: Leadership in
Energy and Environmental Design Platinum Office Building with Innovation

Credit for Acoustic Design." *Journal of the Acoustical Society of America* 124, no. 4: 2546. doi:10.1121/1.4782982.

Royster, Julia D., Larry H. Royster, and Mead C. Killion. 1991. "Sound Exposures and Hearing Thresholds of Symphony Orchestra Musicians." *Journal of the Acoustical Society of America* 85, no. S1: S46. doi:10.1121/1.2026987.

Ryan, Allen F. 2000. "Protection of Auditory Receptors and Neurons: Evidence for Interactive Damage. *Proceedings of the National Academy of Sciences of the United States of America* 97, no. 13: 6939–40. doi:10.1073/pnas.97.13.6939

Rybczynski, Witold. 1986. *Home: A Short History of an Idea.* New York: Viking Press.

Ryberg, J.B. 2009. "A National Project to Evaluate and Reduce High Sound Pressure Levels from Music." *Noise and Health* 11, no. 43: 124–8. doi:10.4103/1463-1741.50698.

Rylander, R., and M. Bjorkman. 2002. "Planning Consequences of the Maximum dB(A) Concept: A Perspective." *Journal of Sound and Vibration* 250, no. 1: 175–9. doi:10.1006/jsvi.2001.3891.

Sadhra, S., C.A. Jackson, T. Ryder, and M.J. Brown. 2002. "Noise Exposure and Hearing Loss among Student Employees Working in University Entertainment Venues." *Annals of Occupational Hygiene* 46, no 55: 455–63. doi:10.1093/annhyg/mefo51.

Sakamoto, H., F. Hayashi, S. Sugiura, and M. Tsujikawa. 2002. "Psycho-Circulatory Responses Caused by Listening to Music, and Exposure to Fluctuating Noise or Steady Noise." *Journal of Sound and Vibration* 250, no. 1: 23–9. doi:10.1006/jsvi.2001.3885.

Salt, Alec N., and Timothy E. Hullar. 2010. "Responses of the Ear to Low Frequency Sounds, Infrasound and Wind Turbines." *Hearing Research* 268, no. 1: 12–21. doi:10.1016/j.heares.2010.06.007.

Salvaterra, Neanda. 2008. "Will Harlem Sway to a New Drumbeat?" *Black Star News*, 11 January. http://www.blackstarnews.com/ny-watch/others/will-harlem-sway-to-a-new-drumbeat.html.

Salvi, Richard, Berthold Langguth, Suzanne Kraus, Michael Landgrebe, Brian Allman, Dalian Ding, and Edward Lobarinas. 2011. "Tinnitus and Hearing Loss and Changes in Hippocampus." *Seminars in Hearing* 32, no. 2: 203–11. doi:10.1055/s-0031-1277243.

Sanborn, Vic. 2009. "London's Street Noises: 'The Enraged Musician' by William Hogarth." *Jane Austen's World.* 3 November. Accessed 7 February 2019. http://janeaustensworld.wordpress.com/2009/11/03/londons-street-noises-the-enraged-musician-by-william-hogarth/.

Sanjuan, Julio, Olga Rivero, Eduardo J. Aguilar, Jose C. González, Maria D. Moltó, Rosa de Frutos, Klaus-Peter Lesch, and Carmen Nájera. 2006. "Serotonin Transporter Gene Polymorphism (5-HTTLPR) and Emotional Response to Auditory Hallucinations in Schizophrenia." *International Journal of Neuropsychopharmacology* 9, no. 1: 131–3. doi:10.1017/S1461145705005559.

Sanyal, Tania, Vivek Kumar, Tapas Chandra Nag, Suman Jain, Vishnu Sreenivas, and Shashi Wadhwa. 2013. "Prenatal Loud Music and Noise: Differential Impact on Physiological Arousal, Hippocampal Synaptogenesis and Spatial Behavior in One Day-Old Chicks." *PLoS One* 8, no. 7: e67347. doi:10.1371/journal.pone.0067347.

Särkämö, Teppo, Mari Tervaniemi, Sari Laitinen, Anita Forsblom, Seppo Soinila, Mikko Mikkonen, Taina Autti, Heli M. Silvennoinen, Jaakko Erkkilä, Matti Laine, et al. 2008. "Music Listening Enhances Cognitive Recovery and Mood after Middle Cerebral Artery Stroke." *Brain* 131, no. 3: 866–76. doi:10.1093/brain/awn013.

Satoh, M., and S. Kuzuhara. 2008. "Training in Mental Singing While Walking Improves Gait Disturbance in Parkinson's Disease Patients." *European Neurology* 60, no. 5: 237–43. doi:10.1159/000151699.

Sayk, Friedhelm, Christoph Becker, Christina Teckentrup, Horst-Lorenz Fehm, Jan Struck, Jens Peter Wellhoener, and Christoph Dodt. 2007. "To Dip or Not to Dip: On the Physiology of Blood Pressure Decrease during Nocturnal Sleep in Healthy Humans." *Hypertension* 49, no. 5: 1070–6. doi:10.1161/HYPERTENSIONAHA.106.084343.

Schafer, R. Murray, ed. 1977a. *European Sound Diary*. Burnaby, BC: World Soundscape Project.

– ed. 1977b. *Five Village Soundscapes*. Burnaby, BC: World Soundscape Project.

– 1977c. *The Tuning of the World*. New York: Knopf.

– 1992. *A Sound Education*. Indian River, ON: Arcana Editions.

– 1993. *Voices of Tyranny, Temples of Silence*. Indian River, ON: Arcana Editions.

– 1994. *Our Sonic Environment and the Soundscape: The Tuning of the World*. Rochester, VT: Destiny Books.

Schlaug, Gottfried, Andrea C. Norton, Sarah Marchina, Lauryn R. Zipse, and Catherine Y. Wan. 2010. "From Singing to Speaking: Facilitating Recovery from Nonfluent Aphasia." *Future Neurology* 5, no. 5: 657–65. https://www.ncbi.nlm.nih.gov/pmc/articles/PMC2982746/.

Schmidt, Jesper H., and Mads Klokker. 2014. "Health Effects Related to Wind Turbine Noise Exposure: A Systematic Review." *PLoS One* 9, no. 12: e114183. doi:10.1371/journal.pone.0114183.

Schmuziger, Nicolas, Jochen Patscheke, and Rudolf Probst. 2006. "Hearing in Nonprofessional Pop/Rock Musicians." *Ear and Hearing* 27, no. 4: 321–30. doi:10.1097/01.aud.0000224737.34907.5e.

Schreckenberg, Dirk, Barbara Griefahn, and Markus Meis. 2010. "The Associations between Noise Sensitivity, Reported Physical and Mental Health, Perceived Environmental Quality, and Noise Annoyance." *Noise and Health* 12, no. 46: 7–16. doi:10.4103/1463-1741.59995.

Schulte-Körne, Gerd, Wolfgang Demiel, Jürgen Bartling, and Helmut Remschmidt. 1998. "Auditory Processing and Dyslexia: Evidence for a Specific Speech Processing Deficit." *NeuroReport* 9, no. 2: 337–40. https://journals-lww-com.ezproxy.lib.ucalgary.ca/neuroreport/Fulltext/1998/01260/Auditory _processing_and_dyslexia__evidence_for_a.29.aspx.

Schwartz, Hillel. 1995. "Realizing the Ideal: The Responsibility of the World's Religions." In *Noise and Silence: The Soundscape and Spirituality*. Noise Pollution Clearinghouse. nonoise.org.

– 2011. *Making Noise: From Babel to the Big Bang and Beyond*. New York: Zone Books.

Schwela, Dietrich. 2001. Scientific Committee on Emerging and Newly Identified Health Risks (SCENIHR). 2008. "Potential Health Risks of Exposure to Noise from Personal Music Players and Mobile Phones Including a Music Playing Function." Preliminary Report. European Commission. Accessed 7 January 2019. http://ec.europa.eu/health/ph_risk/committees/04_scenihr/docs /scenihr_o_017.pdf.

Seep, Benjamin, Robin Glosemeyer, Emily Hulce, Matt Linn, and Pamela Aytar. 2000. *Classroom Noise Booklet*. Melville, NY: Acoustical Society of America. http://www.nonoise.org/library/classroom/.

Segal, S., E. Eviatar, J. Lapinsky, N. Shlamkovitch, and A. Kessler. 2003. "Inner Ear Damage in Children Due to Noise Exposure from Toy Cap Pistols and Firecrackers: A Retrospective Review of 53 Cases." *Noise and Health* 5, no. 18: 13–18. http://www.noiseandhealth.org/text.asp?2003/5/18/13/31823.

Seidler, A., M. Wagner, M. Schubert, P. Dröge, J. Pons-Kühnemann, E. Swart, H. Zeeb, and J. Hegewald. 2016. "Myocardial Infarction Risk Due to Aircraft, Road, and Rail Traffic Noise: Results of a Case: Control Study Based on Secondary Data." *Deutsch Arzteblatt International* 113, no. 24: 407–14. doi:10.3238/arztebl.2016.0407.

Seitz, Aaron R., Robyn Kim, Virginie van Wassenhove, and Ladan Shams. 2007. "Simultaneous and Independent Acquisition of Multisensory and Unisensory Associations." *Perception* 36, no. 10: 1445–53. doi:10.1068/p5843.

Seixas, Noah, Rick Neitzel, Lianne Sheppard, and Bryan Goldman. 2005. "Alternative Metrics for Noise Exposure among Construction Workers." *Annals of Occupational Hygiene* 49, no. 6: 493–502. doi:10.1093/annhyg/mei009.

Sekhar, Deepa L., Sarah J. Clark, Matthew M. Davis, Dianne C. Singer, and Ian M. Paul. 2014. "Parental Perspectives on Adolescent Hearing Loss Risk and Prevention." *Journal of the American Medical Association: Otolaryngological Head and Neck Surgery* 140, no. 1: 22–8. doi:10.1001/jamaoto.2013.5760.

Selander, Jenny, Maria Albin, Ulf Rosenhall, Lars Rylander, Marie Lewné, and Per Gustavsson. 2016. "Maternal Occupational Exposure to Noise during Pregnancy and Hearing Dysfunction in Children: A Nationwide Prospective Cohort Study in Sweden." *Environmental Health Perspectives* 124, no. 6: 855–60. https://www.ncbi.nlm.nih.gov/pmc/articles/PMC4892921/.

Sergeyenko, Yevgeniya, Kumud Lall, M. Charles Liberman, and Sharon G. Kujawa. 2013. "Age-Related Cochlear Synaptopathy: An Early-Onset Contributor to Auditory Functional Decline." *Journal of Neuroscience* 33, no. 34: 13686–94. doi:10.1523/JNEUROSCI.1783-13.2013.

Shahin, Antoine, Larry E. Roberts, and Laurel J. Trainor. 2004. "Enhancement of Auditory Cortical Development by Musical Experience in Children." *Neuro-Report* 15, no. 12: 1917–21. doi:10.1097/00001756-200408260-00017.

Shambo, Lyda, Tony Umadhay, and Alessia Pedoto. 2015. "Music in the Operating Room: Is It a Safety Hazard?" *American Association of Nurse Anaesthetists Journal* 83, no. 1: 43–8. https://www.aana.com/docs/default-source/aana-journal-web-documents-1/music-in-or-0215-pp43-48.pdf.

Shargorodsky, Josef, Sharon G. Curhan, Gary C. Curhan, and Roland Eavey. 2010. "Change in Prevalence of Hearing Loss in US Adolescents." *Journal of the American Medical Association* 304, no. 7: 772–8. doi:10.1001/jama.2010.1124.

Shaywitz, B.A., S.E. Shaywitz, K.R. Pugh, W.E. Mencl, R.K. Fulbright, P. Skudlarski, R.T. Constable, K.E. Marchione, J.M. Fletcher, G.R. Lyon, and J.C. Gore. 2002. "Disruption of Posterior Brain Systems for Reading in Children with Developmental Dyslexia." *Biological Psychiatry* 52, no. 2: 101–10. https://www.ncbi.nlm.nih.gov/pubmed/12114001.

Shek, Vivian, and Emery Schubert. 2009. "Background Music at Work: A Literature Review and Some Hypotheses." Second International Conference on Music Communication Science. Sydney, HCSNet. https://www.researchgate.net/publication/267817702_Background_Music_at_Work_-_A_literature_review_and_some_hypotheses.

Shellenbarger, Sue. 2012. "At Work, Do Headphones Really Help?" *Wall Street Journal*, 7 June. http://www.wsj.com/articles/SB10001424052702303395604577432341782110010.

– 2012. "Indecent Exposure: The Downsides of Working in a Glass Office." *Wall Street Journal*, 4 January. https://www.wsj.com/articles/SB10001424052970203550304577138652786729324.

Shertzer, Kay E., and Juanita Fogel Keck. 2001. "Music and the PACU Environment." *Journal of Perianesthesiology Nursing* 16, no. 2: 90–102. doi:10.1053/jpan.2001.22594.

Shi, Yongbing, and William Hal Martin. 2013. "Noise Induced Hearing Loss in China: A Potentially Costly Public Health Issue." *Journal of Otology* 8, no. 1: 51–6. doi:10.1016/S1672-2930(13)50007-9.

Shield, Bridget M., and Julie E. Dockrell. 2004. "External and Internal Noise Surveys of London Primary Schools." *Journal of the Acoustical Society of America* 115, no. 2: 730–8. doi:10.1121/1.1635837.

Shield, Bridget M., Emma Greenland, and Julie E. Dockrell. 2010. "Noise in Open Plan Classrooms in Primary Schools: A Review." *Noise and Health* 12: 225–34. doi:10.4103/1463-1741.70501.

Shih, Yi-Nuo, Rong-Hwa Huang, and Han-sun Chiang. 2009. "Correlation between Work Concentration Level and Background Music: A Pilot Study." *Work: A Journal of Prevention, Assessment and Rehabilitation* 33, no. 3: 329–33. doi:10.3233/WOR-2009-0880.

Shih, Yi-Nuo, Rong-Hwa Huang, and Hsin-Yu Chiang. 2012. "Background Music: Effects on Attention Performance." *Work: A Journal of Prevention, Assessment and Rehabilitation* 42, no. 4: 573–8. doi:10.3233/WOR-2012-1410.

Shimojo, Shinsuke, and Ladan Shams. 2001. "Sensory Modalities Are Not Separate Modalities: Plasticity and Interactions." *Current Opinion in Neurobiology* 11: 505–9. doi:10.1016/S0959-4388(00)00241-5.

Shu, Hua, Hong Peng, and Catherine McBride-Chang. 2008. "Phonological Awareness in Young Chinese Children: Findings Underscore the Unique Importance of Both Tone and Syllable for Early Character Acquisition in Chinese Children." *Developmental Science* 11, no. 1: 171–81. doi:10.1111/j.1467-7687.2007.00654.x.

Sim, Stuart. 2007. *Manifesto for Silence: Confronting the Politics and Culture of Noise.* Edinburgh: Edinburgh University Press.

Singh, Narendra, and S.C. Davar. 2004. "Noise Pollution: Sources, Effects and Control." *Journal of Human Ecology* 16, no. 3: 181–7. doi:10.1080/09709274 .2004.11905735.

Singhal, Sangeeta, Berendra Yadav, S.F. Hashmi, and Mohammad D. Muzammil. 2009. "Effects of Workplace Noise on Blood Pressure and Heart Rate." *Biomedical Research* 20, no. 2: 122–6. http://www.biomedres.info/biomedical-research/effects-of-workplace-noise-on-blood-pressure-and-heart-rate.pdf.

Siskinds. 2009. "Regulation of Noise." 15 March. https://www.siskinds.com /regulation-of-noise/.

Sizov, Natalia V., John-Paul B. Clarke, Liling Ren, Kevin R. Elmer, and Belur N. Shivashankara. 2005. "Noise Impact Study of a New 2004 Noise Abatement Procedure at the Louisville Airport." *Journal of the Acoustical Society of America* 118, no. 3: 1851. doi:10.1121/1.4778650.

Skånberg, Annbritt. 2004. "Road Traffic Noise-Induced Sleep Disturbances: A Comparison between Laboratory and Field Settings." *Journal of Sound and Vibration* 277, no. 3: 465–7. doi:10.1016/j.jsv.2004.03.006.

Skånberg, Annbritt, and E. Öhrström. 2002. "Adverse Health Effects in Relation to Urban Residential Soundscapes." *Journal of Sound and Vibration* 250, no. 1: 151–5. doi:10.1006/jsvi.2001.3894.

Śliwińska-Kowalska, Mariola, Bartosz Bilski, Ewa Zamysłowska-Szmytke, Piotr Kotyło, Marta Fiszer, Wiktor Wesołowski, Małgorzata Pawlaczyk-łuszczyńska, Małgorzata Kucharska, and Adam Dudarewicz. 2001. "Hearing Impairment in the Plastics Industry Workers Exposed to Styrene and Noise." English abstract of article in Polish. *Medycyna Pracy* 52, no. 5: 297–303. https://www.ncbi.nlm .nih.gov/pubmed/11828842.

Śliwińska-Kowalska, Mariola, and Adrian Davis. 2012. "Noise-Induced Hearing Loss." *Noise and Health* 14, no. 61: 274–80. http://www.noiseandhealth.org /text.asp?2012/14/61/274/104893.

Slocombe, Katie E., Tanja Kaller, Josep Call, and Klaus Zuberbühler. 2010. "Chimpanzees Extract Social Information from Agonistic Screams." *PLoS One* 5, no. 7: e11473. doi:10.1371/journal.pone.0011473.

Smaldino, Joseph J., and Carl C. Crandell. 2000. "Classroom Amplification Technology: Theory and Practice." *Language, Speech, and Hearing Services in Schools* 31: 371–5. doi:10.1044/0161-1461.3104.371.

Smith, Andrew, Beth Waters, and Hywel Jones. 2010. "Effects of Prior Exposure to Office Noise and Music on Aspects of Working Memory." *Noise and Health* 12, no. 49: 235–43. doi:10.4103/1463-1741.70502.

Smith, Bruce R. 1999. *The Acoustic World of Early Modern England*. Chicago: University of Chicago Press.

– 2004. "Listening to the Wild Blue Yonder: The Challenges of Acoustic Ecology." In *Hearing Cultures: Essays on Sound, Listening and Modernity*, edited by Veit Erlmann, 21–41. Oxford: Berg.

Smith, Mark M., ed. 2004. *Hearing History: A Reader*. Athens: University of Georgia Press.

Smith, Pauline A., Adrian Davis, Melanie Ferguson, and Mark E. Lutman. 2000. "The Prevalence and Type of Social Noise Exposure in Young Adults in England." *Noise and Health* 2, no. 6: 41–56. http://www.noiseandhealth.org/text .asp?2000/2/6/41/32650.

Smollett, Tobias G. [1771] 1966. *The Expedition of Humphry Clinker*. New York: Holt, Rinehart and Winston.

Smoorenburg, Guido F. 1993. "Risk of Noise-Induced Hearing Loss Following Exposure to Chinese Firecrackers." *Audiology* 32, no. 6: 333–43. doi:10.3109 /00206099309071864.

Soames Job, Raymond F. 1999. "Noise Sensitivity as a Factor Influencing Human Reaction to Noise." *Noise and Health* 1: 57–68. http://www.noiseandhealth .org/text.asp?1999/1/3/57/31713.

Sobotova, Lubica, Jana Jurkovicova, Zuzana Stefanikova, Ludmila Sevcikova, and

Lubica Aghova. 2010. "Community Response to Environmental Noise and the Impact on Cardiovascular Risk Score." *Science of the Total Environment* 408, no 6: 1264–70. doi:10.1016/j.scitotenv.2009.12.033.

Sobrian, S., V. Vaughn, W. Ashe, B. Markovic, V. Djuric, and B. Jankovic. 1997. "Gestational Exposure to Loud Noise Alters the Development and Postnatal Responsiveness of Humoral and Cellular Components of the Immune System in Offspring." *Environmental Research* 73, no. 1–2: 227–41. doi:10.1006/enrs.1997.3734.

Society for Science and the Public. 1936. "Noise Blamed for Many Evils at Meeting of Deafened." *Science News-Letter* 29, no. 791: 370. doi:10.2307/3912466.

Söderlund, Göran, Sverker Sikström, and Andrew Smart. 2007. "Listen to the Noise: Noise Is Beneficial for Cognitive Performance in ADHD." *Child Psychology and Psychiatry* 48, no. 8: 840–7. doi:10.1111/j.1469-7610.2007.01749.x.

Södersten, Maria, Svante Granqvist, Britta Hammarberg, and Annika Szabo. 2002. "Vocal Behavior and Vocal Loading Factors for Preschool Teachers at Work Studied with Binaural DAT Recordings." *Journal of Voice* 16, no. 3: 356–71. doi:10.1016/S0892-1997(02)00107-8.

Solet, Jo M., and Paul R. Barach. 2012. "Managing Alarm Fatigue in Cardiac Care." *Progress in Pediatric Cardiology* 33, no.1: 85–90. doi:10.1016/j.ppedcard .2011.12.014.

Spaeth, J., L. Klimek, W.H. Döring, A. Rosendahl, and R. Mösges. 1993. "How Badly Does the 'Normal-Hearing' Young Man of 1992 Hear in the High Frequency Range?" English abstract of article in German. *Deutsche Gesellschaft der Hals- Nasen- und Ohrenärzte, Zeitschaft für Nasen- und Ohrenheilkund*e (HNO) 41, no. 8: 385–8. https://www.ncbi.nlm.nih.gov/pubmed/8407380.

Stanchina, Michael L., Muhanned Abu-Hijleh, Bilal K. Chaudhry, Carol C. Carlisle, and Richard P. Millman. 2005. "The Influence of White Noise on Sleep in Subjects Exposed to ICU Noise." *Sleep Medicine* 6, no. 5: 423–8. doi:10.1016/j.sleep.2004.12.004.

Stansfeld, Stephen A. 1992. "Noise, Noise Sensitivity and Psychiatric Disorder: Epidemiological and Psychophysiological Studies." *Psychological Medicine Monograph Supplement* 22: 1–44. doi:10.1017/S0264180100001119.

Stansfeld, Stephen A., Birgitta Berglund, C. Clark, I. Lopez-Barrio, P. Fischer, E. Ohrström, M.M. Haines, J. Head, S. Hygge, I. van Kamp, and B.F. Berry. 2005. "Aircraft and Road Traffic Noise and Children's Cognition and Health: A Cross-National Study." *Lancet* 365, no. 9475: 1942–9. doi:10.1016/S0140-6736(05)66660-3.

Stansfeld, Stephen A., and Rosanna Crombie. 2011. "Cardiovascular Effects of Environmental Noise: Research in the United Kingdom." *Noise and Health* 13, no. 52: 229–33. http://www.noiseandhealth.org/text.asp?2011/13/52/229/80159.

Stauss, Harald M. 2003. "Heart Rate Variability." *American Journal of Physiology: Regulatory, Integrative and Comparative Physiology* 285, no. 5: R927–31. doi:10.1152/ajpregu.00452.2003.

Steelcase. 2006. "The State of the Cubicle: What's Now and What's Next?" Accessed 8 July 2012. http://www.steelcase.com/en/search.aspx?k=cubicle %20size.

Steelcase WorkSpace Futures. 2010. "Benching: An Idea Whose Time Has Come … Again." Accessed 16 February 2019. https://www.steelcase.com/content /uploads/2015/01/whitepaper-benching-v2.3.pdf.

Stefanics, Gabor, Tim Fosker, Martina Huss, Natasha Mead, Denes Szucs, and Usha Goswami. 2011. "Auditory Sensory Deficits in Developmental Dyslexia: A Longitudinal ERP Study." *Neuroimage* 57, no. 3: 723–32. doi:10.1016/j .neuroimage.2011.04.005.

Sternberg, Esther M. 2009. *Healing Spaces: The Science of Place and Well-being.* Cambridge, MA: Belknap Press of Harvard University Press.

Stevens, Anthony C., Janet M. Sharp, and Becky Nelson. 2001. "The Intersection of Two Unlikely Worlds: Ratios and Drums." *Teaching Children Mathematics* 7, no. 6: 376–83. https://www.jstor.org/stable/i40053794.

Stevenson, Ryan, Joseph Schlesinger, and Mark Wallace. 2013. "Effects of Divided Attention and Operating Room Noise on Perception of Pulse Oximeter Pitch Changes: A Laboratory Study." *Anesthesiology* 2, no. 118: 376–81. doi:10.1097/ALN.0b013e31827d417b.

Stiles, William B. 1999. "Signs and Voices in Psychotherapy." *Psychotherapy Research* 9, no. 1: 1–21. doi:10.1080/10503309912331332561.

Stinson, Elizabeth. 2015. "The Future of Wind Turbines? No Blades." *Wired*, 15 May. https://www.wired.com/2015/05/future-wind-turbines-no-blades/.

Stone, Jennifer S., and Douglas A. Cotanche. 2007. "Hair Cell Regeneration in the Avian Auditory Epithelium." *International Journal of Developmental Biology* 51, no. 6–7: 633–47. doi:10.1387/ijdb.072408js.

Størmer, Carl C.L., and Niels C. Stenklev. 2007. "Rock Music and Hearing Disorders." English abstract of article in Norwegian. *Tidsskr Nor Laegeforen* 12, no. 7: 874–7. http://www.ncbi.nlm.nih.gov/pubmed/17435808.

Stosić, Ljiljana, Goran Belojević, and Suzana Milutinović. 2009. "Effects of Traffic Noise on Sleep in an Urban Population." English abstract of article in Serbian. *Arhiv za Higijenu Rada i Toksikologiju* 60, no. 3: 335–42. doi:10.2478 /10004-1254-60-2009-1962.

Sułkowski, Wiesław J. 2009. "Noise-Induced Hearing Loss in Children and Youth: Causes and Prevention." English abstract of article in Polish. *Medycyny Pracy* 60, no. 6: 513–17. https://www.ncbi.nlm.nih.gov/pubmed/20187499.

Sutow, Elliott J., Wayne A. Maillet, James C. Taylor, and Gordon C. Hall. 2004.

"In Vivo Galvanic Currents of Intermittently Contacting Dental Amalgam and Other Metallic Restorations." *Dental Materials* 20, no. 9: 823–31. doi:10.1016/j.dental.2003.10.012.

Svensson, Eva B., Thaís C. Morata, Per Nylén, Edward F. Krieg, and Ann-Christin Johnson. 2004. "Beliefs and Attitudes among Swedish Workers Regarding the Risk of Hearing Loss." *International Journal of Audiology* 43, no. 10: 585–93. doi:10.1080/14992020400050075.

Swaminathan, Nikhil. 2007. "Fact or Fiction? Babies Exposed to Classical Music End Up Smarter." *Scientific American*, 13 September. https://www.scientific american.com/article/fact-or-fiction-babies-ex/.

Tallal, Paula. 2000. "The Science of Literacy: From the Laboratory to the Classroom." *Proceedings of the National Academy of Sciences of the United States of America* 97, no. 6: 2402–4. doi:10.1073/pnas.97.6.2402.

Tallal, Paula, Steve L. Miller, Gail Bedi, Gary Byma, Xiaoqin Wang, Srikantan S. Nagarajan, Christoph Schreiner, William M. Jenkins, Michael M. Merzenich. 1996. "Language Comprehension in Language-Learning Impaired Children Improved with Acoustically Modified Speech." *Science* 271, no. 5245: 81–4. doi:10.1126/science.271.5245.81.

Tangermann, Victor. 2018. "Bring the Noise: Why Electric Vehicles Need to Make More Sound, Right Now." *Futurism*, 10 May. https://futurism.com /electric-vehicles-quiet-dangerous-noise/.

Tatum, Jeremy B. 1996. "The Physics, Physiology, and Psychology of Noise." November. Talk available on the Right to Quiet Society website. http://www.quiet.org/readings/tatum.htm.

Taylor-Ford, Rebecca, Anita Catlin, Mollie LaPlante, and Candace Weinke. 2008. "Effect of a Noise Reduction Program on a Medical-Surgical Unit." *Clinical Nursing Research* 17, no. 2: 74–88. doi:10.1177/1054773807312769.

Teder-Sälejärvi, Wolfgang A., Karen L. Pierce, Eric Courchesne, and Steven A. Hillyard. 2005. "Auditory Spatial Localization and Attention Deficits in Autistic Adults." *Cognitive Brain Research* 23, no. 2–3: 221–34. doi:10.1016/j.cog brainres.2004.10.021.

Teixeira, Cleide F., Lia G. da Silva Augusto, and Thaís C. Morata. 2002. "Occupational Exposure to Insecticides and Their Effects on the Auditory System." *Noise and Health* 4, no. 14: 31–9. http://www.noiseandhealth.org/text.asp ?2002/4/14/31/31811.

– 2003. "Hearing Health of Workers Exposed to Noise and Insecticides." English abstract of article in Portuguese. *Revista Saude Publica* 37, no. 4: 417–23. https://www.ncbi.nlm.nih.gov/pubmed/12937701.

Tempey, Nathan. 2017. "The NYPD Claimed Its LRAD Sound Cannon Isn't a Weapon: A Judge Disagreed." *Gothamist*, 1 June. http://gothamist.com/2017 /06/01/lrad_lawsuit_nypd.php.

Teramoto, Wataru, Souta Hidaka, Yoichi Sugita, Shuichi Sakamoto, Jiro Gyoba, Yukio Iwaya, and Yôiti Suzuki. 2012. "Sounds Can Alter the Perceived Direction of a Moving Visual Object." *Journal of Vision* 12, no. 3: 11. doi:10.1167/12.3.11.

ter Bogt, Tom F.M., Marc J.M.H. Delsing, Maarten van Zalk, Peter G. Christenson, and Wim H.J. Meeus. 2011. "Intergenerational Continuity of Taste: Parental and Adolescent Music Preferences." *Social Forces* 90, no. 1: 297–319. doi:10.1093/sf/90.1.297.

Thaut, Michael H. 2005. "The Future of Music in Therapy and Medicine." *Annals of the New York Academy of Sciences* 1060, no. 1: 303–8. doi:10.1196/annals.1360.023.

Thibaud, Jean-Paul. 2014. "The Sonic Existence of Urban Ambiances." In *Invisible Places / Sounding Cities*. Proceedings, 18–20 July. Viseu, PT: Jardins Efeméros. http://invisibleplaces.org/IP2014.pdf.

Thompson, Emily. 2002. *The Soundscape of Modernity: Architectural Acoustics and the Culture of Listening in America, 1900–1933*. Cambridge, MA: MIT Press.

– 2004. "Wiring the World: Acoustical Engineers and the Empire of Sound in the Motion Picture Industry, 1927–1930." In *Hearing Cultures: Essays on Sound, Listening and Modernity*, edited by Veit Erlmann, 191–209. Oxford: Berg.

Thorne, Peter R., Shanthi N. Ameratunga, Joanna Stewart, Nicolas Reid, Warwick Williams, Suzanne C. Purdy, George Dodd, and John Wallaart. 2008. "Epidemiology of Noise-Induced Hearing Loss in New Zealand." *New Zealand Medical Journal* 121, no. 1280: 33–44. www.nzma.org.nz/__data/assets/pdf_file/0007/17809/Vol-121-No-1280-22-August-2008.pdf.

Thraenhardt, Bettina. 2006–7. "Hearing Voices." *Scientific American Mind* 17, no. 6: 74–9. https://www-jstor-org.ezproxy.lib.ucalgary.ca/stable/pdf/24921633.pdf?refreqid=excelsior%3A153aa6043e4dfe5c6084cc8bb4ddcdf9.

Thurston, Floyd E. 2013. "The Worker's Ear: A History of Noise-Induced Hearing Loss." *American Journal of Industrial Medicine* 56, no. 3: 367–77. doi:10.1002/ajim.22095.

Tomei, Francesco, Bruno Papaleo, Tiziana P. Baccolo, Benedetta Persechino, Giovanni Spanò, and Maria V. Rosati. 1994. "Noise and Gastric Secretion." *American Journal of Industrial Medicine* 26, no. 3: 367–72. doi:10.1002/ajim.4700260310.

Tomei, Francesco, Sergio Fantini, Enrico Tomao, Tiziana Paola Baccolo, and Maria Valeria Rosati. 2000. "Hypertension and Chronic Exposure to Noise." *Archives of Environmental Health: An International Journal* 55, no. 5. doi:10.1080/00039890009604023.

Tomei, Gianfranco, Maria F. Anzani, Teodorico Casale, Federico Piccoli, Daniela Cerratti, Matteo Paolucci, C. Filippelli, M. Fioranti, and Francesco Tomei. 2009. "Extra-Auditory Effects of Noise." English abstract of article in Italian. *Giornale Italiano di Medicina del Lavoro ed Ergonomia* 31, no. 1: 37–48. https://europepmc.org/abstract/med/19558038.

Tomei, Gianfranco, Mario Fioravanti, Daniela Cerratti, Angela Sancini, Enrico Tomao, Maria V. Rosati, Dante F. Vacca, Tania Palitti, Manuela Di Famiani, Roberto Giubilati, et al. 2010. "Occupational Exposure to Noise and the Cardiovascular System: A Meta-Analysis." *Science of the Total Environment* 408, no. 4: 681–9. doi:10.1016/j.scitotenv.2009.10.071.

Toop, David. 2010. *Sinister Resonance: The Mediumship of the Listener*. New York: Continuum Books.

Topf, Margaret. 1992. "Effects of Personal Control over Hospital Noise on Sleep." *Research in Nursing and Health* 15, no. 1: 19–28. doi:10.1002/nur.4770150105.

Trainor, Laurel J., Antoine J. Shahin, and Larry E. Roberts. 2003. "Effects of Musical Training on the Auditory Cortex in Children." *Annals of the New York Academy of Sciences* 999, no. 1: 506–13. doi:10.1196/annals.1284.061.

– 2009. "Understanding the Benefits of Musical Training: Effects on Oscillatory Brain Activity." *Annals of the New York Academy of Sciences* 1169, no. 1: 133–42. doi:10.1111/j.1749-6632.2009.04589.x.

Trask, Douglas K., Bruce Abkas, and Nathan Jous. 2006. "Listening Habits and Noise Exposure of MP3 Player Users." *Otolaryngology: Head and Neck Surgery* 135, no. 2S: 142–3. doi:10.1016/j.otohns.2006.06.865.

Traynor, Robert. 2011. "Hearing Beethoven, Part II: The Medical Conclusion." *Hearing Health and Technology Matters*, 12 May. https://hearinghealthmatters .org/hearinginternational/2011/hearing-beethoven-part-ii-the-medical- conclusion/.

Truax, Barry. 1978. *Handbook for Acoustic Ecology*. Vancouver: A.R.C. Publications.

– 2001. *Acoustic Communication*. 2nd ed. Westport, CT: Ablex.

Tse, Mimi M.Y., Moon F. Chan, and Iris F.F. Benzie. 2005. "The Effect of Music Therapy on Postoperative Pain, Heart Rate, Systolic Blood Pressure and Analgesic Use Following Nasal Surgery." *Journal of Pain and Palliative Care Pharmacotherapy* 19, no. 3: 21–9. doi:10.1080/J354v19n03_05.

Tuomi, Seppo K., and Marlize Jelliman. 2009. "Hear Today, Hearing Loss Tomorrow: A Preliminary Survey of South African First-Year University Students' Personal Audio Player User Habits and Knowledge." *South African Family Practice* 51, no. 2: 166–7. doi:10.1080/20786204.2009.10873835.

Uchida, Yasue, Tsutomu Nakashima, Fujiko Ando, Naoakira Niino, and Hiroshi Shimokata. 2005. "Is There a Relevant Effect of Noise and Smoking on Hearing? A Population-Based Aging Study." *International Journal of Audiology* 44, no. 2: 86–91. doi:10.1080/14992020500031256.

Ullmann, Yehuda, Lucian Fodor, Irena Schwarzberg, Nurit Carmi, Amos Ullmann, and Yitzchak Ramon. 2008. "The Sounds of Music in the Operating Room." *Injury* 39, no. 5: 592–97. doi:10.1016/j.injury.2006.06.021.

United States Census Bureau. 2019. "American Housing Survey (AHS)." Accessed 3 February. https://www.census.gov/programs-surveys/ahs/.

United States Centers for Disease Control and Prevention. 2012. "Summary of 2009 National CDC EHDI." Accessed 3 February 2019. http://www.cdc.gov /ncbddd/hearingloss/2009-data/2009_ehdi_hsfs_summary_508_ok.pdf.

United States Department of Labor. Bureau of Labor Statistics. 2011. Employment of Women, by Industry, 1964–2010 (chart). "Women at Work." March. Accessed 3 February 2019. http://www.bls.gov/spotlight/2011/women/.

Uno, Masaaki, Junko Abe, Chihiro Sawai, Yuhko Sakaue, Atsushi Nishitani, Yuriko Yasuda, Kento Tsuzuki, Tomoyuki Takano, Masaki Ohno, Tsuyoshi Maruyama, et al. 2006. "Effect of Additional Auditory and Visual Stimuli on Continuous Performance Test (Noise-Generated CPT) in AD/HD Children: Usefulness of Noise-Generated CPT." *Brain and Development* 28, no. 3: 162–9. doi:10.1016/j.braindev.2005.06.007.

Vaisman, Daria. 2001–2. "The Acoustics of War." *Cabinet Magazine* 5 (Winter). http://cabinetmagazine.org/issues/5/acousticsofwar.php.

Van Allen, Leslie. 2007. "The Soundscape of New York City in the 1930s." Accessed 7 February 2019. http://ecoear.proscenia.net/wfaelibrary/library /articles/van_allen_NYC1930%27s.pdf.

van Dongen, Diana, Tjabe Smid, and Daniëlle R.M. Timmermans. 2014. "Symptom Attribution and Risk Perception in Individuals with Idiopathic Environmental Intolerance to Electromagnetic Fields and in the General Population." *Perspectives in Public Health* 134, no. 3: 160–8. doi:10.1177/1757913913492931.

van Kempen, Elise E., Hanneke Kruize, Hendriek C. Boshuizen, Caroline B. Ameling, Brigit A. Staatsen, and Augustinus E. de Hollander. 2002. "The Association between Noise Exposure and Blood Pressure and Ischemic Heart Disease: A Meta-Analysis." *Environmental Health Perspectives* 110, no. 3: 307–17. doi:10.1289/ehp.02110307.

Västfjäll, Daniel. 2002. "Influences of Current Mood and Noise Sensitivity on Judgments of Noise Annoyance." *Journal of Psychology* 136, no. 4: 357–70. doi:10.1080/00223980209604163.

Velluti, R.A. 2002. "Procesamiento de la Información Auditiva en Estados de Vigilia y Sueño." *Vigilia-Sueno* 14, no. 2: 87–98. http://psiqu.com/2-5202.

Vidal, Joan, Marie-Hélène Giard, Sylvie Roux, Catherine Barthélémy, and Nicole Bruneau. 2008. "Cross-Modal Processing of Auditory-Visual Stimuli in a No-Task Paradigm: A Topographic Event-Related Potential Study." *Clinical Neurophysiology* 119, no. 4: 763–71. doi:10.1016/j.clinph.2007.11.178.

Viet, Susan Marie, Michael Dellarco, Dorr G. Dearborn, and Richard Neitzel. 2014. "Assessment of Noise Exposure to Children: Considerations for the National Children's Study." *Journal of Pregnancy and Child Health* 1, no. 105. doi:10.4172/2376-127X.1000105.

Vinokur, Roman. 2004. "Acoustic Noise as a Non-Lethal Weapon." *Sound and Vibration* (October). http://www.sandv.com/downloads/0410vino.pdf.

Vitello, Paul. 2007. "That Racket? It's the Sound of Suburbia." *New York Times*, 22 July. https://www.nytimes.com/2007/07/22/nyregion/nyregionspecial2 /22rNoise.html.

Vogel, Ineke, Johannes Brug, Esther J. Hosli, Catharina P.B. van der Ploeg, and Hein Raat. 2008. "MP3 Players and Hearing Loss: Adolescents' Perceptions of Loud Music and Hearing Conservation." *Journal of Pediatrics* 152, no. 3: 400–4.e1. doi:10.1016/j.jpeds.2007.07.009.

Vogel, Ineke, Johannes Brug, Catharina P.B. van der Ploeg, and Hein Raat. 2010. "Discotheques and the Risk of Hearing Loss among Youth: Risky Listening Behavior and Its Psychosocial Correlates." *Health Education Research* 25, no. 5: 737–47. doi:10.1093/her/cyq018.

Vogel, Ineke, Hans Verschuure, Catharina P.B. van der Ploeg, Johannes Brug, and Hein Raat. 2009. "Adolescents and MP3 Players: Too Many Risks, Too Few Precautions." *Pediatrics* 123, no. 6: e953–8. doi:10.1542/peds.2008-3179.

– 2010. "Estimating Adolescent Risk for Hearing Loss Based on Data from a Large School-Based Survey." *American Journal of Public Health* 100, no. 6: 1095–100. doi:10.2105/AJPH.2009.168690.

Vogiatzis, Konstantinos, and Patrick Vanhonacker. 2015. "Noise Reduction in Urban LRT Networks by Combining Track Based Solutions." *Science of the Total Environment* 568: 1344–54. doi:10.1016/j.scitotenv.2015.05.060.

Wachman, Elisha M., and Amir Lahav. 2011. "The Effects of Noise on Preterm Infants in the NICU." *Archives of Disease in Childhood: Fetal and Neonatal Edition* 96, no. 4: F305–9. doi:10.1136/adc.2009.182014.

Walker, Richard S., A. Gardiner Wade, Gelsomina Iazzetti, and Nikhil K. Sarkar. 2003. "Galvanic Interaction between Gold and Amalgam: Effect of Zinc, Time and Surface Treatments." *Journal of the American Dental Association* 134, no. 11: 1463–7. doi:10.14219/jada.archive.2003.0075.

Wallace, Carrie J., Judith Robins, Lynn S. Alvord, and James M. Walker. 1999. "The Effect of Earplugs on Sleep Measures during Exposure to Simulated Intensive Care Unit Noise." *American Journal of Critical Care* 8, no. 4: 210–19. http://ajcc.aacnjournals.org/content/8/4/210.abstract.

Walworth, Darcy, Christopher S. Rumana, Judy Nguyen, and Jennifer Jarred. 2008. "Effects of Live Music Therapy Sessions on Quality of Life Indicators, Medications Administered and Hospital Length of Stay for Patients Undergoing Elective Surgical Procedures for Brain." *Journal of Music Therapy* 45, no. 3: 349–59. doi:10.1093/jmt/45.3.349.

Wang, Bo, and Jian Kang. 2011. "Effects of Urban Morphology on the Traffic Noise Distribution through Noise Mapping: A Comparative Study between UK and China." *Applied Acoustics* 72, no. 8: 556–68. doi:10.1016/j.apacoust.2011.01.011.

Wang, Shu-Ming, Lina Kulkarni, Jackqulin Dolev, and Zeev N. Kain. 2002. "Music and Preoperative Anxiety: A Randomized, Controlled Study." *Anesthesia and Analgesia* 94, no. 6: 1489–94. doi:10.1213/00000539-200206000-00021.

Waugh, Rob. 2012. "Try and Ignore THIS! Inventor Creates £5,000 Bicycle Horn That's 178 Decibels: Louder Than a Concorde Taking Off." *Daily Mail*, 4 May. http://www.dailymail.co.uk/sciencetech/article-2139651/Try-ignore-THIS-Inventor-creates-5-000-bicycle-horn-thats-178-decibels—louder-Concorde-taking-off.html.

Waxman, Sandra. 2016. "Listening to Speech Has Remarkable Effects on a Baby's Brain." Aeon. 7 June. Accessed 7 February 2019. https://aeon.co/ideas/listening-to-speech-has-remarkable-effects-on-a-baby-s-brain.

Webb, Alexandra R., Howard T. Heller, Carol B. Benson, and Amir Lahav. 2015. "Mother's Voice and Heartbeat Sounds Elicit Auditory Plasticity in the Human Brain before Full Gestation." *Proceedings of the National Academy of Sciences of the United States of America* 112, no. 10: 3152–7. doi:10.1073/pnas .1414924112.

Weichenberger, Markus, Robert Kühler, Martin Bauer, Johannes Hensel, Rüdiger Brühl, Albrecht Ihlenfeld, Bernd Ittermann, Jürgen Gallinat, Christian Koch, Tilmann Sander, et al. 2015. "Brief Bursts of Infrasound May Improve Cognitive Function: An fMRI Study." *Hearing Research* 328: 87–93. doi:10.1016 /j.heares.2015.08.001.

Weinberger, Sharon. 2008. "I Was a Sonic Blaster Guinea Pig." *Wired*, 13 February. https://www.wired.com/2008/02/i-was-a-puke-ra/#previouspost

Weissenburger, J.T. 2004. "Room-to-Room Privacy and Acoustical Design Criteria." *Sound and Vibration* (February). http://www.sandv.com/downloads /0402weis.pdf.

Welch, David, and Guy Fremaux. 2017. "Understanding Why People Enjoy Loud Sound." *Seminars in Hearing* 38, no. 4: 348–57. doi: 10.1055/s-0037-1606328

Weldon, Sharon-Marie, Terhi Korkiakangas, Jeff Bezemer, and Roger Kneebone. 2015. "Music and Communication in the Operating Theatre." *Journal of Advanced Nursing* 71, no. 12: 2763–74. doi:10.1111/jan.12744.

West, James, Ilene Busch-Vishniac, Joseph King, and Natalia Levit. 2008. "Noise Reduction in an Operating Room: A Case Study." *Journal of the Acoustical Society of America* 123, no. 5: 3677. doi:10.1121/1.2935033.

Westman, J.C., and J.R. Walters. 1981. "Noise and Stress: A Comprehensive Approach." *Environmental Health Perspectives* 41: 291–309. doi:10.1289/ehp .8141291.

Wharton School of the University of Pennsylvania. 2002. "Telephone Call Centers: The Factory Floors of the 21st Century." Knowledge@Wharton. 10 April.

Accessed 11 February 2019. http://knowledge.wharton.upenn.edu/article/telephone-call-centers-the-factory-floors-of-the-21st-century/.

Widén, Stephen E., and Soly I. Erlandsson. 2004. "Self-Reported Tinnitus and Noise Sensitivity among Adolescents in Sweden." *Noise and Health* 7, no. 25: 29–40. http://www.noiseandhealth.org/text.asp?2004/7/25/29/31649.

Wild, D.C., Marnie J. Brewster, and Anil R. Banerjee. 2005. "Noise-Induced Hearing Loss Is Exacerbated by Long-Term Smoking." *Clinical Otolaryngology* 30, no. 6: 517–20. doi:10.1111/j.1749-4486.2005.01105.x.

Williams, Amber L., Wim van Drongelen, and Robert E. Lasky. 2007. "Noise in Contemporary Neonatal Intensive Care." *Journal of the Acoustical Society of America* 121, no. 5, part 1: 2681–90. doi:10.1121/1.2717500.

Williams, David. 2008. "iPods Blamed for Huge Rise in Accidents." *London Evening Standard*, 18 March. http://www.standard.co.uk/news/ipods-blamed-for-huge-rise-in-accidents-6667633.html.

Williams, Joan A., and Sharon A. Lynch. 2010. "Dyslexia: What Teachers Need to Know." *Kappa Delta Pi Record* 46, no. 2: 66–70. doi:10.1080/00228958.2010.10516696.

Williams, Timothy. 2008. "An Old Sound in Harlem Draws New Neighbors' Ire." *New York Times*, 6 July. http://www.nytimes.com/2008/07/06/nyregion/06drummers.html.

Willich, Stefan N., Karl Wegscheider, Martina Stallmann, and Thomas Keil. 2006. "Noise Burden and the Risk of Myocardial Infarction." *European Heart Journal* 27, no. 3: 276–82. doi:10.1093/eurheartj/ehi658.

Wilson, Wayne J., and Nicole Herbstein. 2003. "The Role of Music Intensity in Aerobics: Implications for Hearing Conservation." *Journal of the American Academy of Audiology* 14, no. 1: 29–38. doi:10.3766/jaaa.14.1.5.

Witterseh, Thomas, David P. Wyon, and Geo Clausen. 2004. "The Effects of Moderate Heat Stress and Open-Plan Office Noise Distraction on SBS Symptoms and on the Performance of Office Work." *Indoor Air* 14, no. s8: 30–40. doi:10.1111/j.1600-0668.2004.00305.x.

Wong, Ann C.Y., and Ryan, Allen F. 2015. "Mechanisms of Sensorineural Cell Damage, Death and Survival in the Cochlea." *Frontiers of Aging Neuroscience* 7, no. 58: 7–21. doi:10.3389/fnagi.2015.00058.

Woodlief, C.B., L.H. Royster, and B.K. Huang. 1969. "Effect of Random Noise on Plant Growth." *Journal of the Acoustical Society of America* 46, no. 2B: 481–2. doi:10.1121/1.1911721.

Work Safe BC. 2019. "OHS Guidelines, Part 7: Noise, Vibration, Radiation and Temperature." Division 1, Noise Exposure. Accessed 11 February. https://www.worksafebc.com/en/law-policy/occupational-health-safety/searchable-ohs-regulation/ohs-guidelines/guidelines-part-07.

Workers' Compensation Board of Nova Scotia. 2017. "Policy Manual on Noise Exposure." http://www.wcb.ns.ca/policymanual/pminoise.html. Document replaced online by Policy Number 1.2.6R1, "Workplace Noise Levels." https://www.wcb.ns.ca/Portals/wcb/Policy%20Manual/Section%201/Section%201.2/1-2-6R1.pdf?ver=2017-04-16-104619-853.

World Health Organization. 2015. "1.1 Billion People at Risk for Hearing Loss." 27 February. https://www.who.int/mediacentre/news/releases/2015/ear-care/en/.

– 2017. "Global Costs of Unaddressed Hearing Loss and Cost-Effectiveness of Interventions." WHO Report. Accessed 11 February 2019. http://apps.who.int/iris/bitstream/handle/10665/254659/9789241512046-eng.pdf.

– 2019. "Noise: Data and Statistics." Accessed 11 February. http://www.euro.who.int/en/health-topics/environment-and-health/noise/data-and-statistics.

World Health Organization, Regional Office for Europe. 2011. *Burden of Disease from Environmental Noise: Quantification of Healthy Life Years Lost in Europe.* Accessed 11 February 2019. https://apps.who.int/iris/bitstream/handle/10665/326424/9789289002295-eng.pdf?sequence=1&isAllowed=y.

Wu, Marco, Isaac Ng, W.K. Szeto, and Maurice Yeung. 2014. "Challenges in Planning against Road Traffic Noise in Hong Kong." In *43rd International Congress and Exposition on Noise Control Engineering (inter.noise 2014): Improving the World through Noise Control – Conference Proceedings*, 2435. Melbourne: Australian Acoustical Society. http://www.acoustics.asn.au/conference_proceedings/INTERNOISE2014/papers/p273.pdf.

Wyness, James A. 2008. "Soundscape as Discursive Practice." In *Proceedings of the Institute of Acoustics* 30, no. 2. https://soundartarchive.net/articles/Wyness-2008.pdf.

Xu, Shanshan, Guangdi Chen, Chunjing Chen, Chuan Sun, Danying Zhang, Manuel Murbach, Niels Kuster, Qunli Zeng, and Zhengping Xu. 2013. "Cell Type-Dependent Induction of DNA Damage by 1800 MHz Radio-frequency Electromagnetic Fields Does Not Result in Significant Cellular Dysfunctions." *PLoS One* 8, no. 1: e54906. doi:10.1371/journal.pone.0054906.

Yang, Jennifer. 2010. "Toronto Police Get 'Sound Cannons' for G20." *Toronto Star*, 27 May. https://www.thestar.com/news/gta/g20/2010/05/27/toronto_police_get_sound_cannons_for_g20.html.

Yaremchuk, Kathleen, Linda Dickson, Kenneth Burk, and Bhagyalakshimi G. Shivapuja. 1997. "Noise Level Analysis of Commercially Available Toys." *International Journal of Pediatric Otorhinolaryngology* 41, no. 2: 187–97. doi:10.1016/S0165-5876(97)00083-9.

Yavas, Mehmet S., and Lakshmi J. Gogate. 1999. "Phoneme Awareness in Children: A Function of Sonority." *Journal of Psycholinguistic Research* 28, no. 3: 245–60. doi:10.1023/A:1023254114696.

Yildirim, Ilhami, Metin Kilinc, Erdogan Okur, Fatma Inanc Tolun, M. Akif Kiliç, Ergul Belge Kurutas, and Hasan Çetin Ekerbiçer. 2007. "The Effects of Noise on Hearing and Oxidative Stress in Textile Workers." *Industrial Health* 45, no. 6: 743–9. doi:10.2486/indhealth.45.743.

Young, Ryan. 2010. "Ancient Noise Ordinances." Inertia Wins! Accessed 8 February 2019. http://inertiawins.com/2010/07/31/ancient-noise-ordinances/.

Yu, Xide, Tao Liu, and Dingguo Gao. 2015. "The Mismatch Negativity: An Indicator of Perception of Regularities in Music." Special Issue, *Behavioral Neurology*. doi:10.1155/2015/469508.

Zenner, Hans P., V. Struwe, G. Schuschke, M. Spreng, G. Stange, P. Plath, W. Babisch, E. Rebentisch, P. Plinkert, K.D. Bachmann, et al. 1999. "Hearing Loss Caused by Leisure Noise." English abstract of article in German. *Deutsche Gesellschaft der Hals- Nasen- und Ohrenärzte (HNO)* 47, no. 4: 236–48. https://www.researchgate.net/publication/12891171_Hearing_loss_caused_by_leisure_noise.

Zhang, Lei, Ji T. Gong, Hu Q. Zhang, Quan H. Song, Guang H. Xu, Lei Cai, Xiao D. Tang, Hai F. Zhang, Fang-E Liu, Zhan S. Jia, et al. 2015. "Melatonin Attenuates Noise Stress-Induced Gastrointestinal Motility Disorder and Gastric Stress Ulcer: Role of Gastrointestinal Hormones and Oxidative Stress in Rats." *Journal of Neurogastroenterology and Motility* 21, no. 2: 189–99. doi:10.5056/jnm14119.

Zhang, Xiaopai, Xiaojun Qiu, and Yan Liu. 2014. "An Integrated Passive and Active Control System for Snoring Noise Cancellation." In *Proceedings of the 21st International Congress on Sound and Vibration*, 1–6. Beijing: International Institute of Acoustics and Vibration (IIAV).

Zheng, Kui-Cheng, and Makoto Ariizumi. 2007. "Modulations of Immune Functions and Oxidative Status Induced by Noise Stress." *Journal of Occupational Health* 49, no. 1: 32–8. doi:10.1539/joh.49.32.

Zirke, Nina, G. Goebel, and B. Mazurek. 2010. "Tinnitus and Psychological Comorbidities." English abstract of article in German. *Deutsche Gesellschaft der Hals- Nasen- und Ohrenärzte (HNO)* 58, no. 7: 726–32. doi:10.1007/s00106-009-2050-9.

Zirke, Nina, Claudia Seydel, Agnieszka J. Szczepek, Heidi Olze, Heidemarie Haupt, and Birgit Mazurek. 2013. "Psychological Comorbidity in Patients with Chronic Tinnitus: Analysis and Comparison with Chronic Pain, Asthma or Atopic Dermatitis Patients." *Quality of Life Research* 22, no. 2: 263–72. doi:10.1007/s11136-012-0156-0.

Zolfagharifard, Ellie. 2015. "Can YOU Hear Wind Farms? Researchers Prove
    Human Hearing Is Better than Thought and 'Turbine Phenomenon' Could
    Alter the Brain." *Daily Mail*, 10 July. http://www.dailymail.co.uk/sciencetech
    /article-3156778/Can-hear-wind-farms-hum-electricity-Researchers-claims-
    human-hearing-better-thought.html.

Index